Rigid-Flexible Coupling Hoisting Robots

Modeling, Analysis, and Control

Bin Zi

Hefei University of Technology
Hefei City, Anhui Province, China

Bin Zhou

Hefei University of Technology
Hefei City, Anhui Province, China

WILEY

Published by John Wiley & Sons, Inc., Hoboken, New Jersey.

Library of Congress Cataloging-in-Publication Data:

Names: Zi, Bin author | Zhou, Bin (Professor of mechanical engineering) author
Title: Rigid–Flexible Coupling Hoisting Robots: Modeling, Analysis, and Control / Bin Zi, Bin Zhou
Description: Hoboken, New Jersey: Wiley, [2026] | Includes bibliographical references and index.
Identifiers: LCCN 2025036243 | ISBN 9781394308941 hardback | ISBN 9781394308965 ePDF | ISBN 9781394308958 epub | ISBN 9781394308972
Subjects: LCSH: Robots–Dynamics | Hoisting machinery | Robots, Industrial
Classification: LCC TJ211.4 Z53 2026
LC record available at https://lccn.loc.gov/2025036243

Cover Design: Wiley
Cover Image: © TongTa/stock.adobe.com

Contents

Preface *vii*
Main Symbol Table *ix*

1 Introduction *1*
1.1 The Evolution of Rigid–Flexible Coupling Robots *1*
1.2 The History and Development of Rigid–Flexible Coupling Hoisting Robots *8*
1.3 The Applications of RFCHRs in Various Fields *11*
1.3.1 Construction *11*
1.3.2 Ocean *13*
1.3.3 Storage *14*
1.4 Scope and Organization of This Book *17*
References *20*

2 Kinematics and Dynamic Modeling of Rigid–Flexible Coupled Hoisting Robots *27*
2.1 Preamble *27*
2.2 Mechanism Design and Kinematic Analysis of RFCHRs *29*
2.2.1 Mechanism Design of RFCHRs *29*
2.2.2 Kinematic Modeling of RFCHRs *32*
2.3 Dynamic Modeling of RFCHRs *42*
2.3.1 Dynamic Modeling of RFCHR Based on the Lagrange Method *42*
2.3.2 Dynamic Modeling of RFCHR Based on the Newton–Euler Method *45*
2.4 Conclusion *49*
References *49*

3 Motion Decoupling, Reconfigurable Design of Rigid–Flexible Coupling Hoisting Robots *51*
3.1 Preamble *51*
3.2 Motion-decoupling Design for a 7-DOF RFCHR *53*
3.2.1 Coupling Characteristic Analysis and Motion-decoupling Method *53*
3.2.2 Mechanical Design of 7-DOF RFCHR *57*
3.3 Modular and Reconfigurable Mechanism Design of RFCHRs *61*
3.3.1 Design Methodology *61*
3.3.2 Mechanical Description *62*
3.3.3 Typical Configuration *64*
3.4 Integrated Mechanism Design of Dual-machine Collaborative RFCHRs *65*
3.4.1 Mechanical Design *66*
3.4.2 Kinematic Modeling *67*
3.4.3 Dynamic Modeling *69*
3.5 Conclusion *72*
References *72*

4 Optimization Design of Rigid–Flexible Coupling Hoisting Robots *75*
4.1 Preamble *75*
4.2 Multi-objective Optimization Design for Workspace and Dexterity *76*
4.2.1 Kinematic Modeling and Static Modeling of RFCHR *76*
4.2.2 Performance Indices of RFCHR *81*
4.2.3 Multi-objective Optimal Design *83*
4.3 Multi-objective Optimization Design for Reliability, Workspace, and Stiffness *89*
4.3.1 Performance Indices of RFCHR *89*
4.3.2 Multi-objective Optimization Design *101*
4.4 Experiment and Verification *109*
4.5 Conclusion *114*
References *114*

5 Kinematic Analysis of Rigid–Flexible Coupling Hoisting Robots with Uncertainty *117*
5.1 Preamble *117*
5.2 Kinematic Uncertainty Analysis with Random Parameters *119*
5.2.1 Mechanism Description *119*
5.2.2 Inverse Kinematics *120*
5.2.3 DACS Equilibrium Equation Under Narrowly Random Model *122*
5.2.4 MHRM for Luffing Angular Response Field of the DACS with Narrow Uncertainty *123*

5.2.5 Numerical Examples *125*
5.2.6 Conclusion *133*
5.3 Kinematic Uncertainty Analysis with Interval Variables *133*
5.3.1 Interval Kinematic Equilibrium Equation *133*
5.3.2 Hybrid Compound Function/Subinterval Perturbation Method *134*
5.3.3 Numerical Examples *137*
5.3.4 Conclusion *143*
5.4 Kinematic Uncertainty Analysis Based on Evidence Theory *143*
5.4.1 Architecture and Kinematics *143*
5.4.2 Error Transfer Model *147*
5.4.3 Uncertainty Analysis Based on Evidence Theory *149*
5.4.4 Simulation and Comparison *152*
5.4.5 Conclusion *160*
5.5 Kinematic Uncertainty Analysis with Hybrid Random and Interval Parameters *160*
5.5.1 Hybrid Uncertain DACS with Random and Interval Parameters *160*
5.5.2 The LAR Analysis of the DACS with Small Uncertainty *162*
5.5.3 Hybrid LAR Field Calculation of the DACS *164*
5.5.4 Numerical Examples *169*
5.5.5 Conclusion *173*
5.6 Conclusion *177*
References *178*

6 Dynamic Analysis of Rigid–Flexible Coupling Hoisting Robots with Uncertainty *181*
6.1 Preamble *181*
6.2 Static Uncertainty Analysis with Fuzzy Parameters *183*
6.2.1 Fuzzy Static Equilibrium Equation *184*
6.2.2 CFFPM *185*
6.2.3 MCFFPM *188*
6.2.4 Numerical Examples *192*
6.2.5 Conclusion *200*
6.3 Dynamic Uncertainty Analysis with Hybrid Random and Interval Parameters *201*
6.3.1 LSOAAC Equilibrium Equation Under the Hybrid Uncertain Model *201*
6.3.2 MHUAM for the Dynamic Response Analysis of LSOAAC *204*
6.3.3 Hybrid LSOAAC Response Field Calculation *207*
6.3.4 Numerical Examples *215*

6.3.5 Conclusion *237*
6.4 Conclusion *238*
References *238*

7 Trajectory Planning and Tracking Control of Rigid–Flexible Coupling Hoisting Robots *241*
7.1 Preamble *241*
7.2 Trajectory Planning for RFCHRs *244*
7.2.1 Inverse Kinematic Modeling *244*
7.2.2 Dynamic Modeling *246*
7.2.3 Point-to-point Trajectory Planning *249*
7.2.4 Numerical Simulation and Experiments *261*
7.2.5 Conclusion *264*
7.3 Fuzzy Control for RFCHRs *264*
7.3.1 Fuzzy Trajectory Tracking Control *265*
7.3.2 Numerical Simulations *270*
7.3.3 Experiment Validation *272*
7.3.4 Conclusion *274*
7.4 Force Control for RFCHRs *275*
7.4.1 Construction of Experimental Platform *275*
7.4.2 Cable Driving Force Control Method *275*
7.4.3 Control and Monitoring Program Design *279*
7.4.4 Test and Verification Experiment *280*
7.4.5 Conclusion *281*
7.5 Conclusion *282*
References *282*

8 Platform Development and Application for Rigid–Flexible Coupling Hoisting Robots *285*
8.1 Preamble *285*
8.2 Platform Development and Performance Verification of an RFCHR for Yard Operations *287*
8.2.1 Physical Prototype Development *287*
8.2.2 Robot Motion Performance Experiment *289*
8.3 Platform Development and Performance Verification for the 7-DOF RFCHR *293*
8.3.1 Physical Prototype Development *293*
8.3.2 Verification of Decoupling Performance *295*
8.3.3 Overall Performance *297*
8.4 Conclusion *298*
References *299*

Index *301*

Preface

This book centers on the novel design, mechanical principles, control strategies, development processes, and practical applications of rigid–flexible coupling hoisting robots, synthesizing the authors' original research and advancements in this emerging field. Unlike traditional rigid or purely flexible systems, the book addresses the unique challenges of integrating rigid structures with flexible cable-driven mechanisms, exploring their potential in precision hoisting, industrial logistics, and complex trajectory execution. The book systematically introduces typical mechanical architectures, such as the 7-DOF rigid–flexible coupling hoisting robot (RFCHR), with discretized pulley decoupling mechanisms, under-constrained cable-suspended robots with variable-angle/height masts, and yard-operation-specific platforms integrating tension sensing and motion capture systems. These designs are analyzed through the lens of practical applications, from radio telescope feed structures to container stacking in logistics. Core mechanical theories are elaborated, including inverse kinematic/dynamic modeling via Lagrange's equations, multiobjective optimization for workspace and stiffness, and uncertainty analysis incorporating fuzzy parameters, random-interval variables, and hybrid uncertain models. Novel control strategies, such as fuzzy PID optimized by adaptive whale algorithms and structural invariance-based force control, are proposed to address nonlinear dynamics and inter-joint interference. Experimental validations, including trajectory planning with polynomial/cycloid methods and real-world motion tracking, bridge theoretical frameworks with engineering practice. Aimed at graduate students and practicing engineers, this book serves as a comprehensive textbook and reference, balancing theoretical rigor with practical insights to advance the design and deployment of rigid–flexible coupling hoisting robots in industrial automation, construction, and precision manipulation.

The primary characteristics of this book are outlined as follows:

- Introduces the development, classification, operating principles, and design methodologies of rigid–flexible coupling hoisting robots, integrating rigid structural mechanics with flexible cable-driven dynamics.
- Focuses on the mechanics, control strategies, and industrial applications of rigid–flexible coupling systems, presenting mechanical architectures like 7-DOF robots with decoupling mechanisms and yard-operation platforms.
- Details kinematic and dynamic modeling and trajectory tracking control for rigid–flexible coupling hoisting robots, emphasizing stiffness optimization and uncertainty analysis.
- Explores integrated mechanical design, control algorithm optimization, and error-performance analysis for rigid–flexible coupling hoisting robots.
- Demonstrates the development of experimental platforms for validating complex trajectories and multirobot coordination, addressing challenges like cable winding variability and sensor integration in real-world logistics and hoisting tasks.

Without the help and participation of many individuals, this book would not have been possible. In particular, the authors would like to thank Prof. Weidong Zhu from the University of Maryland for his assistance and guidance in the relevant research for this book. The authors also would like to thank postgraduate students Sipan Li, Qinghan Wang, Shengsheng Yu, Guojian Shen, Zhi Wang, Xin Chen, Feng Xu, Jiahao Zhao, Tao Zhao, Bingyao Wang, Jun Gao, Yuhang Wang, Zhiyuan Wang, Yang Liu, and so on for their research on rigid–flexible coupling hoisting robots. Meanwhile, the authors would also like to thank all experts and scholars in this field for their contributions to the rigid–flexible coupling hoisting robot. The authors would like to express their sincere gratitude for the ongoing financial support provided by the National Natural Science Foundation of China (NSFC), especially, under the Grants 52205258, 52335002, and 51925502.

Bin Zi
Bin Zhou

Main Symbol Table

Symbols	Description
${}^{O}_{O_1}\boldsymbol{T}$, ${}^{O}_{A}\boldsymbol{T}$, ${}^{O}_{B}\boldsymbol{T}$	Transformation matrices
θ_i	The value of angle
$s\theta_i/c\theta_i$	$\sin\theta_i/\cos\theta_i$
l_{O_1A}	The length between points O_1 and A
l_{AB}	The length between points A and B
l_i	The ith cable length
$\boldsymbol{l}_i$	The ith cable
ABO_{12}	The triangular rigid support arm
B_4B_5	The lifting arm
$O_{11}O_{12}$	The luffing telescopic cylinder
δ_1, δ_2, and δ_3	The intersection angles
Δx	The change value of variable
$\boldsymbol{f}, \boldsymbol{e}$	Rotation matrices
${}^4r_C$, ${}^5r_D$	Homogeneous coordinates
O_i	The base point of the ith coordinate system
R_A, R_B	The radii of the virtual circle
$\boldsymbol{b}_i$	The position vector of the cable outlet points
$\boldsymbol{P}_i$	The position vector of P
$\mathbf{cp}_i$	The position vector of the connection point
${}^{O}\boldsymbol{R}_{\boldsymbol{P}}$	The rotation matrix
$\boldsymbol{u}_i$	The unit direction vector of the cable
ρ	Linear density
g	Gravitational acceleration

(*Continued*)

(Continued)

Symbols	Description
γ_i	The inclination between the tangent at the end point and the horizontal plane
d_i	The sag of the ith cable
ε	Accuracy
${}^{O_1}r_1$	The position vector of center of rotatable base
${}^{A}r_2$	The position vector of center of triangular rigid support arm
${}^{B}r_3$	The position vector of center of lifting arm
${}^{B}r_4$	The position vector of center of payload
${}^{O}\boldsymbol{v}_i$	The velocity vectors
τ_i	The torque of each joint
$\boldsymbol{M}(\boldsymbol{\Theta})$	Inertia matrix
$\boldsymbol{C}(\dot{\boldsymbol{\Theta}}, \boldsymbol{\Theta})$	Centrifugal and Coriolis force matrix
$\boldsymbol{G}(\boldsymbol{\Theta})$	Gravity matrix
K_{out}	The stiffness of the pulley system
K_{in}	The stiffness of the input driving cable
$\boldsymbol{R}$	The attitude rotation matrix
$\boldsymbol{J}$	Jacobian matrix
${}^{O}_{T}\boldsymbol{R}$	The pose matrix
$\boldsymbol{T}_i$	The vector of tension of the ith cable
$\boldsymbol{F}_P$	The net force on P
$\boldsymbol{M}_Q$	The external torque on P
$\boldsymbol{W}$	The wrench on P
η	The global dexterity index
W_G	The workspace index
δ **exf**	The infinitesimal external force
$\boldsymbol{A}^T$	The structural matrix of the system
$\boldsymbol{r}_{O_p}$	The position vector of centroid O_p in the base frame
$\boldsymbol{T}_i(\boldsymbol{K}_i(\boldsymbol{X}))$	The composite function matrix
$\boldsymbol{y}^I$	The interval vector
$\boldsymbol{A}_k$	The joint focal element
Δp_x	The calculation function of a response
$\boldsymbol{g}_0$	The threshold vector
$\boldsymbol{X}^R$	The property of the random vector
$\boldsymbol{W}_{\boldsymbol{A}_i\boldsymbol{B}_i}$	The first virtual work from lifting arms A_iB_i

Symbols	Description
$\boldsymbol{W}_{C_1C_2}$	The second virtual work from the payload C_1C_2
f_m	The property of the fuzzy variable
α_m	The cut level corresponding to f_m
f^I_{m,α_m}	The interval parameter with respect to α_m
$\boldsymbol{F}^I_u$	The uth interval vector
$\boldsymbol{\alpha}_u$	The uth cut level vector
$\boldsymbol{I}$	Identity matrix
DD	The stochastic model
K_1	The kinetic energy of the end-effector
U_1	The potential energy of the end-effector
K_2	The kinetic energy of cable mast rotating along the circular base
U_2	The potential energy of cable mast rotating along the circular base
K_3	The kinetic energy of cable mast driven by the hydraulic cylinder in the Z-direction
U_3	The potential energy of cable mast driven by the hydraulic cylinder in the Z-direction
e_j	The error of each joint angle
e_{cj}	The error rate of each joint angle
ΔK_p, ΔK_i, and ΔK_d	PID parameters
T_{fA}, T_{fD}, T_{fP_i}	The static friction torques
T_{BA}, T_{BD}, T_{BP_i}	The viscous friction torques
T_{JA}, T_{JD}, T_{JA_i}	The moments of inertia

1

Introduction

1.1 The Evolution of Rigid–Flexible Coupling Robots

Rigid–flexible coupling robot (RFCR) is a class of robotic systems that integrate the characteristics of rigid structures and flexible materials. The core lies in the organic combination of rigid and flexible components to achieve coordinated regulation of stiffness, flexibility, and adaptability [1]. Based on the complementarity of rigid and flexible characteristics, RFCRs overcome the limitations of single rigid or flexible systems and serve as an important development direction in the robotics field to address the needs of complex tasks [2]. From the perspective of actuation and control methods, RFCRs can be roughly classified into the following categories: the "rigid actuation-flexible mechanism," the "flexible actuation-rigid support," the "smart material driven," and so on. These types of RFCRs have their respective advantages in different application scenarios.

For the "rigid actuation-flexible mechanism" type of RFCRs, they are the most mature in application and have the longest development history at the present stage. Such robots combine the advantages of traditional rigid robots and soft robots, typically driven by rigid drive units (such as motors, reducers, tracks, or gears) to actuate flexible or continuously deformable mechanical structures [3]. Traditional rigid robots use rigid components like metal skeletons and gear hinges to form joints, with power provided by rigid actuators such as motors. In contrast, soft robots employ flexible materials like silicone and elastomers, relying on pneumatic or solid-state actuators to achieve continuous deformation. The structural design of rigid-drive flexible-mechanism robots lies between the two extremes. Built on the high output force and precise control capabilities of rigid drives, they introduce flexible backbones or joints to achieve high redundancy and compliance. These robots typically feature more degrees of freedom and significant deformation capabilities, with structures that can bend or stretch arbitrarily.

They inherently possess compliance and environmental adaptability while maintaining the high output torque and speed of rigid drives, thus ensuring excellent interactive safety and high load-bearing capacity [4].

The mechanical structure of this type can be traced back to the mid-twentieth century. Its technological evolution runs through the interdisciplinary integration of mechanical engineering, control theory, materials science, and other disciplines, forming a complete context from theoretical exploration to engineering practice. RFCRs have been used widely in fields such as scientific research, industry, and people's livelihood [5].

In the 1970s, the team of Professor Hirose from the Tokyo Institute of Technology in Japan proposed the "Active Cord Mechanism (ACM)" and manufactured the world's first snake-like robot prototype, ACM-III [6].

In the 1980s, Shigeo Hirose and others developed the cable-driven continuum manipulator ELASTOR with an eight-segment flexible spring backbone [7]. Additionally, Wilson at Duke University in the United States proposed research on fluid-driven flexible arms [8].

In 1995, Pratt et al. [9] proposed the concept of series elastic actuators, introducing adjustable elastic components between rigid motors and flexible springs. In 2014, the team of Chen Xiaoping from the University of Science and Technology of China proposed the "pneumatic honeycomb network" structure [10]. By combining the honeycomb-type rigid–flexible structure with airbags, the structure achieves high compliance and high load-carrying capacity. Furthermore, Meng et al. [11] designed a flexible robotic arm. Utilizing a pneumatic honeycomb mechanism and airbags, this arm is capable of performing complex tasks such as opening doors, drawers, and twisting off bottle caps in daily life environments.

In addition, significant progress has also been made in the research of cable-driven RFCRs. In the 1980s, as shown in Figure 1.1, a video tape recorder system named SkyCam with four cables is designed, which is widely used in large-space shooting operations. Meanwhile, its maximum speed can reach 44.8 km/h [12]. In addition, in the famous Five-hundred-meter Aperture Spherical radio Telescope (FAST) project in China led by academician Nan Rendong and others, six steel cables are used to drive the 30-ton feed cabin, and in combination with the Stewart platform, a positioning accuracy of the millimeter level (3 mm) is achieved, as shown in Figure 1.2. This is a typical RFCR of the "rigid actuation-flexible mechanism" type [13]. As shown in Figure 1.3, Zhao et al. [14] proposed an under-constrained RFCR with variable angle and height cable mast. Hamida et al. [15] designed a cable-driven RFCR, which can be intended for upper-limb rehabilitation exercises. Meanwhile, four different topology solutions were also considered. Chen et al. [16] proposed a cable-driven parallel waist rehabilitation robot with two-level control algorithm to assist the patients with waist injuries to do some rehabilitation training. Tho et al. [17] designed an RFCR structure for the

Figure 1.1 A video tape recorder system for large space.

Figure 1.2 Large spherical radio telescope. *Source*: Chinese Academy of Sciences, https://english.nao.cas.cn/research/researchdivisions/radioastronomy/202103/t20210323_266393.html, last accessed Nov. 21, 2025.

3D printing in a large workspace. Overall, in future applications, "rigid actuation-flexible mechanism" robots will increasingly serve in fields such as intelligent manufacturing, flexible production lines, and human–robot collaboration, as well as explore new scenarios like extraterrestrial planetary surface exploration and medical diagnostics [18].

The relevant research for the "flexible actuation-rigid support" type of RFCRs can be traced back to bionic explorations and breakthroughs in material technology in the late twentieth century [19]. The core idea is to achieve highly adaptive motion through flexible driving components while maintaining system stability using rigid supporting structures [2]. Such robots possess both soft and hard characteristics, ensuring load support while maintaining good compliance. For example, in the field of exoskeleton robots, actuation can be divided into electric,

Figure 1.3 Under-constrained RFCR with variable angle and height cable mast. *Source*: [14]/Springer Nature/CC BY 4.0.

pneumatic, hydraulic, or hybrid types, while structures can be categorized into rigid exoskeletons and flexible exoskeletons. Rigid exoskeletons use rigid metal structures for load-bearing, often being bulky and complex; flexible exoskeletons replace some rigid skeleton components with soft, lightweight materials, making the structure more portable and comfortable [20]. In continuous bending manipulators, similar classifications exist: cable-driven continuous arms can be divided into three categories: "pure flexible segmental drive," "rigid–flexible hybrid segmental drive," and "discrete rigid full drive." In short, "flexible drive-rigid support" robots refer to robotic systems that use flexible actuators to transmit power and rigid components to form load-bearing skeletons, belonging to the design paradigm of soft-hard coupled robotics [21].

As early as the 1950s, McKibben and others developed braided pneumatic artificial muscles, laying the foundation for flexible actuators [22]. In the 1960s, the first continuous bending robot prototypes emerged: Anderson and Horn's "Tensor Arm," designed with a cable-driven diving arm structure, is recognized as the first conceptual continuous robot prototype [23]. Over the following decades, continuous backbone robots and flexible drive technologies gradually diversified; in 1989, the US National Institute of Standards and Technology (NIST) introduced the first large-scale cable-driven parallel robot, "RoboCrane," for large-range handling tasks [24, 25].

During the mid-to-late 1990s, researchers made advancements in kinematic modeling and design innovation: Chirikjian [26] proposed the kinematic theory of continuous backbones, while Hirose and colleagues developed bionic snake robot prototypes [27]. In 1999, Robinson and Davies formally defined the concept of "continuous robots" and completed their classification, marking the conceptual development of the flexible robotics field [28].

Entering the twenty-first century, "soft robotics" research became a hotspot—the term "soft robotics" was internationally adopted for the first time in 2008 to describe robots with compliant joints or flexible components [29]. In recent years, with breakthroughs in bionics and structural innovation, multiple key milestones have emerged: for example, China launched the major scientific research project "Tri-Co Robot" (flexible collaborative robot) in 2016 [30]; in 2023, Southern University of Science and Technology proposed an anisotropic stiffness structure (OASS) based on single-sided origami mechanisms, achieving an innovative structure with both rigidity and flexibility in different directions [31]. Harvard University's Wyss Institute (led by Conor Walsh) has made outstanding contributions in the field of flexible-driven exoskeletons (soft exoskeletons), with its developed soft cable-driven exoskeletons significantly improving gait [32–34]. Universities such as Carnegie Mellon University and the Georgia Institute of Technology have long engaged in deep research on soft robotics theory and control [35, 36]. In Europe, the Fraunhofer Institute for Production Technology (Fraunhofer IPA) team has led research on cable-driven parallel robots (such as the IPAnema series), proposing design concepts that significantly enhance load capacity, workspace, and dynamic performance in industrial scenarios [37, 38]. The team led by Liu Hongbin at the Institute of Automation, Chinese Academy of Sciences, has achieved groundbreaking results in minimally invasive flexible surgical robot technology (such as the development of minimally invasive flexible neurosurgical robots) [39]; the team led by Chen Xuefeng at Xi'an Jiaotong University has also made new progress in the design and modeling of flexible continuum robots, as shown in Figure 1.4 [40]. Additionally, the industrialization process of soft robotics is also accelerating further.

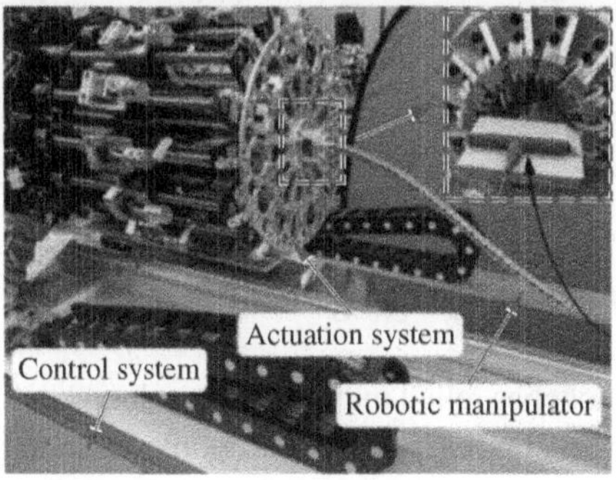

Figure 1.4 Continuum robotic system prototype. *Source:* [40]/IEEE/CC BY 4.0.

In terms of applications, continuous bending robots leverage their advantages of being long, narrow, and bendable, making them widely used in minimally invasive surgery and intraluminal operations [41]. For example, superconducting spiral continuous core tube robots are frequently employed in endoscopic and cardiac surgeries, as they can smoothly pass through the natural orifices of body cavities to achieve precise operations [42]. Additionally, variable-stiffness soft surgical robots enable surgeons to switch more flexibly between rigid and soft states during procedures [43]. Traditional rigid exoskeletons mostly use electric drive, but they suffer from heavy weight and poor comfort. Emerging flexible exoskeletons adopt lightweight flexible materials and cable-driven systems to provide close-fitting assistance for hand or lower limb movements. For instance, Harvard's soft exoskeleton "Soft Exosuit" uses flexible cables to generate auxiliary torque at the hip and ankle joints, which can improve foot drop and reduce walking burden [34]. Meanwhile, due to their high compliance, these robots can be applied in scenarios such as pipeline maintenance, marine exploration, or post-disaster search and rescue [44]. Japan's Land Equipment Development Agency has developed cable-driven snake robots for search and rescue in earthquake debris [27]; NASA and CMU teams in the United States have also researched flexible robots capable of moving within pipelines [45]. Evidently, robots that combine flexible drive and rigid support are gradually expanding from single manufacturing scenarios to a broader range of intelligent equipment and service fields. In summary, the future of "flexible actuation-rigid support" robots will evolve toward higher compliance, stronger load capacity, and greater intelligence, enabling them to accomplish complex tasks that traditional rigid robots struggle to handle [46–48].

For the "smart material-driven" type of RFCRs, this type of robot refers to robotic systems that utilize smart materials (such as shape memory alloys [SMAs], dielectric elastomers, and hydrogels) as driving units while incorporating rigid structures as skeletons or load-bearing frameworks [49]. This structure combines flexible deformation capabilities with certain mechanical stability, making it suitable for scenarios such as bionic execution, precise manipulation, wearable devices, and minimally invasive medical procedures.

With the development of material systems, typical manufacturing processes have also evolved continuously: early methods mostly employed compression molding and fiber winding (such as helical wound actuators), while recent advancements have introduced new techniques like layer-by-layer 3D/4D printing, soft lithography, and micro–nano-processing, enabling high integration of structures and materials [50, 51].

In terms of control strategies, early approaches relied mainly on open-loop heating or simple voltage driving [52]. As the technology has matured, model-based closed-loop control, fuzzy control, and machine learning algorithms have emerged to compensate for the nonlinear hysteresis of intelligent materials and achieve precise positioning [53, 54].

As early as the 1980s, smart materials were initially applied to driving mechanisms, with SMAs being the main type, and rigid structures typically consisting of metal or plastic skeletons. In 1988, a prototype of a bionic gripper driven by an SMA spring emerged. In 1993, MIT developed a bionic finger structure based on SMA [55]. In addition, Nguyen et al. [56] show a 30-mg crawling robot driven by a high-frequency flexible SMA microactuator.

In the early twenty-first century, researchers began to systematically construct rigid–flexible coupled actuation structures, attempting to integrate precision rigid structures with flexible actuation materials in microrobots and medical devices. For example, Harvard University has developed microdevices with SMA and rigid hinges [57]. Teams such as those from Stanford University have integrated hydrogels into microfluidic rigid platforms to achieve soft operation ends [58].

Since the 2010s, various smart materials (dielectric elastomers, liquid crystal elastomers, carbon nanotubes, etc.) have been integrated into complex rigid frameworks, simultaneously improving control precision and load capacity; some have entered the practical application stage in medical, wearable, and bionic robotics. For example, in 2016, Harvard introduced the Octobot: some regions feature rigid casings, with fuel circuits embedded in flexible bodies, as shown in Figure 1.5 [59]. In 2019, MIT proposed a robotic arm with rigid connecting rods and flexible torsional muscles, enabling complex path grasping [60].

Since the 2020s, task-scenario-oriented development has flourished, including medical catheters, bionic soft arms, pipeline detection robots, and more [61]. Through the integration of structural design, material customization, and intelligent control, some rigid–flexible coupling structures have become commercially

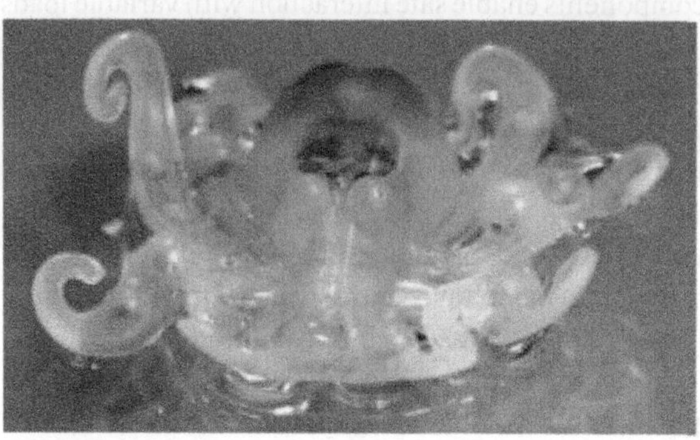

Figure 1.5 Octobot. *Source*: [59]/with permission of Springer Nature.

viable or entered mass testing phases. In 2021, the team led by Zhao Huichan at Tsinghua University proposed a rigid–flexible coupled soft pipeline robot, which combines rigid segments (control circuits and sensing modules) with dielectric elastomer actuators [62]. In 2023, the team led by Su Songjin utilized thermoresponsive hydrogels as driving units, with rigid frameworks restricting their deformation directions to achieve high-speed underwater swimming [63].

To sum up, "smart material-driven" type of RFCRs is rapidly evolving toward integration, miniaturization, and intelligence. In the future, such robots will further integrate multiple smart materials with complex rigid structures to achieve higher actuation performance and structural stability. Meanwhile, relying on advanced manufacturing technologies and intelligent control algorithms, robots will possess stronger environmental adaptability, autonomous sensing, and precise manipulation capabilities, with broad applications in cutting-edge fields such as minimally invasive medicine, wearable devices, soft bionics, and extreme environment exploration. This will drive the evolution of robotic systems toward high performance, multifunctionality, and autonomous intelligence [64].

1.2 The History and Development of Rigid–Flexible Coupling Hoisting Robots

Rigid–flexible coupling hoisting robot (RFCHR) is a special type of RFCR [65]. It is specifically developed to perform hoisting or lifting operations that involve complex dynamic interactions between rigid and flexible components. Unlike conventional rigid-body hoisting systems, the RFCHR incorporates flexible elements—such as compliant links, elastic joints, or cable-driven segments—which allow for enhanced adaptability, vibration suppression, and energy efficiency. The rigid components provide structural support and positional accuracy, while the flexible components enable safe interaction with variable loads and dynamic environments.

RFCHRs are particularly suited for tasks in construction, warehousing, and automated manufacturing where traditional rigid mechanisms may struggle with payload variations, dynamic disturbances, or constrained spaces. By leveraging the advantages of both rigidity and flexibility, RFCHRs offer improved performance in terms of precision, safety, and operational versatility. Research on RFCHRs typically focuses on dynamic modeling, control strategies, and structural optimization to ensure stable hoisting under changing conditions and external disturbances [66].

RFCHR is a large manipulator that combines rigid structures with flexible (cable/rope) actuation, which can be traced back to the 1980s. Early concepts replaced rigid links with cables to achieve very large workspaces and high payload-to-weight ratios. For example, the SkyCam sports camera (late 1980s), as shown in Figure 1.1 [12]. In parallel, as shown in Figure 1.6, NIST in the United States began the RoboCrane project in 1989: a Stewart-platform-like crane using six cables to control a suspended end-effector. The RoboCrane enabled precise 6-DOF load placement under teleoperation or CNC, supporting tasks like welding, grinding, sawing, and assembly on construction sites [67, 68]. For instance, a scaled RoboCrane was used to assemble modular bridge sections with full 6-DOF control. Meanwhile, Xidian University pioneered cable cranes for astronomy. In 1999, they proposed the feed-cabin cable-driven support for the 500 m FAST radio telescope. Today FAST's movable feed is guided by multiple winches and cables, a celebrated early RFCHR application, as shown in Figure 1.2 [13].

Since entering the twenty-first century, research on RFCHR has flourished globally. Research institutions such as the NIST, Fraunhofer Institute for Applied Physic, French National Centre for Scientific Research, Xidian University, Hefei University of Technology, and China University of Mining and Technology have published numerous research achievements on the design, dynamic modeling, and control of RFCHRs [25, 69].

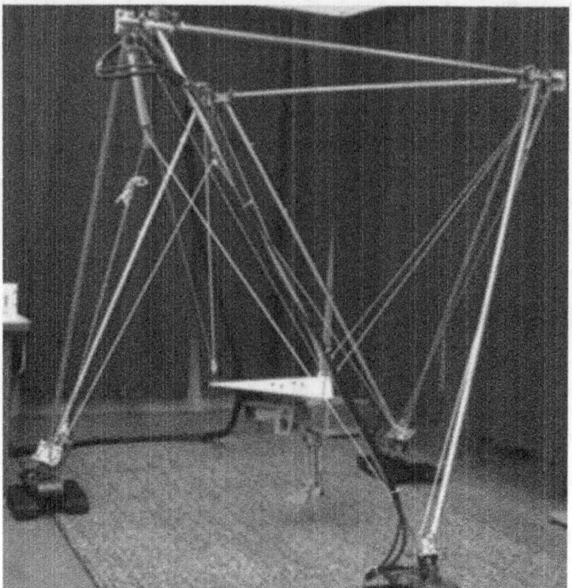

Figure 1.6 NIST RoboCrane model. *Source*: [67]/National Institute of Standards and Technology (NIST)/Public domain.

For instance, Japan's FALCON project and WARP system conduct high-speed motion research using cable-driven parallel robots and propose the system design principle that a cable-driven parallel robot with n degrees of freedom requires at n+1 cables [71]. In 2004, France's ONERA has developed the SACSO wind tunnel model, which uses a parallel robot with nine cables for model positioning and attitude control in wind tunnel tests [72]. In 2009, researchers at the University of Duisburg-Essen in Germany have established the SEGESTA experimental platform to study kinematic modeling, workspace calculation, tension distribution, and trajectory planning of cable robots [73]. In 2012, the Fraunhofer IPA Institute in Germany has developed the IPAnema series of cable-driven parallel robots for industrial applications such as construction and maintenance, as shown in Figure 1.7 [70].

In recent years, the high-strength steel cables or synthetic ropes as actuators are used in the RFCHR. Compared to rigid links, cables offer very light weight and very large reachable volumes. Some designs even incorporate spring elements or compliant joints for additional flexibility. In 2020, Tang and Chen demonstrate hybrid wave-compensation rigs for offshore hoisting: a 4-DOF cable-driven mechanism that flexibly adjusts to sea motion [74]. In addition, significant advancements in intelligent control and multirobot collaboration have further propelled the development of RFCHR. These breakthroughs have enhanced the robots' adaptability to dynamic environments, improved coordination accuracy in complex tasks, and expanded their application scenarios in fields such as intelligent manufacturing, large-scale structure construction, and emergency rescue [75]. For example, Shen et al. [76] present a model-based optimal boundary control design

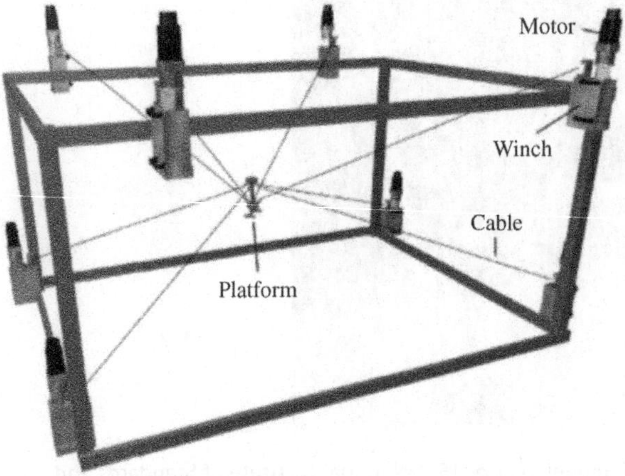

Figure 1.7 IPAnema series of cable-driven parallel robots. *Source*: [70]/National Institute of Standards and Technology (NIST)/Public domain.

for an aerial robotic system composed of a quadrotor carrying a flexible cable. Javier [77] proposed a new analytical model for the drum–cable interaction in mechanical systems actuated through cables, which can capture the extensibility of the cable and the drum–cable contact interaction. Taheri et al. [78] present a reinforcement learning-based policy optimization workflow for training and evaluating controllers of RFCHR, which can inspect the stability and performance in the whole operational envelope of the RFCHR for safe deployment on real machines. Zhu et al. [79] designed an intelligent RFCHR with path planning function, which has the characteristics of foldable assembly, wireless intelligent remote control, intelligent anticollision, dynamic path planning, laser setting out, and payload real-time monitoring. Teerapat et al. [80] proposed a motion compensation algorithm for multistep speed control to achieve precise displacement of RFCHR by dynamically optimizing the duration at each planned speed.

In addition, this field is also supported by dedicated conferences and literature. For example, in 2018, the "Cable-Driven Parallel Robots" symposium is held. Bin Zi's books such as *Design, Analysis and Control of Cable-Suspended Parallel Robots and Its Applications* have collected many research advancements [75]. Meanwhile, numerous patents reflect the maturing technology, including patents on multicable geometries, hybrid drives, and interference-avoiding mechanisms.

In summary, from early cable-camera systems and the RoboCrane in the 1990s, RFCHRs have grown into a broad field. Today, they embody a mix of crane-like hardware and advanced robotics: flexible structures for long reach, multicable parallel architectures, and complex control of coupled rigid–flexible dynamics. Applications span construction, logistics, and large-structure assembly. As of 2025, the trend continues toward integration of smart control (AI, vision feedback), lighter materials, and cooperative multirobot deployments [81].

1.3 The Applications of RFCHRs in Various Fields

1.3.1 Construction

Iturralde et al. [82] designed a new RFCHR, which can be used for the installation of curtain wall modules, as shown in Figure 1.8. Compared with manual installation, the adoption of RFCHR has significantly improved the efficiency of curtain wall installation, greatly shortened the construction period, reduced installation costs, and decreased the incidence of safety accidents. This is a typical example of RFCHR's application in the construction industry. Based on workspace and kinematic sensitivity, McDonald et al. [83] proposed a 3-DOF RFCHR for automated construction.

In addition, as an indispensable tool in the construction industry, tower cranes also represent a typical application of RFCHR, as shown in Figure 1.9 [84].

Figure 1.8 A RFCHR for the installation of curtain wall modules. *Source*: [82]/with permission of Elsevier.

Figure 1.9 Schematic diagram of tower crane. *Source*: [84]/with permission of Elsevier.

At present, the industrial application of tower cranes has been quite mature, and the current research focus lies in effectively solving the problem of swing suppression for tower crane loads in engineering applications. For instances, Sun et al. [85] designed an adaptive control scheme for underactuated tower cranes to achieve simultaneous slew/translation positioning and swing suppression. Besides, Wang et al. [86] proposed a new adaptive controller based on the neural network for tower cranes with obstacle avoidance, and the controller can simultaneously achieve obstacle avoidance for tower crane lifting payload and swing suppression of the payload. Tian et al. [87] considered the change in cable length of RFCHR and proposed a real-time modified trajectory planning method for tower cranes with heavy-load lifting and lowering movements. In addition, exciting results have also been achieved in the research of the double-pendulum system.

In recent years, RFCHR has achieved remarkable breakthroughs in the construction field through technology integration and scenario expansion, especially in intelligent control and efficiency improvement.

1.3.2 Ocean

As shown in Figure 1.10, the RFCHR structure has garnered extensive and versatile applications in the domain of marine shipboard cranes. This innovative structural design, characterized by its unique combination of rotational flexibility and cable-driven hoisting mechanisms, has seamlessly integrated into the operational frameworks of various marine vessels, ranging from cargo ships to specialized-offshore engineering platforms. In this field, its mature applications are vividly demonstrated through its remarkable capabilities in efficiently handling heavy lifting tasks with high precision and reliability in the harsh marine environment. For instance, it enables shipboard cranes to smoothly execute complex operations such as the loading and unloading of oversized cargo containers, the precise positioning of marine engineering equipment during offshore construction, and the

Figure 1.10 Offshore ship-mounted tower crane. *Source*: CSSC South China Marine Machinery Co. http://www.scmmc.com.cn/men/productshow.php?cid=285&id=395, last accessed July 25, 2025.

stable hoisting of supplies in challenging sea conditions with strong winds and waves. These practical implementations not only highlight the advanced technical advantages of the RFCHR structure but also solidify its indispensable position in modern marine logistics and offshore engineering, driving the continuous improvement of marine crane operation efficiency and safety standards.

At present, both academia and industry have achieved good results in the design and application of RFCHR controllers. For example, Qian et al. [88] proposed an observer-based nonlinear feedback controller for offshore shipboard RFCHRs, which is used to achieve precise boom/trolley positioning of RFCHRs and fast load swing suppression against complex unknown external disturbances. Li et al. [89] proposed a self-adapting trajectory tracking controller, which can compensate the dead zone constraint, unknown gravity forces, and external disturbances. Wang et al. [90] designed a new triple-tagline system, which can suppress the payload swing in arbitrary directions.

In summary, RFCHR controllers are transforming marine and shipbuilding operations through improved precision, efficiency, and adaptability. Ongoing research and industrial collaboration continue to push the boundaries of these technologies, addressing both technical challenges and sustainability goals.

1.3.3 Storage

Warehousing is currently one of the most important and widely applied fields for rigid–flexible coupling lifting robots, which benefits from their unique advantages and compatibility with warehousing operations. In the warehousing environment, tasks such as cargo handling and stacking are arduous and require high efficiency and precision. Rigid–flexible coupling lifting robots combine the high load-bearing capacity of rigid structures with the flexibility of flexible structures, making them capable of easily meeting the handling needs of various types of cargo. Their rigid components ensure stability when lifting heavy goods, preventing shaking or imbalance caused by excessive weight; the flexible components allow the robot to freely navigate through complex warehouse spaces, accurately placing goods in designated storage locations and effectively improving the utilization rate of warehouse space.

Zargari et al. [91] designed a single RFCHR system that integrates actuators, cameras, and computer vision, which is used for fast and accurate scanning of storage racks in warehouses. Zhou et al. [65] designed an RFCHR for warehouse cargo hoisting and transportation and introduced a performance-optimized design approach for RFCHRs, taking into account cable deformation and stochastic loads.

An et al. [92] proposed a proposed a novel RFCHR with movable anchor points and enabled the RFCHR to achieve obstacle avoidance without changing the predetermined trajectory under different conditions through a self-designed collision detection method and obstacle avoidance algorithm. Ha et al. [93] designed a new RFCHR tailored for logistics warehouses with high-level but short rack intervals, characterized by large working spaces, high operational repetition rates, and rapid task processing capabilities. This robot has been successfully tested and deployed in logistics warehouses, as shown in Figure 1.11.

In addition to the construction, ocean, and storage, RFCHRs are also widely applied in industrial manufacturing. On automotive production lines, they can accurately hoist various components to achieve efficient assembly, significantly enhancing production efficiency. In the medical field, they can assist in transporting medical equipment and even aid in the precise positioning of surgical instruments in some special scenarios. In logistics and transportation, they are used for rapid loading/unloading and transshipment of goods to optimize logistics processes. In power inspection, they help workers hoist inspection equipment to reach inaccessible high-altitude power facilities.

With continuous technological advancements, RFCHRs are expected to emerge in more emerging fields in the future. Their performance will continue to be optimized, and their intelligence level will gradually improve, enabling more precise, efficient, and safe operations. Through deep integration with other cutting-edge technologies, they will bring more innovative changes to various industries, drive

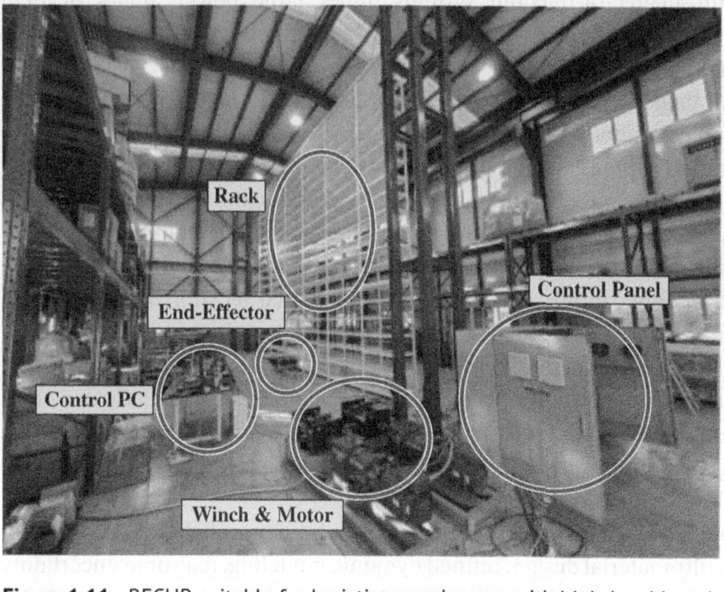

Figure 1.11 RFCHR suitable for logistics warehouses with high-level but short rack intervals. *Source*: [93]/with permission of Springer Nature.

industrial upgrading, and play an increasingly important role in global economic development.

Currently, research on RFCHRs still faces notable challenges and inadequacies across multiple domains, including mechanism design, dynamic analysis, uncertainty analysis, motion decoupling, control strategies, and industrial applications. In mechanism design, material and structural optimization is insufficiently addressed, with systematic methods for selecting flexible joint materials and controlling stiffness often lacking, leading to significant precision losses due to structural deformation and durability issues. Modular and reconfigurable designs are also underdeveloped, limiting adaptability to diverse scenarios and increasing maintenance costs due to poor replaceability of flexible components.

In dynamic analysis, simplified models such as beam elements or lumped mass approaches are frequently employed, overlooking high-order vibration modes of flexible bodies and causing inaccuracies in dynamic response predictions. In addition, regarding uncertainty analysis, existing models are often focused narrowly on parametric perturbations, while external disturbances and unmodeled dynamics are inadequately quantified. Real-time estimation methods for dynamic uncertainties, such as online deformation and disturbance compensation, are largely absent, with reliance instead placed on offline parameter identification.

In motion decoupling, traditional methods are rendered ineffective by strongly coupled multi-degree-of-freedom systems under large power angles or high impedance ratios, and dynamic decoupling strategies based on static models are lacking in adaptability to changing conditions.

Control strategies are challenged by the contradiction between real-time performance and computational burden, as high-frequency control requirements for flexible vibration suppression and trajectory tracking are exceeded by the capabilities of embedded systems. Traditional control methods exhibited poor robustness against uncertainties, while emerging techniques like deep reinforcement learning suffered from limited training data and generalization capabilities.

In industrial applications, reliability is compromised by short service lives of flexible components and frequent maintenance needs, which are compounded by challenges in fault diagnosis and the absence of online health monitoring. Environmental adaptability is affected by issues such as thermal deformation in high-temperature environments and sensor failures in harsh conditions, while engineering validation is insufficient, with limited long-term reliability data provided for lab-developed systems.

Overall, addressing these gaps requires interdisciplinary collaboration and innovations in multi-material design, refined dynamic modeling, real-time uncertainty estimation, adaptive control algorithms, and robust industrial validation to facilitate the transition from theoretical research to large-scale practical application.

1.4 Scope and Organization of This Book

This book centers on the novel design, mechanical principles, control strategies, development processes, and practical applications of RFCHRs, summarizing and extending the authors' work in this field. It can provide inspiration and directions for addressing the existing issues and shortcomings of RFCHRs. It introduces typical mechanical architectures and their practical applications, covering feed cable-suspended structures used in large radio telescopes, as well as under-constrained cable-suspended parallel robots with adjustable parameters, and cable-driven parallel robots for yard operations. The fundamental mechanics are systematically addressed, covering inverse kinematic and dynamic modeling of cable-suspended systems using Lagrange's equations, multi-objective optimization design for workspace and stiffness, and uncertainty analysis incorporating fuzzy parameters, random-interval variables, and hybrid uncertain models. These theoretical frameworks are validated through numerical simulations and experiments, such as trajectory planning using polynomial and cycloid methods, and fuzzy PID control strategies optimized by adaptive whale algorithms to enhance tracking accuracy. For performance improvement, novel designs like mechanical decoupling mechanisms (discretized pulleys and 3D-printed components) and hybrid-driven systems are proposed to mitigate inter-joint interference and nonlinear dynamics. Experimental platforms are developed, integrating high-precision motion capture, tension sensing, and servo-driven cable systems to verify complex trajectories (linear, spiral, and stacking) and validate force control strategies based on structural invariance principles. The book bridges theory and practice by addressing key challenges such as cable winding variability, measurement limitations, and prototype tolerances, providing case studies with numerical analyses and experimental validations. It encompasses advanced concepts like motion decoupling, adaptive control, and sensor integration, offering a comprehensive guide for both theoretical understanding and engineering implementation. The purpose of this book is to function as a textbook for postgraduate courses and a reference material for engineers in the field of RFCHRs. To achieve this goal, the book is structured as follows:

In Chapter 2, the mechanism designs and mechanics modeling of the three new RFCHRs are presented. The mechanism designs of three robots are carried out, and structural models are built to verify their effectiveness. The kinematic modeling is conducted, including the forward and inverse kinematic analysis, and the transformation matrices and vector equations of cable length are derived based on the Denavit–Hartenberg theory and geometric method. Moreover, the dynamic modeling is established based on the Lagrange method and Newton–Euler method, and the dynamic equations considering the inertia matrix, the matrices of centrifugal and Coriolis forces, and gravity matrix are derived.

In Chapter 3, motion decoupling and reconfigurable design of RFCHRs are presented. The coupling characteristics during joint rotation are analyzed, followed by the introduction of motion decoupling methods based on translational and rotational compensation, and the design of a 7-DOF cable-driven serial spray-painting robot featuring these decoupling mechanisms is presented, including a decoupling mechanism for yaw motion and rotational compensation mechanisms for roll motion, along with a tension-amplification mechanism using movable pulleys. Modular and reconfigurable mechanism design is introduced, using modular branches as basic components to reconfigure topological structures by adjusting the number of branches and connection points, enabling configurations from 2-DOF to 6-DOF. A dual-machine collaborative RFCHR mechanical prototype is developed, with kinematic and dynamic modeling conducted, including pose transformation matrices, Jacobian matrix derivation, and Lagrangian dynamic equations considering payload and component masses. The design enhances flexibility, adaptability, and task versatility for diverse applications.

In Chapter 4, the optimization design of RFCHRs is systematically presented, focusing on two distinct mechanisms and their performance enhancements. A multi-objective optimization is introduced for workspace and dexterity, establishing kinematic, static models, and performance indices. An adaptive adjustment inertia weight particle swarm optimization algorithm is proposed, outperforming traditional PSO and genetic algorithms. Then, based on the RFCHR in Chapter 2, the dynamic uncertainty and reliability under random/interval parameters are studied. A hybrid random-interval perturbation method is developed to quantify kinematic errors, validated through Monte Carlo simulations and experiments. The physical prototype demonstrates reduced response errors and improved robustness. Finally, the effectiveness of the optimized design is verified.

In Chapter 5, kinematic analysis of RFCHRs with uncertainty is presented. Kinematic uncertainty analysis with random parameters for a dual automobile cranes system is conducted using the modified hybrid random method, combining perturbation theory and Monte Carlo simulations to predict luffing angular response fields under narrowly distributed random payload parameters, demonstrating its efficiency and accuracy. For interval variables, the hybrid compound function/subinterval perturbation method is proposed to handle large bounded uncertainties by dividing interval variables into subintervals, improving accuracy compared to vertex methods and Monte Carlo methods with reduced relative errors. An evidence theory-based uncertainty analysis method is introduced for cable-driven parallel robots, quantifying epistemic uncertainties in cable lengths to calculate credibility and plausibility functions efficiently. Additionally, a hybrid perturbation compound function-based moment method is developed to address hybrid random and interval parameters, modeling kinematic error propagation in collaborative systems. Numerical examples and comparisons validate the robustness of these methods in enhancing reliability and design accuracy for RFCHRs.

In Chapter 6, dynamic analysis of RFCHRs with uncertainty is presented. Static uncertainty analysis with fuzzy parameters is conducted using the compound function/fuzzy perturbation method (CFFPM) and modified CFFPM, transforming fuzzy parameters into interval variables via level-cut techniques to improve calculation accuracy and efficiency compared to the Monte Carlo method. For dynamic uncertainty analysis with hybrid random and interval parameters, a modified hybrid uncertain analysis method (MHUAM) is proposed, combining random interval perturbation and moment methods to model uncertainties in loads and structural parameters. Numerical examples validate that MHUAM efficiently predicts dynamic response fields with narrow uncertainties, outperforming traditional methods in computational cost. The chapter provides robust frameworks for uncertainty quantification, enhancing reliability in engineering applications.

In Chapter 7, trajectory planning and tracking control of RFCHRs are presented. Point-to-point trajectory planning for an under-constrained cable-suspended parallel robot is conducted using three algebraic methods, with the fifth-degree polynomial method proving optimal for smooth motion. A fuzzy PID control strategy based on an adaptive whale optimization algorithm is proposed, improving trajectory tracking accuracy for cable-driven parallel robots compared to traditional methods. A cable force control experimental platform is built, adopting a PID controller with structural invariance principle to achieve precise force control, validated by real-time monitoring and experiments. The chapter integrates planning, control, and experimentation to enhance robot performance in engineering applications.

In Chapter 8, the development and application of RFCHR platforms are presented. A 7-DOF RFCHR is designed with mechanical decoupling mechanisms like discretized pulleys and 3D-printed components, complemented by a fuzzy PID control system to address nonlinear dynamics. Experimental validation focuses on motion decoupling performance and trajectory tracking, demonstrating effective reduction of interjoint interference. A yard operations-oriented robot is also developed, integrating high-precision motion capture and tension sensors to execute linear, spiral, and task-space stacking trajectories. The chapter highlights the integration of mechanical design and control strategies, providing practical platforms for industrial hoisting and logistics tasks, and verifying their feasibility in real-world applications (Figure 1.12).

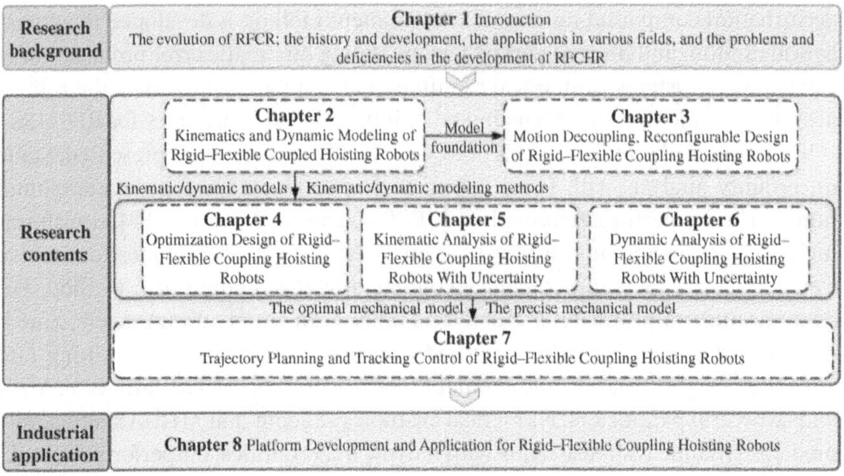

Figure 1.12 The main contents and chapter arrangement for this book.

References

1 Nandihal P V, Mohan A, Saha S K. Dynamics of Rigid-Flexible Robots and Multibody Systems. Singapore: Springer, 2022.
2 Li B, Li X, Gao H, et al. Advances in flexible robotic manipulator systems—Part II: planning, control, applications, and perspectives. IEEE/ASME Transactions on Mechatronics, 2024, 29(3): 1680–1689.
3 Yang Y, Wu Y, Li C, et al. Flexible actuators for soft robotics. Advanced Intelligent Systems, 2020, 2(1): 1900077.
4 Chen H, Zhu J Q, Cao Y, et al. Soft-rigid coupling grippers: collaboration strategies and integrated fabrication methods. Science China Technological Sciences, 2023, 66(11): 3051–3069.
5 Jiang X, Zhang J, Zhang Y, et al. Rigid and flexible structures coupled system study and enhanced vision-aided control for a versatile cable winding robot. IEEE/ASME Transactions on Mechatronics, 2024, 29(6): 4618–4629.
6 Hirose S, Umetani Y. Kinematic control of active cord mechanism with tactile sensors. Transactions of the Society of Instrument and Control Engineers, 1976, 12(5): 543–547.
7 Seeja G, Doss A S A, Hency V B. A survey on snake robot locomotion. IEEE Access, 2022, 10: 112100–112116.
8 Di Lallo A, Yu S, Slightam J E, et al. Untethered fluidic engine for high-force soft wearable robots. Advanced Intelligent Systems, 2024, 6(11): 2400171.

9 Pratt G A, Williamson M M. Series elastic actuators. Proceedings of the 1995 IEEE/RSJ International Conference on Intelligent Robots and Systems. Human Robot Interaction and Cooperative Robots. IEEE, 1995, 1: 399–406.

10 Sun H, Chen X P. Towards honeycomb PneuNets robots. Robot Intelligence Technology and Applications 2: Results from the 2nd International Conference on Robot Intelligence Technology and Applications. Springer International Publishing, 2014, 274: 331–340.

11 Hongjun M, Shupeng Z, Wei Z, et al. Design and control of a new pneumatic quadruped soft robot based on honeycomb structure. IEEE Access, 2024, 12: 98882–98899.

12 Tanaka M, Seguchi Y, Shimada S. Kineto-statics of SkyCam-type wire transport system. Proceedings of USA-Japan Symposium on Flexible Automation, Crossing Bridges: Advances in Flexible Automation and Robotics, 1988: 689–694.

13 Nan R. Five hundred meter aperture spherical radio telescope (FAST). Science in China Series G, 2006, 49: 129–148.

14 Zhao T, Zi B, Qian S, et al. Algebraic method-based point-to-point trajectory planning of an under-constrained cable-suspended parallel robot with variable angle and height cable mast. Chinese Journal of Mechanical Engineering, 2020, 33: 1–18.

15 Hamida I B, Laribi M A, Mlika A, et al. Multi-objective optimal design of a cable driven parallel robot for rehabilitation tasks. Mechanism and Machine Theory, 2021, 156: 104141.

16 Chen Q, Zi B, Sun Z, et al. Design and development of a new cable-driven parallel robot for waist rehabilitation. IEEE/ASME Transactions on Mechatronics, 2019, 24(4): 1497–1507.

17 Tho T P, Thinh N T. Using a cable-driven parallel robot with applications in 3D concrete printing. Applied Sciences, 2021, 11(2): 563.

18 Zhang Y, Li P, Quan J, et al. Progress, challenges, and prospects of soft robotics for space applications. Advanced Intelligent Systems, 2023, 5(3): 2200071.

19 Ma C, Song W, Zhao X, et al. Bioinspired soft robots with integrated biological motion mechanisms and rigid–flexible coupling systems. Small Methods, 2025, 9(7): 2402264.

20 Lee D, Song B, Park S Y, et al. Development and control of an electro-hydraulic actuator system for an exoskeleton robot. Applied Sciences, 2019, 9(20): 4295.

21 Calisti M, Picardi G, Laschi C. Fundamentals of soft robot locomotion. Journal of the Royal Society Interface, 2017, 14(130): 20170101.

22 Salahuddin B B. Hydrogel Based Braided Artificial Muscles. University of Wollongong, 2020.

23 Anderson V C, Horn R C. Tensor arm manipulator. United State Patent, 1970 (3,497,083).

24 James A, Roger B, Nicholas D. The NIST RoboCrane. Journal of Robotic Systems, 1993, 10: 709–724.

25 Qian S, Zi B, Shang W W, et al. A review on cable-driven parallel robots. Chinese Journal of Mechanical Engineering, 2018, 31(1): 1–11.

26 Chirikjian G S. Hyper-redundant manipulator dynamics: a continuum approximation. Advanced Robotics, 1994, 9(3): 217–243.

27 Hirose S, Fukushima E F. Snakes and strings: new robotic components for rescue operations. The International Journal of Robotics Research, 2004, 23 (4–5): 341–349.

28 Robinson G, Davies J B C. Continuum robots-a state of the art. Proceedings of the 1999 IEEE International Conference on Robotics and Automation (Cat. No. 99CH36288C). IEEE, 1999, 4: 2849–2854.

29 Mo X, Ouyang L, Xiong Z, et al. Advances in digital light processing of hydrogels. Biomedical Materials, 2022, 17(4): 042002.

30 Ding H, Yang X, Zheng N, et al. Tri-Co Robot: a Chinese robotic research initiative for enhanced robot interaction capabilities. National Science Review, 2018, 5(6): 799–801.

31 Zhu R, Fan D, Wu W, et al. Soft robots for cluttered environments based on origami anisotropic stiffness structure (OASS) inspired by desert iguana. Advanced Intelligent Systems, 2023, 5(6): 2200301.

32 Polygerinos P, Wang Z, Galloway K C, et al. Soft robotic glove for combined assistance and at-home rehabilitation. Robotics and Autonomous Systems, 2015, 73: 135–143.

33 Mosadegh B, Polygerinos P, Keplinger C, et al. Pneumatic networks for soft robotics that actuate rapidly. Advanced Functional Materials, 2014, 24(15): 2163–2170.

34 Awad L N, Bae J, O'donnell K, et al. A soft robotic exosuit improves walking in patients after stroke. Science Translational Medicine, 2017, 9(400): eaai9084.

35 Bao G, Fang H, Chen L, et al. Soft robotics: academic insights and perspectives through bibliometric analysis. Soft Robotics, 2018, 5(3): 229–241.

36 Zhao W, Zhang Y, Wang N. Soft robotics: research, challenges, and prospects. Journal of Robotics and Mechatronics, 2021, 33(1): 45–68.

37 Pott A, Mütherich H, Kraus W, et al. IPAnema: a family of cable-driven parallel robots for industrial applications. Cable-Driven Parallel Robots. Berlin: Springer, 2012: 119–134.

38 Miermeister P, Pott A. Modelling and real-time dynamic simulation of the cable-driven parallel robot IPAnema. New Trends in Mechanism Science: Analysis and Design. Dordrecht: Springer, 2010: 353–360.

39 Wang S, Zhao Q, Chen J, et al. Co-axial slender tubular robot (CAST): towards robotized operation for transorbital neurosurgery with minimal invasiveness. Proceedings of the 2024 IEEE International Conference on Robotics and Automation (ICRA). IEEE, 2024: 11571–11577.

40 Yang Z, Yang L, Sun Y, et al. A novel contact-aided continuum robotic system: design, modeling, and validation. IEEE Transactions on Robotics, 2024, 40: 3024–3043.

41 Omisore O M, Han S, Xiong J, et al. A review on flexible robotic systems for minimally invasive surgery. IEEE Transactions on Systems, Man, and Cybernetics: Systems, 2020, 52(1): 631–644.

42 Yang Z, Yang H, Cao Y, et al. Magnetically actuated continuum medical robots: a review. Advanced Intelligent Systems, 2023, 5(6): 2200416.

43 Lin B, Song S, Wang J. Variable stiffness methods of flexible robots for minimally invasive surgery: a review. Biomimetic Intelligence and Robotics, 2024, 4(3): 100168.

44 Tian Y, Chen C, Sagoe-Crentsil K, et al. Intelligent robotic systems for structural health monitoring: applications and future trends. Automation in Construction, 2022, 139: 104273.

45 Skibniewski M J. New directions and developments in robotics and site automation in the USA. Proceedings of the 17th International Symposium on Automation and Robotics in Construction. International Association for Automation and Robotics in Construction (IAARC), 2000.

46 Majidi C. Soft robotics: a perspective—current trends and prospects for the future. Soft Robotics, 2014, 1(1): 5–11.

47 Hartmann F, Baumgartner M, Kaltenbrunner M. Becoming sustainable, the new frontier in soft robotics. Advanced Materials, 2021, 33(19): 2004413.

48 Banerjee H, Tse Z T H, Ren H. Soft robotics with compliance and adaptation for biomedical applications and forthcoming challenges. International Journal of Robotics and Automation, 2018, 33(1): 68–80.

49 Li Y, Wang H, Li X, et al. Recent progress in soft robots: principles, designs, and applications. Smart Materials and Structures, 2024, 33(11): 115014.

50 Bhatt A T, Gohil P P, Chaudhary V. Primary manufacturing processes for fiber reinforced composites: history, development & future research trends. IOP Conference Series: Materials Science and Engineering. IOP Publishing, 2018, 330: 012107.

51 Rajabasadi F, Schwarz L, Medina-Sánchez M, et al. 3D and 4D lithography of untethered microrobots. Progress in Materials Science, 2021, 120: 100808.

52 Freire N M A, Estima J O, Cardoso A J M. A voltage-based approach without extra hardware for open-circuit fault diagnosis in closed-loop PWM AC regenerative drives. IEEE Transactions on Industrial Electronics, 2013, 61(9): 4960–4970.

53 Baziyad A G, Ahmad I, Salamah Y B, et al. Robust tracking control of piezo-actuated nanopositioning stage using improved inverse LSSVM hysteresis model and RST controller. Actuators, 2022, 11(11): 324.
54 Yuan Z, Zhou S, Zhang Z, et al. Piezo-actuated smart mechatronic systems: nonlinear modeling, identification, and control. Mechanical Systems and Signal Processing, 2024, 221: 111715.
55 Xia Y, He Y, Zhang F, et al. A review of shape memory polymers and composites: mechanisms, materials, and applications. Advanced Materials, 2021, 33(6): 2000713.
56 Nguyen X T, Calderon A A, Rigo A, et al. SMALLBug: a 30-mg crawling robot driven by a high-frequency flexible SMA microactuator. IEEE Robotics and Automation Letters, 2020, 5(4): 6796–6803.
57 Hussein H, Damdam A, Ren L, et al. Actuation of mobile microbots: a review. Advanced Intelligent Systems, 2023, 5(9): 2300168.
58 Wilson R E. Microfabricated Hydrogel Platforms to Compare Single Cells in 2D Vs 3D Engineered Microenvironments. Stanford University, 2019.
59 Wehner M, Truby R L, Fitzgerald D J, et al. An integrated design and fabrication strategy for entirely soft, autonomous robots. Nature, 2016, 536(7617): 451–455.
60 Huang H, Wang H, Fang C, et al. Grasping by Spiraling: Reproducing Elephant Movements with Rigid-Soft Robot Synergy. arXiv preprint arXiv:2504.01507, 2025.
61 Dheerthi N, Kumar A K, Sarveswaran S, et al. A comprehensive analysis of bio-inspired soft robotic arm for healthcare applications. Multidisciplinary Applications of AI Robotics and Autonomous Systems. IGI Global, 2024: 1–23.
62 Shao Q, Dong X, Lin Z, et al. Untethered robotic millipede driven by low-pressure microfluidic actuators for multi-terrain exploration. IEEE Robotics and Automation Letters, 2022, 7(4): 12142–12149.
63 Zhang D, Xu J, Liu X, et al. Advanced bionic attachment equipment inspired by the attachment performance of aquatic organisms: a review. Biomimetics, 2023, 8(1): 85.
64 Soori M, Dastres R, Arezoo B, et al. Intelligent robotic systems in Industry 4.0: a review. Journal of Advanced Manufacturing Science and Technology, 2024, 4: 2024007.
65 Zhou B, Li S, Zi B, et al. Performance analysis and optimization design of a rigid-flexible coupling hoisting robot considering cable deformation and random loading. Mechanical Systems and Signal Processing, 2025, 223: 111903.
66 Wang Z, Chen T, Zhang Y, et al. A robotic design framework for phase verification of overhead power transmission lines. Journal of Robotics, 2025, 2025(1): 7301970.
67 Bostelman R, Albus J, Dagalakis N, et al. Applications of the NIST RoboCrane. Proceedings of the 5th International Symposium on Robotics and Manufacturing. Maui, HI: NIST, 1994, 5: 1.

68 Lytle A M, Saidi K S, Bostelman R V, et al. Adapting a teleoperated device for autonomous control using three-dimensional positioning sensors: experiences with the NIST RoboCrane. Automation in Construction, 2004, 13(1): 101–118.

69 Sugahara Y, Chen G, Atsumi N, et al. A suspended cable-driven parallel robot for human-cooperative object transportation. ROMANSY 23-Robot Design, Dynamics and Control: Proceedings of the 23rd CISM IFToMM Symposium 23. Springer International Publishing, 2021: 109–117.

70 Schmidt V L. Modeling Techniques and Reliable Real-time Implementation of Kinematics for Cable-driven Parallel Robots Using Polymer Fiber Cables. Stuttgart: Fraunhofer Verlag, 2017.

71 Ming A. Study on multiple degree-of-freedom positioning mechanism using wires (part 1). International Journal of the Japan Society for Precision Engineering, 1994, 28: 131–138.

72 Bousquet J M. Survey of engine integration testing in ONERA wind tunnels. 41st AIAA/ASME/SAE/ASEE Joint Propulsion Conference & Exhibit, 2005: 3705.

73 Shang W, Zhang B, Zhang B, et al. Synchronization control in the cable space for cable-driven parallel robots. IEEE Transactions on Industrial Electronics, 2018, 66(6): 4544–4554.

74 Tang L, Chen Y. Dynamic model of four degree of freedom rope-driven rigid-flexible hybrid wave compensation device. Journal of Mechanical Science and Technology, 2020, 34: 1937–1948.

75 Zi B, Qian S. Design, Analysis and Control of Cable-Suspended Parallel Robots and Its applications. Singapore: Springer, 2017.

76 Shen Y, Franchi A, Gabellieri C. Aerial robots carrying flexible cables: dynamic shape optimal control via spectral method model. IEEE Transactions on Robotics, 2025, 41: 3162–3182.

77 González-Carbajal J. An analytical model for the drum-rope interaction in hoisting mechanisms and cable-driven systems. Mechanism and Machine Theory, 2025, 207: 105939.

78 Taheri A, Rankka A, Gustafsson P, et al. End-effector Cartesian velocity control for redundant loader cranes using reinforcement learning. IEEE Transactions on Robotics, 2024, 41: 484–496.

79 Zhu Y, Zhang X, Xu K, et al. Intelligent switch cabinet hoisting robot with path planning function. Proceedings of the 2022 IEEE International Conference on High Voltage Engineering and Applications (ICHVE). IEEE, 2022: 1–4.

80 Ku T K X, Zuo B, Ang W T. Robotic tower cranes with hardware-in-the-loop: enhancing construction safety and efficiency. Automation in Construction, 2024, 168: 105765.

81 Bouaouda A, Pannequin R, Charpillet F, et al. Dynamic modeling and AI-based control of a cable-driven parallel robot. IFAC-PapersOnLine, 2023, 56(2): 10021–10026.

82 Iturralde K, Feucht M, Illner D, et al. Cable-driven parallel robot for curtain wall module installation. Automation in Construction, 2022, 138: 104235.

83 McDonald E, Arsenault M, Beites S. Design of a 3-degrees-of-freedom cable-driven parallel robot for automated construction based on workspace and kinematic sensitivity. Journal of Mechanisms and Robotics, 2024, 16(2): 021006.

84 Zhang Z, Wong M O, Pan W. Virtual reality enhanced multi-role collaboration in crane-lift training for modular construction. Automation in Construction, 2023, 150: 104848.

85 Sun N, Fang Y, Chen H, et al. Slew/translation positioning and swing suppression for 4-DOF tower cranes with parametric uncertainties: design and hardware experimentation. IEEE Transactions on Industrial Electronics, 2016, 63(10): 6407–6418.

86 Wang K, Ma X, Li J. Neural network-based adaptive swing suppression control for tower cranes with obstacle avoidance. IEEE/ASME Transactions on Mechatronics, 2024, 30: 2177–2188.

87 Tian Z, Yu L, Ouyang H, et al. Swing suppression control in tower cranes with time-varying rope length using real-time modified trajectory planning. Automation in Construction, 2021, 132: 103954.

88 Qian Y, Fang Y. Switching logic-based nonlinear feedback control of offshore ship-mounted tower cranes: a disturbance observer-based approach. IEEE Transactions on Automation Science and Engineering, 2018, 16(3): 1125–1136.

89 Li F, Qian Y, Wu S. A self-adapting trajectory tracking control for double-pendulum marine tower crane considering constraint dead zone against seawave disturbances. Ocean Engineering, 2024, 309: 118358.

90 Wang S, Ren Z, Du J, et al. Payload swing suppression for offshore cranes using a novel triple-tagline system: theory and experiment. Proceedings of the Institution of Mechanical Engineers, Part M: Journal of Engineering for the Maritime Environment, 2020, 234(2): 547–557.

91 Zargari A, Castrejon Z A, Kim D, et al. Cable-driven parallel robot for warehouse monitoring tasks. Proceedings of the 2023 IEEE 13th Annual Computing and Communication Workshop and Conference (CCWC). IEEE, 2023: 1085–1090.

92 An H, Yuan H, Tang K, et al. A novel cable-driven parallel robot with movable anchor points capable for obstacle environments. IEEE/ASME Transactions on Mechatronics, 2022, 27(6): 5472–5483.

93 Ha S H, Woo S W, Kim M C, et al. A cable-driven parallel robot for high-rack logistics automation. International Journal of Control, Automation and Systems, 2024, 22(11): 3329–3340.

2

Kinematics and Dynamic Modeling of Rigid–Flexible Coupled Hoisting Robots

2.1 Preamble

In the 1980s of the twentieth century, rigid–flexible coupling robots came out, and in recent years, scholars have continued to study them. Traditional robots use rigid connecting rods as the transmission unit, and the working space is small. Moreover, the flexible robot tends to undergo vibration and deformation during dynamic operation due to its unique structural characteristics, resulting in its poor stability. The rigid–flexible coupling robot is a new type of parallel robot mechanism after the rigid manipulator arm and the flexible parallel mechanism, which inherits the advantages of small inertia, rapid response, high bearing capacity, and excellent dynamic performance of the flexible cable drive mechanism [1]. It plays an important role in aerospace, medical rehabilitation, construction, logistics, and warehousing [2]. After the birth of the rigid–flexible coupling robot, with continuous research and development, it has thus emerged as a significant research direction in the field of robotics at home and abroad [3]. The application of rigid–flexible coupling robot in the field of hoisting and handling is an important direction in its various applications, the traditional hoisting or handling mode adopts a single lifting robot structure, its hoisting quality is limited by the structure, and here a single girder overhead crane is commonly used in a large space; although it has a large-load capacity, the cost and space requirements are higher, and the tandem manipulator structure is common in production and processing, which is shown in Figure 2.1, but its work quality and scope are relatively limited.

The hoisting equipment mainly realizes the handling of materials and assists in adjusting the attitude of objects, or assists in the installation of some large-weight equipment. With the continuous upgrading of hoisting and handling needs, the structure and function of hoisting equipment are constantly updated and developed. The emergence of rigid–flexible coupling hoisting robots (RFCHRs) makes up for the shortcomings of weak carrying capacity, small working space, and

Grispb/Adobe Stock Photos

Figure 2.1 Series manipulator in production line.

poor flexibility of hoisting robots. However, with the advancement of science and technology and the growing demand for high-end intelligent robots, more stringent requirements have been imposed on the high-precision control of robots. Due to the relatively complex configuration of the rigid–flexible coupling robot, the characteristics of the flexible cord that is easy to deform and can only transmit tensile force, the kinematics and dynamics modeling of the RFCHR pose a challenge. Therefore, it is of great significance to reasonably simplify the model and ensure the accuracy of the simplified model to minimize the error in the process of robot movement.

As you know, dynamics is an important area of study that mainly studies the relationship between the forces acting on an object and the motion of an object. For general parallel mechanisms, the commonly used methods for dynamic modeling are the Newton–Euler method [4], the Lagrange method [5], the virtual work principle method [6], the Kane method [7], and other combination methods [8], and the modeling methods of rigid–flexible coupled robots are generally based on the modeling methods of parallel robots. The Lagrange method is currently one of the most widely used methods in rigid–flexible coupled robots, which does not need to consider the complex interrelationship between structures, calculates the robot from the perspective of energy, and needs to obtain dynamic results through differential calculations. However, the literature on the dynamics modeling of rigid–flexible coupled robots is scarce. Below, we develop dynamic equations for rigid–flexible coupled robot systems, which can be used to control and model the motion of rigid–flexible coupled robots.

This chapter introduces the design and experiment of the next generation of RFCHR structure. Firstly, the structure was optimized to improve the payload of the hoisting robot and build a physical model to verify its effectiveness.

2.2 Mechanism Design and Kinematic Analysis of RFCHRs

2.2.1 Mechanism Design of RFCHRs

2.2.1.1 Mechanism Design of an RFCHR in Large-load Environment

As illustrated in Figure 2.2, a mechanical model of the RFCHR designed for lifting payloads in heavy-load environments is presented. This heavy-load RFCHR comprises a lifting arm, a triangular rigid support arm, a luffing telescopic cylinder, a rotatable base, four driving units, four cables, two bevel gear sets, a cable reel, thrust ball bearings, and a timing belt. The robot features four degrees of freedom, which are furnished by the four driving units and the luffing telescopic cylinder. The rotatable base, connected with thrust ball bearings, is driven by driving unit 1. Its rotational function is realized through the motor torque transmitted via bevel gear set 1. Driving units 2 and 3, the triangular rigid support arm, and the luffing telescopic cylinder are all mounted on the rotatable base. The triangular rigid support arm achieves rotation through the telescopic movement of the luffing telescopic cylinder. The lifting arm is connected to the triangular rigid support arm and is driven by driving units 2 and 3 via cables 1, 2, and 3. Driving unit 4 is

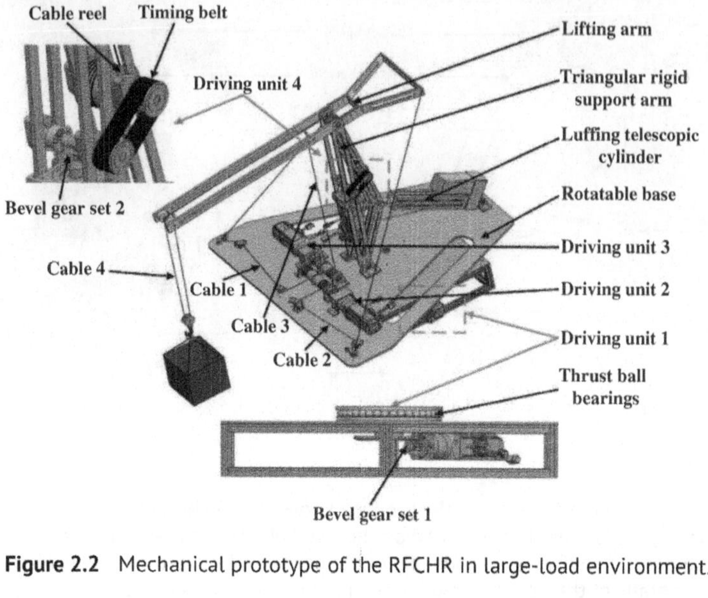

Figure 2.2 Mechanical prototype of the RFCHR in large-load environment.

fixed on the triangular rigid support arm; the retraction of cable 4 is accomplished by transmitting the motor torque to the cable reel through bevel gear set 2 and the timing belt.

2.2.1.2 Mechanism Design of an RFCHR in Low-load Environment

The simplified CAD model of the proposed novel 6-DOF hybrid cable-actuated parallel palletizing robot is shown in Figure 2.3. Actually, the proposed palletizing robot can be seen as a spatial hybrid serial–parallel mechanisms, which are divided into the lower structure and upper structure by the moving platform.

Actuated by four cables, the lower structure, wherein the moving platform is constrained by four cables and the back arm, is a redundant 3-DOF hybrid cable-actuated parallel mechanism. The three DOFs are, respectively, the rotations around two axes (1 Axis and 2 Axis) of lower universal joint and the axis (3 Axis) of upper swivel plate. To constrain the back arm which connects the upper swivel plate and the base by means of two universal joints, the lengths of supporting cables and the back arm should be equal and parallel with each other; moreover, the swivel plate and the cross shafts of universal joint should be concentric. Unlike the mechanisms in [9], the supporting cable not only constrains the upper swivel plate parallel to the lower swivel plate but also transfers moderate torque from the

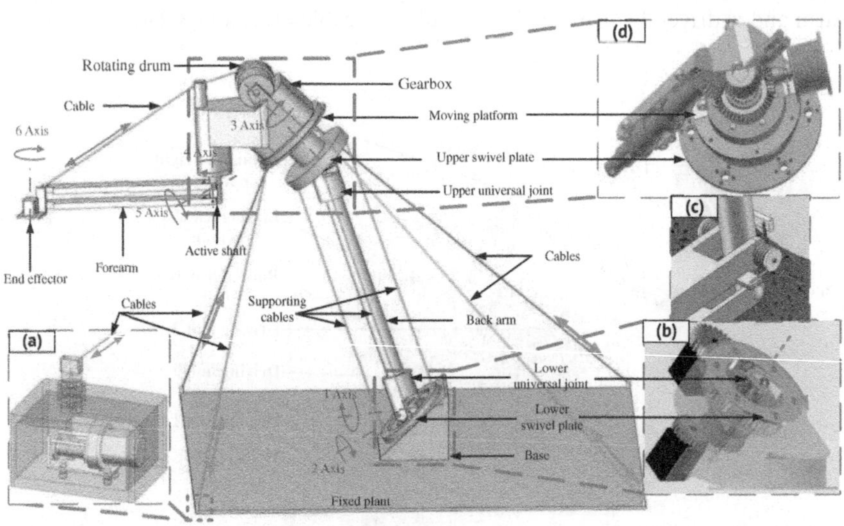

Figure 2.3 The simplified CAD model of the proposed novel 6-DOF hybrid palletizing robot: (a) details of windlass module installed on the fixed plant, (b) details of gear reducer system mounted on the base, (c) details of the gear box mounted on the moving platform, and (d) details of the gear box mounted on the moving platform.

lower swivel plate to the upper swivel plate. As seen in Figure 2.3b, two motors and gear reducer system mounted on the base actuate the lower universal joint and swivel plate to rotate around the same axis. Moreover, to maximize the intersection of workspace and pallets, the axis of base was leaned to pallets. The distribution of the windlass module (see Figure 2.3a) can be easily adjusted with the change of the environment.

The upper structure is a 3-DOF hybrid arm. Fixed on the moving platform, the gearbox transfers the input rotations of upper universal joint and upper swivel plate into rotations of drum and shaft to actuate the forearm to rotate around 4 Axis and 5 Axis by using two pairs of bevel gears (see Figure 2.3d). Utilizing parallelogram mechanisms, the forearm can constrain the axis of wrist (6 Axis) parallel to the axis of active shaft (4 Axis). Without considering the DOF of the wrist, the upper structure actually is a nonredundant and under-constrained 2-DOF hybrid cable-actuated parallel mechanism; hence, cable tensioning can be provided by the gravity when end effector carries goods. However, when there is a load on the end effector, the structure cannot bear the upward force at the moment of grasping; here, we can add a pair of torsional springs (see Figure 2.3c) to prevent the relaxation of cable. Due to the parallelogram mechanisms employed in the upper and lower structures, in general, the axis of wrist is vertical, which makes the proposed robot more suitable for palletizing than the structure in [10]. Particularly, the rotational DOF of the moving platform gives the upper structure a better adaptability to cope with complex situations.

The novel features of hybrid palletizing robot can be listed as follows: (1) the inertia of the structure is further reduced by transferring the actuator to ground and utilizing cable; (2) the modular windlass can be easily rearranged to accommodate different tasks; (3) the employment of parallelogram mechanisms makes the structure more suitable for palletizing.

2.2.1.3 Mechanism Design of an RFCHR for Container Yard Operations

The mechanical configuration of RFCHR for container yard operations is designed as shown in [11]. The robot can be divided into three main parts: the first part comprises the operating arm module utilized for strapping containers in container yards. This module is capable of longitudinal movement along the left and right vertical frames. Additionally, an internal mechanism within the operating arm module enables limited lateral movement.

The second part primarily consists of the drive units for the motion platform and the track vehicle component responsible for the overall movement of the rigid–flexible hoisting robots. The drive units are fixed onto the track vehicle, and the left and right vertical frames are also mounted on it. The height of the vertical frames can be adjusted reasonably according to the stacking requirements.

The third part is the motion platform of the robot, which is driven by eight cables. One end of each cable is connected to the bottom drive mechanism, while the other end is guided through a series of pulleys and connected to the top end point of the motion platform. The bottom of the motion platform is equipped with standard container lifting frames, fulfilling the requirements for stacking empty containers in the container yard. Additionally, an internal robot arm is installed for strapping containers at the top of container stacks in the yard.

During strapping operations, the robot arm can be extended over a large range via a telescopic mechanism (coarse adjustment) and precisely positioned using a precision adjustment mechanism (fine adjustment). This "coarse-fine" dual-level adjustment mechanism, in conjunction with the multi-degree-of-freedom attitude adjustment of the motion platform, ensures accurate alignment with the strapping positions at the top of container stacks.

2.2.2 Kinematic Modeling of RFCHRs

2.2.2.1 Kinematic Modeling of RFCHR in Large-load Environment

The kinematic model of the RFCHR is depicted in Figure 2.4. A global coordinate system O–XYZ is established at the rotation center of the rotatable base. A moving coordinate system O_1–$X_1Y_1Z_1$ is also positioned at the rotation center of the rotatable base. Additionally, moving coordinate systems A–$X_AY_AZ_A$ and B–$X_BY_BZ_B$ are placed at the pivot centers of the triangular rigid support arm and the lifting arm, respectively.

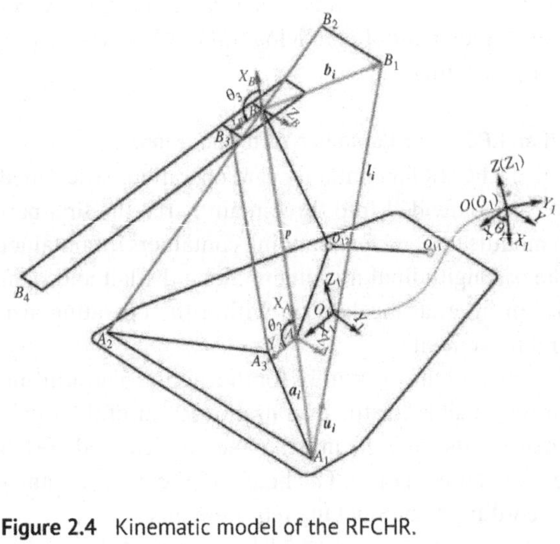

Figure 2.4 Kinematic model of the RFCHR.

In accordance with the Denavit–Hartenberg theory, the transformation matrices ${}^{O}_{O_1}\boldsymbol{T}$, ${}^{O}_{A}\boldsymbol{T}$, and ${}^{O}_{B}\boldsymbol{T}$ are expressed as

$$
{}^{O}_{O_1}\boldsymbol{T} = \begin{bmatrix} c\theta_1 & -s\theta_1 & 0 & 0 \\ s\theta_1 & c\theta_1 & 0 & 0 \\ 0 & 0 & 1 & 0 \\ 0 & 0 & 0 & 1 \end{bmatrix} \tag{2.1}
$$

$$
{}^{O}_{A}\boldsymbol{T} = {}^{O}_{O_1}\boldsymbol{T} \cdot {}^{O_1}_{A}\boldsymbol{T} = \begin{bmatrix} c\theta_1 c\theta_2 & -c\theta_1 s\theta_2 & -s\theta_1 & l_{O_1A} c\theta_1 \\ s\theta_1 c\theta_2 & -s\theta_1 c\theta_1 & c\theta_1 & l_{O_1A} s\theta_1 \\ -s\theta_2 & -c\theta_2 & 0 & 0 \\ 0 & 0 & 0 & 1 \end{bmatrix} \tag{2.2}
$$

$$
{}^{O}_{B}\boldsymbol{T} = {}^{O}_{O_1}\boldsymbol{T} \cdot {}^{O_1}_{A}\boldsymbol{T} \cdot {}^{A}_{B}\boldsymbol{T} \tag{2.3}
$$

$$
\begin{bmatrix} c\theta_1 c\theta_2 c\theta_3 - c\theta_1 s\theta_2 s\theta_3 & -c\theta_1 c\theta_2 s\theta_3 - c\theta_1 s\theta_2 c\theta_3 & -s\theta_1 & l_{O_1A} c\theta_1 + l_{AB} c\theta_1 c\theta_2 \\ s\theta_1 c\theta_2 c\theta_3 - s\theta_1 s\theta_2 s\theta_3 & s\theta_1 c\theta_2 s\theta_3 - s\theta_1 s\theta_2 c\theta_3 & c\theta_1 & l_{O_1A} s\theta_1 + l_{AB} s\theta_1 c\theta_2 \\ -c\theta_2 s\theta_3 - s\theta_2 c\theta_3 & s\theta_2 s\theta_3 - c\theta_2 c\theta_3 & 0 & -l_{AB} s\theta_2 \\ 0 & 0 & 0 & 1 \end{bmatrix} \tag{2.4}
$$

where θ_1, θ_2, and θ_3 denote rotational angles, l_{O1A} stands for the distance between points O_1 and A, and l_{AB} represents the length between points A and B. It should be noted that s and c are abbreviations for "sin" (sine) and "cos" (cosine), respectively.

The vector diagram illustrating the cable lengths of the RFCHR in heavy-load environments is presented in Figure 2.5. Here, $\boldsymbol{a}_i$ denotes the position vector of point A_i within the moving coordinate system A–$X_A Y_A Z_A$, $\boldsymbol{p}$ represents the position vector of the triangular rigid support arm in the same coordinate system A–$X_A Y_A Z_A$, $\boldsymbol{b}_i$ stands for the position vector of point B_i in the moving coordinate system B–$X_B Y_B Z_B$, and $\boldsymbol{u}_i$ is the unit direction vector along the cable. Based on Eqs. (2.1)–(2.3), the following relationship is derived:

$$
l_i = \|\boldsymbol{l}_i\| = \sqrt{(\boldsymbol{p} + \boldsymbol{b}_i - \boldsymbol{a}_i)^T (\boldsymbol{p} + \boldsymbol{b}_i - \boldsymbol{a}_i)},\ i = 1, 2, 3 \tag{2.5}
$$

where l_i denotes the length of the ith cable; $\boldsymbol{l}_i$ represents the vector of the ith cable.

Assuming the position coordinates of point B_4 in the global coordinate system O–XYZ are given as (x, y, z). The rotational angle θ_1 can be calculated by

$$\theta_1 = \arctan\left(\frac{y}{x}\right) \tag{2.6}$$

To determine other poses of the RFCHR, the triangular rigid support arm and the lifting arm can be treated as a two-link mechanism with two degrees of freedom. As illustrated in Figure 2.6, ABO_{12} denotes the triangular rigid support arm, B_4B_5

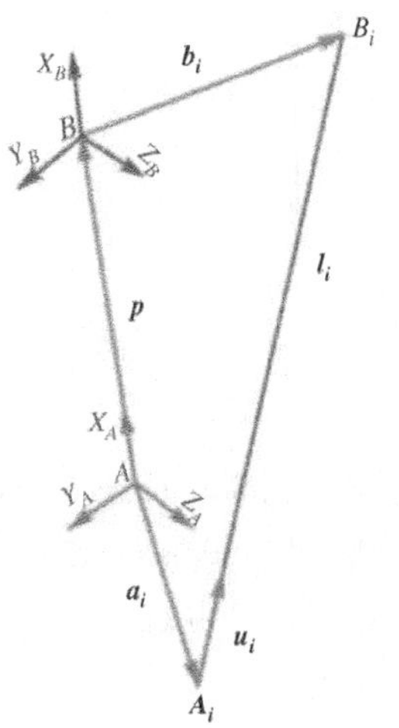

Figure 2.5 Vector diagram of cable length.

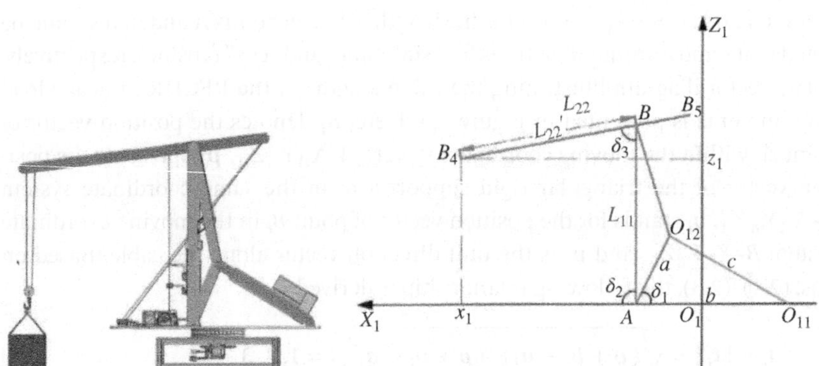

(a) (b)

Figure 2.6 Side view and geometrical relationship of the RFCHR: (a) side view and (b) geometrical relationship.

represents the lifting arm, and $O_{11}O_{12}$ stands for the luffing telescopic cylinder. Using the geometric method and the cosine theorem [12], the intersection angles δ_1, δ_2, and δ_3 are expressed as

$$\delta_1 = \cos^{-1}\left(\frac{a^2 + b^2 - c^2}{2 \cdot a \cdot b}\right) \tag{2.7}$$

$$\delta_2 = \cos^{-1}\left(\frac{L_{11}^2 + \left(x_1 - l_{AO_1}\right)^2 + z_1^2 - {L'_{22}}^2}{2 \cdot L_{11} \cdot \sqrt{\left(x_1 - l_{AO_1}\right)^2 + z_1^2}}\right) + \tan^{-1}\left(\frac{z_1}{x_1 - l_{AO_1}}\right) \tag{2.8}$$

$$\delta_3 = \cos^{-1}\left(\frac{L_{11}^2 + {L'_{22}}^2 - \left(x_1 - l_{AO_1}\right)^2 - z_1^2}{2 \cdot L_{11} \cdot L'_{22}}\right) \tag{2.9}$$

Here, x_1 and z_1 are the coordinates of point B_4 along the X_1-axis and Z_1-axis in the O_1–$X_1Y_1Z_1$ coordinate system, with $x_1 = \sqrt{x^2 + y^2}$, $z_1 = z$; L_{11} denotes the length of AB; L_{22} represents the length of B_4B_5; L'_{22} is the length of BB_4; and l_{AO_1} stands for the length of AO_1. Using Eqs. (2.8) and (2.9), the angles θ_2 and θ_3 can be calculated as

$$\theta_2 = -\delta_2 \tag{2.10}$$

$$\theta_3 = \frac{\pi}{2} - \delta_3 \tag{2.11}$$

Here, δ_2 refers to the intersection angle between AB and X_1-axis, while δ_3 denotes the intersection angle between AB and BB_4. The expansion amount Δx of the luffing telescopic cylinder can be calculated by

$$\begin{aligned} \Delta x &= c - c_0 \\ &= \sqrt{a^2 + b^2 - 2 \cdot a \cdot b \cdot \cos \delta_1} - \sqrt{a^2 + b^2 - 2 \cdot a \cdot b \cdot \cos \delta'_1} \end{aligned} \tag{2.12}$$

Here, δ_1 denotes the intersection angle between AO_{11} and AO_{12}; δ'_1 the value of δ_1 when the triangular rigid support arm is vertical; a the length of AO_{12}; b the length of AO_{11}; c the length of $O_{11}O_{12}$; and c_0 the length of $O_{11}O_{12}$ when the triangular rigid support arm is vertical.

The above is the positive kinematic modeling of the RFCHR, and we can obtain the specific position of the end according to the length of the flexible cable and the angle of each joint angle.

2.2.2.2 Inverse Kinematic Analysis of RFCHR in Low-load Environment

This subsection analyzes the kinematic models of the proposed RFCHR in low-load environment and the RFCHR for container yard operations. For the former, two passive DOFs such that rotations of the lower universal joint and swivel plate are fixed, as seen in Figure 2.7, make either of two axes of lower universal joint horizontal and set as 1 Axis, the other axis is set as 2 Axis. The ranges of motion of six axes are specified in Table 2.1.

Schematic diagrams of the lower and upper structures are shown in Figures 2.8 and 2.10, respectively. The forward kinematic transformation from the base to the moving platform can be written as

$$ {}^{0}_{4}T = {}^{0}_{1}T\,{}^{1}_{2}T\,{}^{2}_{3}T\,{}^{3}_{4}T \tag{2.13}$$

Similarly, the forward kinematic transformation from the moving platform to the end effector can be written as

$$ {}^{4}_{6}T = {}^{4}_{5}T\,{}^{5}_{6}T \tag{2.14}$$

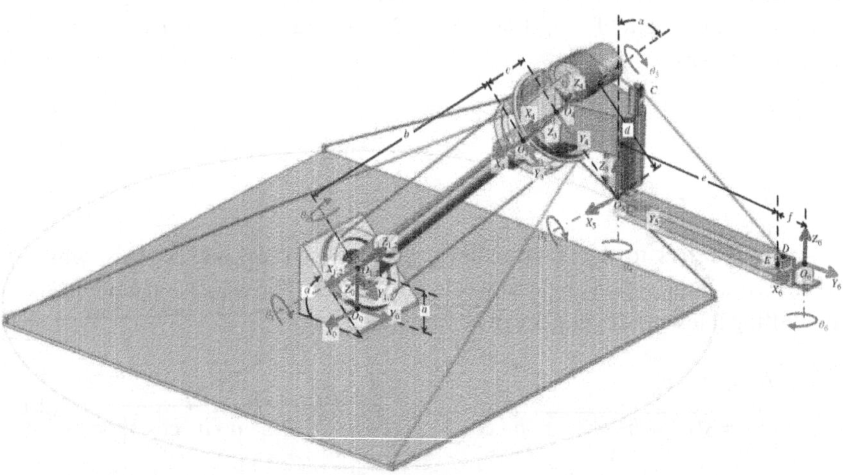

Figure 2.7 The initial configuration and coordinate frames system of the proposed robot.

Table 2.1 The ranges of motion of six axes.

Axis	Range	Axis	Range
1 Axis	$-90° \leq \theta_1 \leq 90°$	4 Axis	$-90° \leq \theta_4 \leq 90°$
2 Axis	$-90° \leq \theta_2 \leq 90°$	5 Axis	$-60° \leq \theta_5 \leq 45°$
3 Axis	$-45° \leq \theta_3 \leq 45°$	6 Axis	$-\infty \leq \theta_6 \leq \infty$

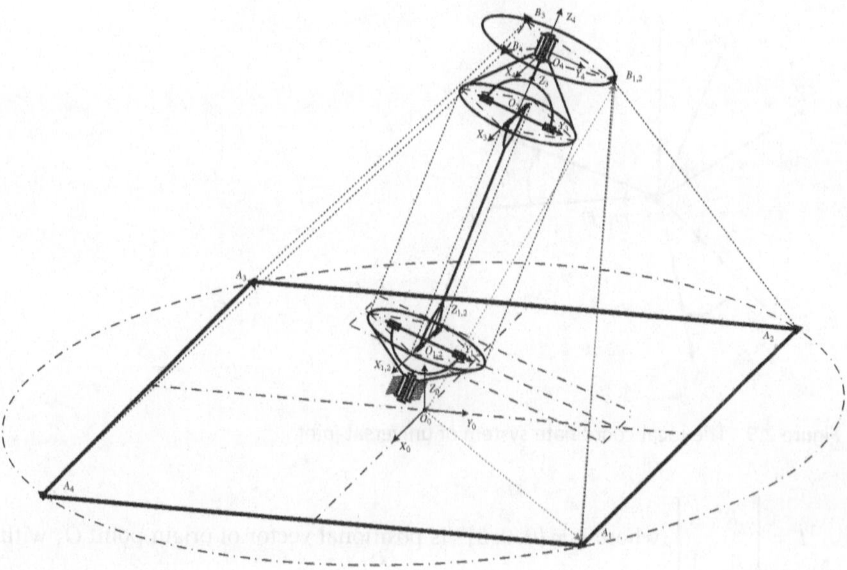

Figure 2.8 Schematic diagrams of the lower structure, the O_1A_1, $A_1B_{1,2}$, $O_0B_{1,2}$ vector loop diagrams the cable kinematic chain.

where ${}^0_4T$, ${}^0_1T$, ${}^1_2T$, ${}^2_3T$, ${}^3_4T$, ${}^4_6T$, ${}^4_5T$, and ${}^5_6T$ represent the transformations from frame O_4 to frame O_0, from frame O_1 to O_0, from frame O_2 to frame O_1, from frame O_3 to frame O_2, from frame O_4 to frame O_3, from frame O_6 to frame O_4, from frame O_5 to frame O_6, and from frame O_6 to frame O_5, respectively.

Based on the arrangement of the local frames and parameters shown in Figures 2.7 and 2.8, we can get ${}^0_1T = \begin{bmatrix} {}^0_1R & a \\ 0 & 1 \end{bmatrix}$, where ${}^0_1R = R_{X_0}(-\alpha_1)$ is the rotation matrix gotten by a rotation $-\alpha_1$ about the fixed X_0-axis; $\boldsymbol{a} = [0, 0, a]^T$ is positional vector of origin point O_1 with respect to frame O_0; ${}^1_2T = \begin{bmatrix} {}^2_1R & 0 \\ 0 & 1 \end{bmatrix}$, where the rotation matrix ${}^2_1R$ can be described by two Euler angles. As shown in Figure 2.9, the local coordinate frame O_2 attached to the back arm can be thought of as a rotation of θ_1 about the X_1-axis, resulting in the intermediate frame O_1'–$X_1'Y_1'Z_1'$ followed by another rotation of θ_2 about the Y_1'-axis. So, the rotation matrix ${}^2_1R$ can be written as

$$
{}^2_1R = R_{X_1Y_1'}(\theta_1, \theta_2) = \begin{bmatrix} c\theta_2 & 0 & s\theta_2 \\ s\theta_1 s\theta_2 & c\theta_1 & -s\theta_1 c\theta_2 \\ -c\theta_1 s\theta_2 & s\theta_1 & c\theta_1 c\theta_2 \end{bmatrix} \tag{2.15}
$$

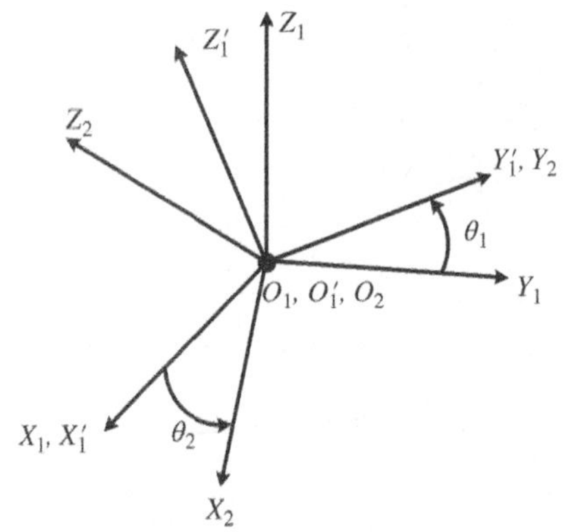

Figure 2.9 The local coordinate system of universal joint.

${}_2^3T = \begin{bmatrix} {}_2^3R & b \\ 0 & 1 \end{bmatrix}$, where $\boldsymbol{b} = [0, 0, b]^T$ is positional vector of origin point O_3 with respect to frame O_2. Due to the constraints of supporting cables, the local frames O_3 and O_1 have same orientations. So, the rotation matrix ${}_2^3R$ can be obtained by

$$
{}_2^3R = \mathrm{inv}\left({}_1^2R\right) = \begin{bmatrix} c\theta_1 & s\theta_1 s\theta_2 & -s\theta_1 c\theta_2 \\ 0 & c\theta_2 & s\theta_2 \\ s\theta_1 & -c\theta_1 s\theta_2 & c\theta_1 c\theta_2 \end{bmatrix} \tag{2.16}
$$

${}_3^4T = \begin{bmatrix} {}_3^4R & c \\ 0 & 1 \end{bmatrix}$, where $\boldsymbol{c} = [0, 0, c]^T$ is positional vector of origin point O_4 with respect to frame O_3, and the rotation matrix ${}_1^2R$ can be gotten by a rotation θ_3 about the Z_3-axis.

${}_4^5T = \begin{bmatrix} {}_4^5R & d \\ 0 & 1 \end{bmatrix}$, where $\boldsymbol{d} = [0, d, 0]^T$ is positional vector of origin point O_5 with respect to frame O_4, and the rotation matrix ${}_4^5R$ can be gotten by X–Z–X Euler angles. So, the rotation matrix ${}_4^5R$ can be written as

$$
{}_4^5R = R_{X_4Z_4'X_4''}(\alpha, \theta_4, \theta_5) = \begin{bmatrix} c\theta_4 & -s\theta_4 s\theta_5 & s\theta_4 c\theta_5 \\ s\alpha s\theta_4 & -s\alpha c\theta_4 s\theta_5 + c\alpha c\theta_5 & -s\alpha c\theta_4 c\theta_5 - c\alpha s\theta_5 \\ -c\alpha s\theta_4 & c\alpha c\theta_4 s\theta_5 + s\alpha c\theta_5 & c\alpha c\theta_4 c\theta_5 - s\alpha s\theta_5 \end{bmatrix} \tag{2.17}
$$

Due to the constraints of parallelogram mechanisms in forearm, 6 Axes are always parallel to 4 Axes. ${}_5^6T$ can be obtained by a translation along the axis Y_5 by a distance e, followed by a rotation $-\theta_5$ about the intermediate X_5'-axis, and another translation along the intermediate Y_5'' by a distance f. So, the transformation from frame O_6 to frame O_5 can be written as

$$ {}_5^6T = \begin{bmatrix} {}_5^6R & {}_5^6Rf + e \\ 0 & 1 \end{bmatrix} \tag{2.18}$$

where ${}_5^6R$, f, and e denote the rotation matrix, and $\boldsymbol{f} = [0, f, 0]^T$ and $\boldsymbol{e} = [0, e, 0]^T$ are the transformation vectors.

Assume that cable attachment points are located on virtual circles on the fixed plant and moving platform and the radii of virtual circles are represented by R_A and R_B, respectively. As shown in Figure 2.8, the cable attachment points are represented with A_i $(i = 1, 2, 3, 4)$ and B_i $(i = 1, 2, 3, 4)$, respectively.

As seen in Figure 2.8, the vector loop equation associated with the ith cable can be written as

$$\overrightarrow{B_iA_i} = \overrightarrow{O_0A_i} - \overrightarrow{O_0B_i} \ (i = 1, 2, 3, 4) \tag{2.19}$$

where the vector $\overrightarrow{B_iA_i}$ is the ith cable vector from point B_i to point A_i. $\overrightarrow{O_0B_i}$ and $\overrightarrow{O_0A_i}$ are the positional vectors of cable attachment points. It should be noted that these three vectors are written in terms of local frame O_2. If the positional vector of the point B_i in the local frame O_4 is represented by homogeneous coordinates ${}^4r_{Bi} = (x_{Bi}, y_{Bi}, z_{Bi}, 1)^T$ and the positional vector of the point A_i in the global coordinate frame is represented by homogeneous coordinates ${}^0r_{Ai} = (x_{Ai}, y_{Ai}, z_{Ai}, 1)^T$, changing the description of r_{Bi} into frame O_0 by using Eq. (2.16), then Eq. (2.19) can be written as

$$ {}^0l_i = {}^0r_{Ai} - {}_0^4T\,{}^0r_{Bi} \ (i = 1, 2, 3, 4) \tag{2.20}$$

Then, by taking the dot product of both sides of Eq. (2.20), we get

$$ {}^0l_i^T\,{}^0l_i = \left({}^0r_{Ai} - {}_0^4T\,{}^4r_{Bi}\right)^T \left({}^0r_{Ai} - {}_0^4T\,{}^4r_{Bi}\right) = l_i^2 \ (i = 1, 2, 3, 4) \tag{2.21}$$

where l_i denotes the length of the ith cable. The inverse kinematic problem for the 3-DOF lower structure is completed.

As for the inverse kinematic problem of the 3-DOF lower structure, the forward kinematic transformation from the moving platform to the end effector is gotten by Eq. (2.14). As seen in Figure 2.10, the vector loop equation associated with the cable actuating forearm can be written as

$$\overrightarrow{DC} = \overrightarrow{O_4D} - \overrightarrow{O_4C} \tag{2.22}$$

where the vector $\overrightarrow{DC}$ is the cable vector from point D to point C. $\overrightarrow{O_4C}$ and $\overrightarrow{O_4D}$ are the positional vectors of cable attachment points and the center of lead pulley. If the positional vectors of the point C in the local frame O_4 are represented

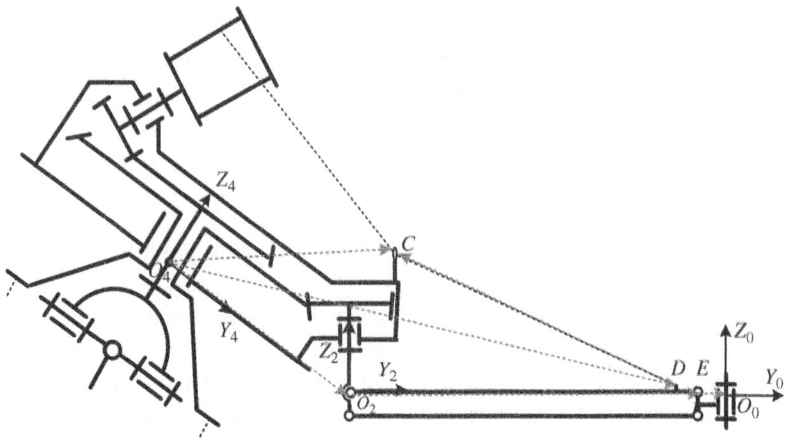

Figure 2.10 Schematic diagrams of the upper structure.

by homogeneous coordinates ${}^4r_C = (x_C, y_C, z_C, 1)^T$ and the positional vectors of the point D in the local frame O_5 are represented by homogeneous coordinates ${}^5r_D = (x_D, y_D, z_D, 1)^T$, changing the description of r_D into frame O_4 by using ${}^5_4T$, then Eq. (2.22) can be written as

$$ {}^4l_{CD} = {}^5_4T\,{}^5r_D - {}^4r_C \tag{2.23} $$

Then, by taking the dot product of both sides of Eq. (2.23), we get

$$ {}^4l_{CD}^T\,{}^4l_{CD} = \left({}^5_4T\,{}^5r_D - {}^4r_C\right)^T \left({}^5_4T\,{}^5r_D - {}^4r_C\right) = l_{CD}^2 \tag{2.24} $$

where l_{CD} denotes the length of cable.

The above is the NI kinematic modeling of the RFCHR, and we can find the length of each flexible rope and the angle of each joint according to the specific posture of the end.

For the RFCHR for container yard operations, the kinematic model of the robot is shown in Figure 2.11. The motion platform is driven by eight cables, with a "4 + 4" layout at the connection points. Specifically, there are four sets of cables located at the four vertices of the upper and lower surfaces of the motion platform. Both upper and lower tiers of cables are connected to the bottom drive unit. In special environments where the typical upper and lower attachment points cannot fully control the robot's movement, the robot's own gravity is utilized as an additional "cable" to constrain the motion range, thereby expanding the workspace.

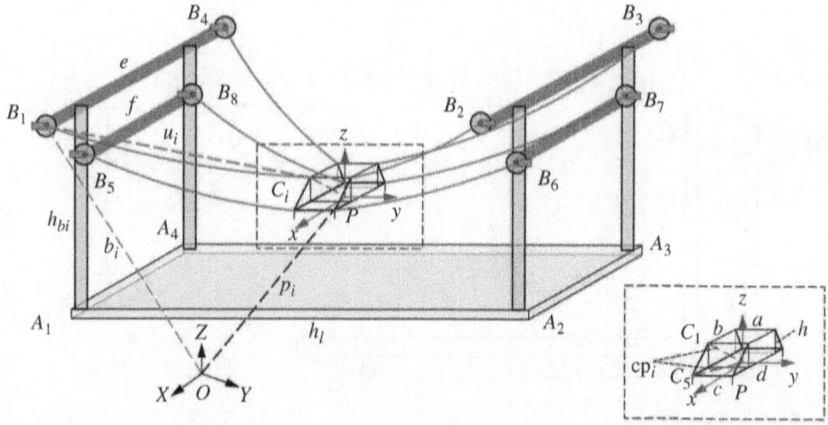

Figure 2.11 The kinematic model of the RFCHR for container yard operations.

As shown in Figure 2.11, point P is the origin of the local coordinate system. B_i and C_i ($i = 1$–8) represent the cable outlet and connection points on the moving platform, respectively. $\boldsymbol{b_i}$ represents the position vector of the cable outlet point in the global coordinate system, $\boldsymbol{p_i}$ is the position vector of the P in the global coordinate system, $\mathbf{cp_i}$ is the position vector of the connection point on the moving platform in the local coordinate system, and ${}^{O}\boldsymbol{R_p}$ denotes the rotation matrix from local to global coordinate systems.

The unit direction vector of the cable can be expressed as

$$\boldsymbol{u_i} = \frac{\boldsymbol{p_i} - \boldsymbol{b_i} - {}^{O}\boldsymbol{R_p}\mathbf{cp_i}}{\left\| \boldsymbol{p_i} - \boldsymbol{b_i} - {}^{O}\boldsymbol{R_p}\mathbf{cp_i} \right\|} \quad (\boldsymbol{i} = 1\text{–}8) \tag{2.25}$$

In fact, due to its own weight, cable will sag in the Z-direction, leading $\boldsymbol{u_i}$ to inaccuracy and thereby affecting modeling precision. To address this, we have established the catenary model shown in Figure 2.12 to accurately describe the kinematics of the robot.

In the local coordinate system, cable B_iC_i can be expressed as follows:

$$z_i = \frac{H_i}{\rho}\left[\text{ch}(\alpha_i) - \text{ch}\left(\frac{2\beta_i y_i}{l_i} - \alpha_i\right)\right] \tag{2.26}$$

where $\alpha_i = \text{sh}^{-1}\left[\frac{\beta_i\,(h_i/l_i)}{\text{sh}\,(\beta_i)}\right] + \beta_i, \beta_i = \frac{\rho l_i}{2H_i}$, and z_i and y_i are z- and y-direction coordinates; l_i, h_i, H_i, V_i, ρ, and g can be represented by the following parameters: horizontal span, vertical span, horizontal component of tension, vertical component of tension, linear density, and gravitational acceleration.

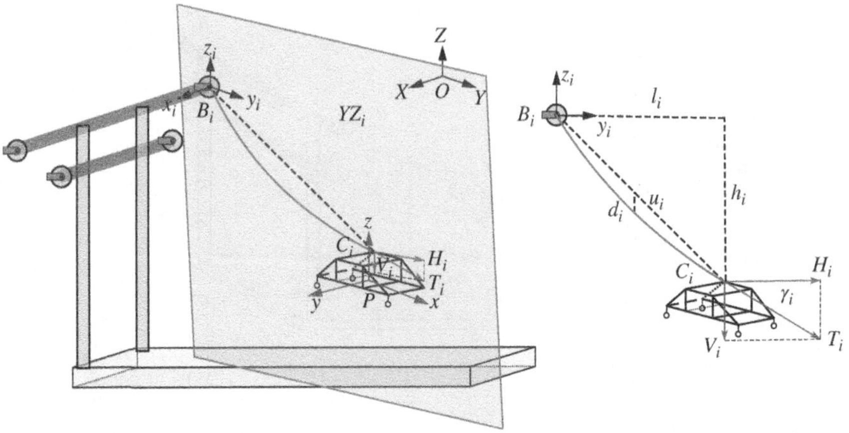

Figure 2.12 Catenary model.

The slope at the end point C_i is

$$\tan \gamma_i = \frac{\partial z_i}{\partial y_i} = -\text{sh}(2\beta_i - \alpha_i) \tag{2.27}$$

where γ_i represents the inclination between the tangent at the end point and the horizontal plane. Additionally, the length of the cable L_i and the corresponding sag d_i are provided as

$$L_i = \sqrt{h^2 + \left[\frac{\text{sh}(\beta_i)}{\beta_i} l_i\right]^2} \tag{2.28}$$

$$d_i = \frac{8H_i\text{sh}(\beta_i)\,\text{sh}^{-1}\left[\frac{\rho g l_i/(2H_i)}{\text{sh}(\beta_i)}\right] - l_i\rho g}{2\rho g} \tag{2.29}$$

Clearly, d_i is highly correlated with H_i. When $d_i = 0$, the catenary curve becomes a straight line. In the iterative algorithm, we use the difference between the sag values of two consecutive iterations being less than the desired accuracy ε as the termination condition

$$\left|d_{i\_new} - d_{i\_old}\right| \leq \varepsilon \tag{2.30}$$

2.3 Dynamic Modeling of RFCHRs

2.3.1 Dynamic Modeling of RFCHR Based on the Lagrange Method

It is assumed that the robot is adequately lubricated, with frictional forces across the entire system being negligible. For the RFCHR operating in heavy-load environments, its rotatable base, triangular rigid support arm, and lifting arm are simplified into a three-link mechanism. As depicted in Figure 2.4, the coordinates of the rotatable base's center along the X-axis, Y-axis, and Z-axis within the moving coordinate system O_1–$X_1Y_1Z_1$ are marked as (x_1, y_1, z_1); the coordinates of the triangular rigid support arm's center along the X-axis, Y-axis, and Z-axis in the moving coordinate system A–$X_AY_AZ_A$ are denoted as (x_2, y_2, z_2); the coordinates of the lifting arm's center along the X-axis, Y-axis, and Z-axis in the moving coordinate system B–$X_BY_BZ_B$ are specified as (x_3, y_3, z_3), and the coordinates of the payload's center along X-axis, Y-axis, and Z-axis in the moving coordinate system B–$X_BY_BZ_B$ are indicated as (x_4, y_4, z_4). Therefore, the position vector of the rotatable base's center ${}^{O_1}\boldsymbol{r}_1$ is expressed as

$$^{O_1}\boldsymbol{r}_1 = \begin{pmatrix} x_1 & y_1 & z_1 & 1 \end{pmatrix}^T \tag{2.31}$$

The position vector representing the center of the triangular rigid support arm, denoted as ${}^{A}\boldsymbol{r}_2$, can be expressed as

$$^{A}\boldsymbol{r}_2 = \begin{pmatrix} x_2 & y_2 & z_2 & 1 \end{pmatrix}^T \tag{2.32}$$

The position vector corresponding to the center of the lifting arm, labeled ${}^{B}\boldsymbol{r}_3$, is expressed as

$$^{B}\boldsymbol{r}_3 = \begin{pmatrix} x_3 & y_3 & z_3 & 1 \end{pmatrix}^T \tag{2.33}$$

The position vector denoting the center of the payload, designated as ${}^{B}\boldsymbol{r}_4$, is expressed as

$$^{B}\boldsymbol{r}_4 = \begin{pmatrix} x_4 & y_4 & z_4 & 1 \end{pmatrix}^T \tag{2.34}$$

To streamline the notation, ${}^{O_1}\boldsymbol{r}_1$, ${}^{A}\boldsymbol{r}_2$, ${}^{B}\boldsymbol{r}_3$, and ${}^{B}\boldsymbol{r}_4$ are respectively simplified to ${}^{1}\boldsymbol{r}_1$, ${}^{2}\boldsymbol{r}_2$, ${}^{3}\boldsymbol{r}_3$, and ${}^{4}\boldsymbol{r}_4$. By taking derivatives of Eqs. (2.31)–(2.34), the velocity vectors ${}^{0}\boldsymbol{v}_i$ (relative to the global coordinate system O–XYZ) for the centers of the rotatable base, triangular rigid support arm, lifting arm, and payload are expressed as

$$\begin{gathered} {}^{O}\boldsymbol{v}_i = \frac{d}{dt}\left({}^{O}\boldsymbol{r}_i\right) = \frac{d}{dt}\left({}^{O}_{i}\boldsymbol{T} \cdot {}^{i}\boldsymbol{r}_i\right) = \left(\sum_{j=1}^{i} \frac{\partial\, {}^{O}_{i}\boldsymbol{T}}{\partial \theta_j} \cdot \dot{\theta}_j\right) \cdot {}^{i}\boldsymbol{r}_i \\ {}^{O}_{i}\boldsymbol{T} = {}^{O}_{O_1}\boldsymbol{T}, i = 1;\ {}^{O}_{i}\boldsymbol{T} = {}^{O}_{A}\boldsymbol{T}, i = 2;\ {}^{O}_{i}\boldsymbol{T} = {}^{O}_{B}\boldsymbol{T}, i = 3, 4 \end{gathered} \tag{2.35}$$

Here, ${}^{O}\boldsymbol{v}_i$ denotes the velocity vectors of the centers of the rotatable base, triangular rigid support arm, lifting arm, and payload relative to the global coordinate system O–XYZ; ${}^{O}\boldsymbol{r}_i$ stands for the position vectors of the centers of these components (rotatable base, triangular rigid support arm, lifting arm, and payload) with respect to the global coordinate system $O\text{–}XYZ$. On the basis of Eq. (2.35), the kinetic energy K_i of each of these centers (rotatable base, triangular rigid support arm, lifting arm, and payload) can be calculated as

$$\begin{aligned} K_i &= \frac{1}{2}m_i \cdot \mathrm{tr}\left({}^O\boldsymbol{v}_i\,, {}^O\boldsymbol{v}_i^T\right) = \frac{1}{2}m_i \cdot \mathrm{tr}\left[\sum_{j=1}^{i}\sum_{k=1}^{i} \frac{\partial\left({}_i^O\boldsymbol{T}\right)}{\partial\theta_j}{}^i\boldsymbol{r}_i \cdot \left({}^i\boldsymbol{r}_i\right)^T \frac{\partial\left({}_i^O\boldsymbol{T}\right)^T}{\partial\theta_k}\dot{\theta}_j\dot{\theta}_k\right] \\ &= \frac{1}{2}m_i \cdot \mathrm{tr}\left[\sum_{j=1}^{i}\sum_{k=1}^{i} \frac{\partial\left({}_i^O\boldsymbol{T}\right)}{\partial\theta_j}{}^i\boldsymbol{r}_i \cdot \left({}^i\boldsymbol{r}_i\right)^T \frac{\partial\left({}_i^O\boldsymbol{T}\right)^T}{\partial\theta_k}\dot{\theta}_j\dot{\theta}_k\right] \end{aligned} \tag{2.36}$$

Here, K_1, K_2, K_3, and K_4 denote the kinetic energy associated with the centers of the rotatable base, triangular rigid support arm, lifting arm, and payload, in that order, and

$$\begin{cases} K_1 = \dfrac{1}{2}m_1{}^O v_1^2 \\ K_2 = \dfrac{1}{2}m_2{}^O v_2^2 \\ K_3 = \dfrac{1}{2}m_3{}^O v_3^2 \\ K_4 = \dfrac{1}{2}m_4{}^O v_4^2 \end{cases} \tag{2.37}$$

Here, m_1, m_2, m_3, and m_4 denote the masses of the rotatable base, triangular rigid support arm, lifting arm, and payload, respectively; ${}^O\boldsymbol{v}_1$, ${}^O\boldsymbol{v}_2$, ${}^O\boldsymbol{v}_3$, and ${}^O\boldsymbol{v}_4$ represent the linear velocity vectors of the centers of the rotatable base, triangular rigid support arm, lifting arm, and payload relative to the global coordinate system $O\text{–}XYZ$. Therefore, according to Eq. (2.37), the total kinetic energy of the RFCHR under heavy-load conditions can be calculated as

$$K = \sum_{i=1}^{4} K_i \tag{2.38}$$

Next, the $(O\text{–}XY)$ -plane is selected as the reference plane for zero potential energy. Accordingly, the potential energy of the rotatable base, triangular rigid support arm, lifting arm, and payload can be computed as

$$\begin{cases} U_1 = 0 \\ U_2 = m_2 g\,{}^0 h_{z_2} \\ U_3 = m_3 g\,{}^0 h_{z_3} \\ U_4 = m_4 g\,{}^0 h_{z_4} \end{cases} \tag{2.39}$$

Here, g denotes the gravitational acceleration; ${}^0h_{z_2}$, ${}^0h_{z_3}$, and ${}^0h_{z_4}$ respectively represent the Z-axis components of the centroid coordinates of the triangular rigid support arm, lifting arm, and payload relative to the global coordinate system O–XYZ. Therefore, based on Eq. (2.39), the total potential energy of the RFCHR in heavy-load environments can be calculated as

$$U = \sum_{i=1}^{4} U_i \tag{2.40}$$

In numerous control systems, cables are frequently simplified to facilitate analysis and the design of control strategies. The core emphasis lies on the primary dynamic characteristics of the system, with minor deformations induced by flexible cables being disregarded. Furthermore, the deformation of flexible cables exerts a negligible impact on control performance; their relatively low flexibility means that such deformations do not notably alter the overall behavior or stability of the controlled system. For this reason, cable deformation is not considered.

Using the Lagrangian dynamic formulation [13] along with Eqs. (2.38) and (2.40), we obtain

$$L = K - U = \sum_{i=1}^{4} (K_i - U_i) \tag{2.41}$$

By differentiating of Eq. (2.41), the dynamic equation of the RFCHR operating in heavy-load conditions can be expressed as

$$\tau_i = \frac{d}{dt}\frac{\partial L}{\partial \dot{\theta}_i} - \frac{\partial L}{\partial \theta_i} = \frac{d}{dt}\frac{\partial K}{\partial \dot{\theta}_i} - \frac{\partial K}{\partial \theta_i} + \frac{\partial U}{\partial \theta_i} \tag{2.42}$$

Here, τ_i represents the torque of each joint. For the sake of simplification, Eq. (2.42) can be rewritten as

$$\boldsymbol{\tau} = \boldsymbol{M}(\boldsymbol{\Theta})\,\ddot{\boldsymbol{\Theta}} + \boldsymbol{C}(\dot{\boldsymbol{\Theta}}, \boldsymbol{\Theta})\,\dot{\boldsymbol{\Theta}} + \boldsymbol{G}(\boldsymbol{\Theta}) \tag{2.43}$$

where $\boldsymbol{\Theta} = \begin{bmatrix} \theta_1 & \theta_2 & \theta_3 \end{bmatrix}^T$ denotes the joint angle vector; $\boldsymbol{M}(\boldsymbol{\Theta}) \in R^{3\times3}$ represents the inertia matrix; $\boldsymbol{C}(\dot{\boldsymbol{\Theta}}, \boldsymbol{\Theta}) \in R^3$ stands for the centrifugal and Coriolis force vector; $\boldsymbol{G}(\boldsymbol{\Theta}) \in R^3$ is the gravity matrix.

2.3.2 Dynamic Modeling of RFCHR Based on the Newton–Euler Method

Given the structural features of the RFCHR, the robot is capable of attaining three degrees of rotational freedom in space. To start with, an analysis is conducted on the dynamics of the RFCHMS. Subsequently, the overall dynamics model is constructed using the Newton–Euler method.

The joint torque transmission mechanism of the RFCHR differs from that of other comparable mechanisms. The lifting arm is driven by a combination of cables and a luffing rigid rod. The torque generated by the lifting arm around the luffing rigid rod does not get transmitted to the next joint [14]. Nevertheless, it should be emphasized that all forces acting on the lifting arm are transferred to the luffing rigid rod. Prior to proceeding with the subsequent analysis, the joint located at the coordinate system O_1–$X_1Y_1Z_1$ is designated as joint 1; the joint at the coordinate system A–$X_AY_AZ_A$ is referred to as joint 2; and the joint at the coordinate system B–$X_BY_BZ_B$ is named joint 3. Furthermore, a schematic illustration of the force distribution in the robot structure is presented in Figure 2.13.

Using the Newton–Euler method, the driving torque of the lifting arm can be calculated as follows:

$$^{3}\boldsymbol{n}_{3} = {}^{3}\boldsymbol{N}_3 + {}^{B}_{O}\boldsymbol{R}\boldsymbol{N}_{\mathrm{ou}} + {}^{3}\boldsymbol{P}_{c_3} \times {}^{3}\boldsymbol{F}_3 + {}^{3}\boldsymbol{P}_{\mathrm{ou}} \times {}^{B}_{O}\boldsymbol{R}\boldsymbol{F}_{\mathrm{ou}} \tag{2.44}$$

where $^{3}\boldsymbol{N}_3$ represents the torque acting on the center of mass of the lifting arm; $\boldsymbol{n}_{\mathrm{ou}}$ denotes the external hoisting torque; ${}^{B}_{O}\boldsymbol{R}$ stands for the rotation matrix; $\boldsymbol{N}_{\mathrm{ou}}$ is the external hoisting torque; $^{3}\boldsymbol{F}_3$ refers to the force exerted on the center of mass of the lifting arm; $\boldsymbol{F}_{\mathrm{ou}}$ indicates the external hoisting force; $^{3}\boldsymbol{P}_{c_3}$ and $^{3}\boldsymbol{P}_{\mathrm{ou}}$ are the position vectors corresponding to $^{3}\boldsymbol{F}_3$ and $\boldsymbol{F}_{\mathrm{ou}}$, respectively. The torque applied to joint 3 can be expressed as

$$\tau_3 = {}^{3}\boldsymbol{n}_3^{T} \cdot {}^{3}\hat{\boldsymbol{Z}}_3 \tag{2.45}$$

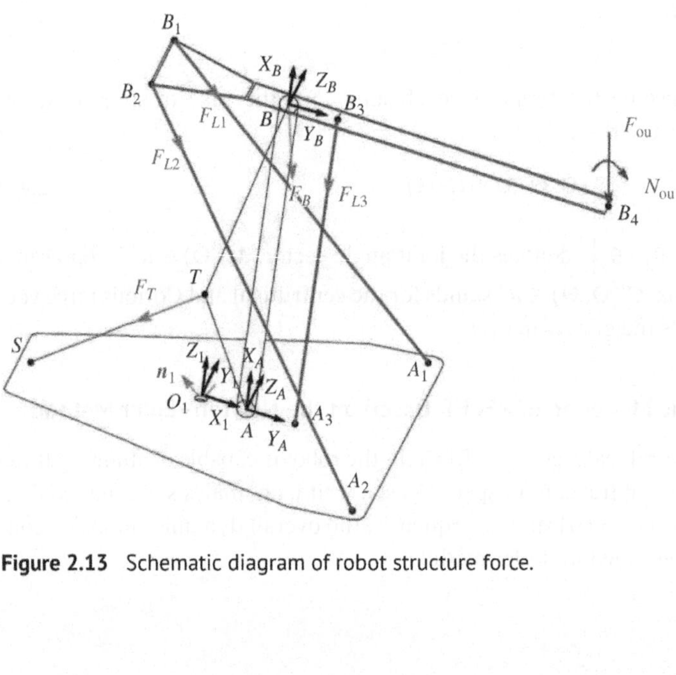

Figure 2.13 Schematic diagram of robot structure force.

where ${}^{3}\hat{Z}_3 = \begin{bmatrix} 0 & 0 & 1 \end{bmatrix}^T$ represents the unit vector along the Z_B-axis. As indicated by the parallel characteristics of the robot illustrated in Figure 2.13, the torque transmitted from the lifting arm to joint 3 is achieved through the cable force $\boldsymbol{F}_{Li}$. It can be expressed by

$$ {}^{3}\boldsymbol{n}_3 = \boldsymbol{b}_i \times \sum_{O}^{B} \boldsymbol{R}\boldsymbol{F}_{Li}, i = 1, 2, 3 \tag{2.46} $$

where $\boldsymbol{F}_{Li} = F_{Li}\boldsymbol{u}_{Li}$, and $\boldsymbol{u}_{Li}$ represents the direction vector of the tension force in the ith cable. Moreover, the force acting on joint 3 can be expressed as

$$ {}^{3}\boldsymbol{f}_3 = {}^{B}_{O}\boldsymbol{R}\boldsymbol{F}_{\text{ou}} + {}^{3}\boldsymbol{F}_3 + \sum \boldsymbol{F}_{Li} \tag{2.47} $$

In the same vein, the force and torque associated with joint 2 can be expressed as follows

$$ {}^{2}\boldsymbol{f}_2 = {}^{A}_{B}\boldsymbol{R}^{3}\boldsymbol{f}_3 + {}^{2}\boldsymbol{F}_2 \tag{2.48} $$

$$ {}^{2}\boldsymbol{n}_2 = {}^{2}\boldsymbol{N}_2 + {}^{A}_{B}\boldsymbol{R}^{3}\boldsymbol{n}_3 + {}^{2}\boldsymbol{P}_{C2} \times {}^{2}\boldsymbol{F}_2 + {}^{2}\boldsymbol{P}_3 \times {}^{A}_{B}\boldsymbol{R}^{3}\boldsymbol{f}_3 \tag{2.49} $$

$$ \tau_2 = {}^{2}\boldsymbol{n}_2^T \cdot {}^{2}\hat{Z}_2 \tag{2.50} $$

Here, ${}^{2}\boldsymbol{N}_2$ is the torque acting on the center of mass of the luffing rigid rod; ${}^{2}\boldsymbol{P}_{C_2}$ denotes the position vector of the center of mass of the luffing rigid rod; ${}^{2}\boldsymbol{F}_2$ represents the force exerted on the center of mass of the luffing rigid rod; ${}^{2}\boldsymbol{P}_3$ is the position vector of joint 3 relative to joint 2; and ${}^{2}\hat{Z}_2 = \begin{bmatrix} 0 & 0 & 1 \end{bmatrix}^T$ is the unit vector along the Z_A-axis.

Furthermore, an analysis of the force acting on the luffing rigid rod is conducted. Given the structural features of the robot, it is evident that the luffing rigid rod serves as the rigid branch chain of the lifting arm. It not only offers support to the robot along the Z_1-axis, but also provides a constraint force to the robot in the X_1-axis. As depicted in Figure 2.13, the connection point between the electric telescopic rod and the luffing rigid rod is defined as point T. $\boldsymbol{F}_T$ denotes the force applied by the electric telescopic rod to the luffing rigid rod. Thus, we can draw the following conclusion:

$$ \boldsymbol{l}_{TA} \times \boldsymbol{F}_T + \boldsymbol{l}_{AB} \times \boldsymbol{F}_B + \boldsymbol{l}_{BG2} \times \boldsymbol{G}_2 = 0 \tag{2.51} $$

Here, $\boldsymbol{l}_{TA}$ denotes the vector from point T to point A in the base coordinate system; $\boldsymbol{l}_{AB}$ is the vector from point A to point B in the base coordinate system; $\boldsymbol{l}_{BG2}$ represents the vector from ${}^{2}\boldsymbol{P}_{C_2}$ to point A in the base coordinate system; $\boldsymbol{G}_2$ is the gravity vector of the luffing rigid rod; and $\boldsymbol{F}_B$ stands for the force exerted on the luffing rigid rod at point B.

Additionally, the known torque, which excludes the driving torque produced by the electric telescopic rod, is expressed as

$$\boldsymbol{N}_T = \boldsymbol{l}_{AB} \times \boldsymbol{F}_B + \boldsymbol{l}_{BG2} \times \boldsymbol{G}_2 \tag{2.52}$$

By associating with Eq. (2.52), Eq. (2.51) can be rephrased as

$$\boldsymbol{l}_{TA} \times \boldsymbol{F}_T = -\boldsymbol{N}_T \tag{2.53}$$

Therefore, when combined with Eq. (2.50), the following result can be derived:

$$\tau_2 = (\boldsymbol{l}_{TA} \times \boldsymbol{F}_T)^T \cdot {}^2\hat{\boldsymbol{Z}}_2 \tag{2.54}$$

assuming $\boldsymbol{u}_T$ is the unit vector in the direction of the driving force of the electric telescopic rod. $\boldsymbol{F}_T$ can thus be expressed as

$$\boldsymbol{F}_T = \boldsymbol{u}_T F_T \tag{2.55}$$

$\boldsymbol{u}_T$ can be expressed as

$$\boldsymbol{u}_T = \frac{\boldsymbol{l}_{TA} - \boldsymbol{l}_{SA}}{\|\boldsymbol{l}_{TA} - \boldsymbol{l}_{SA}\|} \tag{2.56}$$

Here, $\boldsymbol{l}_{SA}$ represents the position vector from point S to point A.

Taking the joint angle into account, the comprehensive dynamic model of the RFCHR can be expressed as

$$\boldsymbol{\tau} = \boldsymbol{M}(\boldsymbol{\theta})\,\ddot{\boldsymbol{\theta}} + \boldsymbol{C}\left(\boldsymbol{\theta}, \dot{\boldsymbol{\theta}}\right) + \boldsymbol{G}\left(\boldsymbol{\theta}\right) + \tau_o \tag{2.57}$$

where $\boldsymbol{\tau}$ denotes the driving torque vector of each joint; $\boldsymbol{M}(\boldsymbol{\theta})$ represents the system inertia matrix; $C(\boldsymbol{\theta}, \dot{\boldsymbol{\theta}})$ stands for the centripetal–Coriolis matrix; $\boldsymbol{G}(\boldsymbol{\theta})$ is the gravity matrix; $\boldsymbol{\tau}_o$ refers to the input external torque vector.

Furthermore, an analysis is conducted on the relationship between the motion of the lifting arm (end-effector) and both the forces acting on the luffing rigid rod and the cable tensions. With the input parameters $\boldsymbol{q} = \begin{bmatrix} x_{O_1} & z_{O_1} & \beta \end{bmatrix}^T$, the angle of joint 2 can be expressed as

$$\theta_2 = -\arccos\left(\frac{z_{O_1}}{x_{O_1}}\right) \tag{2.58}$$

When Eq. (2.58) and $\beta = \theta_2 + \theta_3$ are substituted into Eqs. (2.46) and (2.55), the dynamic model of the RFCHMS is

$$\boldsymbol{F}_S = \begin{bmatrix} F_{L1} \\ F_{L2} \\ F_{L3} \\ F_T \end{bmatrix} = \boldsymbol{K}\left(\ddot{\boldsymbol{q}}, \dot{\boldsymbol{q}}, \boldsymbol{q}\right) \tag{2.59}$$

Here, $\boldsymbol{K}$ is the column vector representing the forces of the branch chains. Based on Eqs. (2.45) and (2.59), the relationship between the robot's input parameters $\boldsymbol{q}$ and the driving forces or torques can be derived.

2.4 Conclusion

In this chapter, three kinds of RFCHRs are proposed, which are suitable for large-load environment, high-dexterity environment, and container yard operations environment, respectively; compared with the traditional structure, the RFCHRs designed in this chapter are driven by flexible cables instead of motor drives, so that the self-weight is reduced, and the degree of freedom of the mechanism is improved. At the same time, the kinematic modeling of the three RFCHRs was carried out, and the motion laws of the RFCHRs were revealed. Finally, the dynamic modeling of RFCHR in large-load environment was carried out respectively by using the Lagrange method and the Newton–Euler method.

References

1 Hong H, Ali J, Ren L. A review on topological architecture and design methods of cable-driven mechanism. Advances in Mechanical Engineering, 2018, 10(5): 1687814018774186.

2 Alamdari A, Krovi V. Design and analysis of a cable-driven articulated rehabilitation system for gait training. Journal of Mechanisms and Robotics, 2016, 8(5): 051018.

3 Zhang Z, Shao Z, You Z, et al. State-of-the-art on theories and applications of cable-driven parallel robots. Frontiers of Mechanical Engineering, 2022, 17(3): 37.

4 Luh J Y S, Walker M W, Paul R P C. On-line computational scheme for mechanical manipulators. Journal of Dynamic Systems, Measurement, and Control, 1980, 102(2): 69–76.

5 Shang W, Xie F, Zhang B, et al. Adaptive cross-coupled control of cable-driven parallel robots with model uncertainties. IEEE Robotics and Automation Letters, 2020, 5(3): 4110–4117.

6 Xu P, Cheung C F, Li B, et al. Design, dynamic analysis, and experimental evaluation of a hybrid parallel–serial polishing machine with decoupled motions. Journal of Mechanisms and Robotics, 2021, 13(6): 061008.

7 Ginsberg J H. Advanced Engineering Dynamics (2nd ed.). Cambridge University Press, 2008.

8 Jamshidifar H, Rushton M, Khajepour A. A reaction-based stabilizer for nonmodel-based vibration control of cable-driven parallel robots. IEEE Transactions on Robotics, 2020, 37(2): 667–674.

9 Khajepour A, Behzadipour S, Dekker R, et al. Light weight parallel manipulators using active/passive cables: U.S. Patent 7,367,771 [P]. 2008-5-6.

10 Yang G, Lin W, Pham C, et al. Kinematic design of a 7-DOF cable-driven humanoid arm: a solution-in-nature approach. Proceedings of the 2005 IEEE/ASME International Conference on Advanced Intelligent Mechatronics. IEEE, 2005: 444–449.

11 Yang L, Sen Q, Bin Z. Mechanical design and tension optimization of rigid-flexible hybrid robots for container yard operations. Proceedings of the 2024 IEEE 2nd International Conference on Sensors, Electronics and Computer Engineering (ICSECE). IEEE, 2024: 1131–1138.

12 Bouchard S, Gosselin C, Moore B. On the ability of a cable-driven robot to generate a prescribed set of wrenches. Journal of Mechanisms and Robotics, 2010, 2(1): 011010.

13 Sun G, Wu L, Kuang Z, et al. Practical tracking control of linear motor via fractional-order sliding mode. Automatica, 2018, 94: 221–235.

14 Zhou B, Li S, Zi B, et al. Performance analysis and optimization design of a rigid-flexible coupling hoisting robot considering cable deformation and random loading. Mechanical Systems and Signal Processing, 2025, 223: 111903.

3
Motion Decoupling, Reconfigurable Design of Rigid–Flexible Coupling Hoisting Robots

3.1 Preamble

At present, there are two primary types of cable coupling problems that occur when the intermediate joint rotates: (1) when the joint's rotation axis is perpendicular to the connecting rod, the rotation of the intermediate joint leads to changes in the length of the driving cables at the rear joint, which in turn causes unnecessary extra movements; (2) when the joint's rotation axis is parallel to the connecting rod, the rotation of the intermediate joint results in the entanglement of the driving cables at the rear joint, leading to cable tangling and jamming. As a consequence, motion coupling can give rise to unnecessary additional motions and even cause cable entanglement and jamming. This not only increases the difficulty of robot modeling and computational complexity but also reduces the robot's motion accuracy and control efficiency. To tackle the aforementioned motion coupling issues in cable-driven robots, researchers have carried out extensive studies. Currently, there are two main approaches to motion decoupling in such robots: algorithmic decoupling and mechanical decoupling [1]. Algorithmic decoupling involves taking into account the changes in cable length during the robot's movement and establishing a motion compensation algorithm for the driving cables. By integrating closed-loop feedback control of cable length variations, it realizes decoupled motion in cable-driven robots. The rigid–flexible hybrid-driven low-cost compliant robot developed by Quigley et al. [2] comprises seven revolute joints. The three-degree-of-freedom (DOF) wrist of the end-effector is rigidly actuated using traditional stepper motors, while the shoulder and elbow joints are driven by cables. Through software compensation for the coupling of the driving cables, the first four revolute joints can achieve decoupled rotation and pitching motions. Liang et al. [3] analyzed and modeled the motion coupling in cable-driven robots and put forward a feedforward compensation control strategy based on a feedforward neural network, which realized the motion decoupling of the end-effector for a laparoscopic cable-driven surgical robot. The above-mentioned active decoupling methods require a large amount of computation, placing high demands on the computing power of the host computer, and their real-time performance and stability still need further improvement. Meanwhile, algorithmic decoupling mainly addresses the motion coupling that occurs when joints perform pitching or yawing rotations. However, it fails to solve the problem of driving cable entanglement and jamming when joints perform rolling rotations. Lee et al. [4–6] developed a cable-driven master–slave servo robot with decoupled joints based on the principle of motion compensation. Yuan et al. [7] designed a planetary cable routing mechanism, which simultaneously realizes joint rotation and compensates for changes in cable length. Jiang et al. [1, 8] designed a modular driving cable decoupling mechanism based on the planetary gear principle, achieving motion decoupling for joint pitching and rolling rotations. Nevertheless, the radial arrangement of multiple sets of planetary gears and driving cables is not beneficial to the lightweight design and scalability of the decoupling mechanism. Zhao et al. [9] also designed a motion-decoupling mechanism for surgical grippers based on the planetary gear principle. Shah et al. [10] designed a novel winding mechanism that ensures the cable length remains constant by decoupling the rotation of the pulley around the central axis, thereby achieving motion decoupling of the joints. However, when multiple cables need to be decoupled simultaneously, the complexity of the mechanism increases. In addition to the aforementioned decoupling mechanisms, Chen et al. [11] proposed a brake cable structure, and Liu et al. [12] proposed a method of centrally threading cables through the axis of the connecting rods. However, both of these methods only achieved partial decoupling.

To facilitate the control of the end-effector position and the global configuration of the robot, this section adopts position-level inverse kinematics. The commonly used solution methods for this include numerical methods and analytical methods. Numerical methods utilize an iterative approach, relying on computers to perform extensive calculations to obtain approximate solutions for inverse kinematics. While this method boasts strong universality, its high computational requirements make it unable to meet the real-time solving needs of the control system. This chapter introduces dual-arm angle and configuration control parameters and proposes an inverse kinematics solution method that simultaneously satisfies global configuration control and non-singularity. Additionally, it presents the modular and reconfigurable structural design of a rigid–flexible coupled hoisting robot and conducts an analysis of several classical configurations.

3.2 Motion-decoupling Design for a 7-DOF RFCHR

3.2.1 Coupling Characteristic Analysis and Motion-decoupling Method

To reduce the overall weight and minimize the motion inertia of cable-driven robots, motors and reduction mechanisms are usually placed close to the base. This means that when driving forces are transmitted from the base to the end-effector, the cables must be routed around intermediate joints along specific paths before being attached to the rear joints. The driving cables connected to the rear joints are thus influenced by the rotation of the intermediate joints, causing relative movement or entanglement of the cables and giving rise to motion coupling problems in cable-driven robots. In the field of robotics, the spatial rotation of a revolute joint can be described in terms of roll, pitch, and yaw, as shown in Figure 3.1 [13].

When the intermediate joint rotates in pitch or yaw, the driving cable connected to the rear joint will undergo changes, leading to coupled motion, as depicted in Figure 3.2. Driving cable i is fixed to the driving pulley of joint i, while driving cable $i + 1$ runs through joint i and is fixed to the driving pulley of joint $i + 1$. Taking driving cable $i + 1$ as a case in point, when joint i rotates clockwise by $\Delta\theta$, link i shifts to the dashed position. At this time, the lengths of cable segments AB and CD stay the same. However, due to the rotation of joint i, the length of cable segment BC wrapped around it increases by $\Delta\theta \cdot r$. Given that the total length of driving

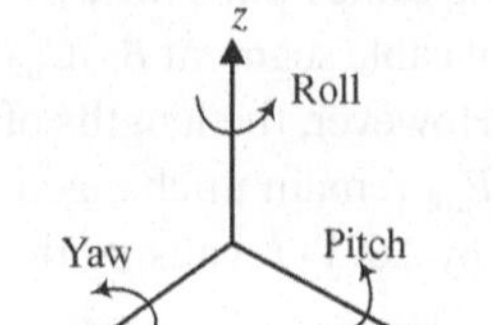

Figure 3.1 Rotation of joints.

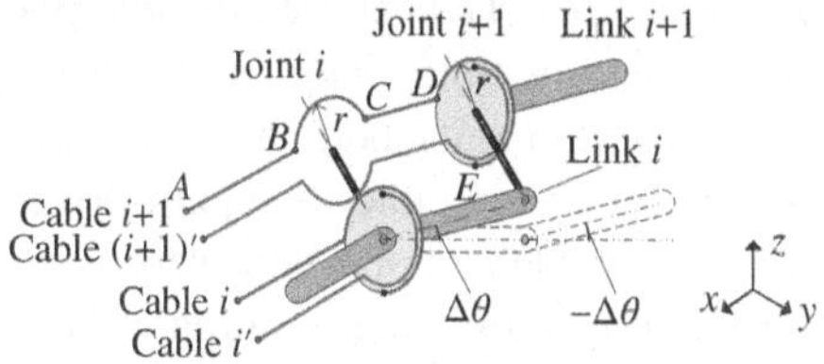

Figure 3.2 Motion coupling when joint yaw or pitch.

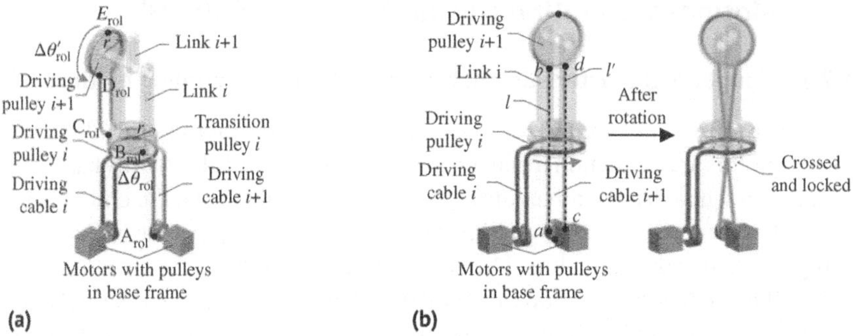

Figure 3.3 The motion coupling and entanglement during roll rotation. (a) The motion coupling during lateral cable routing. (b) The entanglement during intermediate cable routing.

cable $i + 1$ remains unchanged, the length of cable segment DE wrapped around joint $i + 1$ decreases by $\Delta\theta \cdot r$, which causes joint $i + 1$ to rotate counterclockwise by $\Delta\theta$. Eventually, link $i + 1$ rotates to the dashed position.

When the intermediate joint rotates in roll, the driving cable connected to the rear joint will change, leading to motion coupling and even cable entanglement, as illustrated in Figure 3.3. The driving cable $i + 1$ is routed from the side of the transition pulley to the rear joint, with the driving unit fixed to the base. Moreover, one end of cable $i + 1$ is attached to the base's driving pulley, and the other end is secured to the joint's driving pulley. When the driving cable i pulls the driving pulley to rotate counterclockwise by $\Delta\theta_{rol}$, the length of cable segment $B_{rol}C_{rol}$ wrapped around transition pulley i increases by $\Delta\theta_{rol} \cdot r$. However, the lengths of cable segments $A_{rol}B_{rol}$, $C_{rol}D_{rol}$, and the total length $A_{rol}E_{rol}$ remain unchanged. As a result, the length of cable segment $D_{rol}E_{rol}$ decreases by $\Delta\theta_{rol} \cdot r$, causing the link $i + 1$ to rotate counterclockwise by $\Delta\theta'_{rol}$.

As shown in Figure 3.3b, the driving cable $i + 1$ runs through joint i and extends to the rear joint, with the driving unit fixed to the base. When joint i rotates counterclockwise, it drives both link i and driving pulley $i + 1$ to rotate counterclockwise. In the absence of an intermediate transition pulley, segments ac and bd are no longer parallel. This causes the free segments ab and cd between the two pulleys to intersect and lock during rotation. At the same time, as segments ab and cd rotate along with driving pulley $i + 1$, their lengths l and l' both increase. However, the lengths of the driving cable on the motor-driven pulley and joint-driven pulley $i + 1$ remain unchanged. As a result, the total length of driving cable $i + 1$ increases during movement, making it prone to rupture when rotating at large angles.

To tackle the aforementioned motion coupling problems in cable-driven joints, this subsection proposes a motion-decoupling method based on the translation and rotation of movable pulleys. This method is specifically designed to accommodate different types of rotations, including roll, pitch, and yaw.

3.2.1.1 Motion Decoupling Based on Translational Compensation

Figure 3.4 demonstrates the principle of motion decoupling through translational compensation. As depicted in Figure 3.4a,b, the red cable acts as the driving cable for joint $i + 1$. One end of this red cable is fixed to the driving pulley of the motor, then routed through the pulleys on joint i and link i, and ultimately fastened to joint $i + 1$. A key requirement is that the rotation axes of all pulleys must be parallel to one another. The fixed pulley on joint $i + 1$, fixed pulley on link i, and motor with pulley for joint $i + 1$ function as the transmission pulleys, while the decoupling

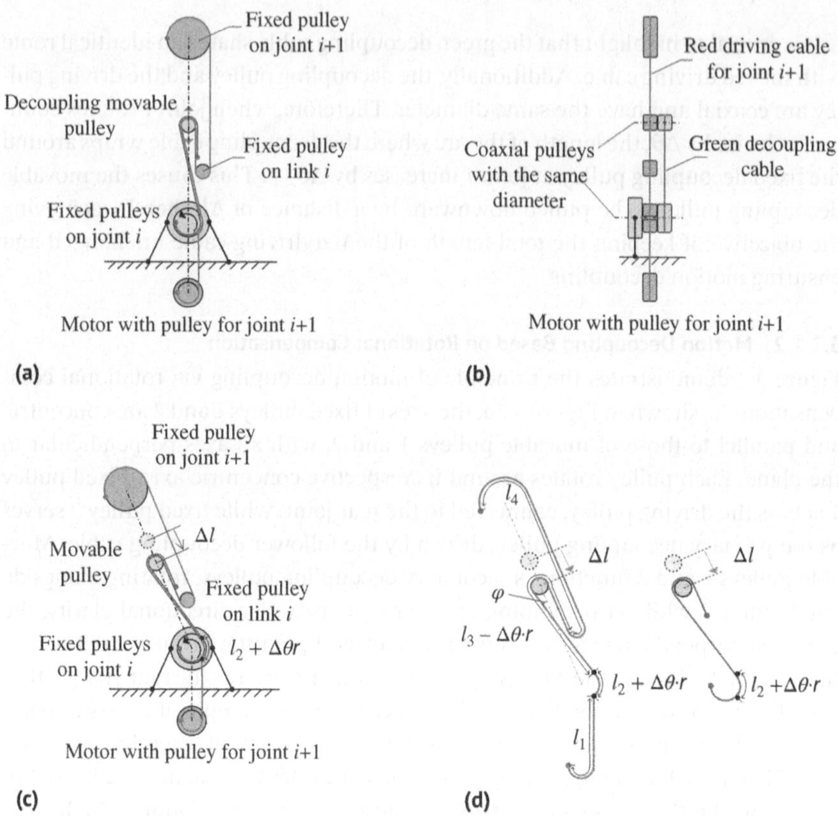

Figure 3.4 The principle of motion decoupling based on translational compensation. (a) Front view before the rotation of joint i; (b) side view before the rotation of joint i; (c) after the rotation of joint i; (d) changes in cable lengths after the rotation of joint i.

movable pulley and fixed pulleys on joint i serve as the decoupling pulleys. It is worth noting that the fixed pulley on joint i is attached to the front link via a cable, ensuring that it does not rotate relative to the front link. The green decoupling cable is fixed to the fixed pulley on joint i, passes through the movable pulley, and is finally secured to link i. It is important to highlight that the green decoupling cable and the red driving cable are parallel to each other.

As shown in Figure 3.4c,d, both the red driving cable and the green decoupling cable are wrapped around the fixed pulley at joint i. When joint i rotates counterclockwise, the lengths of the red and green cables increase simultaneously to $l_2 + \Delta\theta \cdot r$, while the lengths l_1 and l_4 stay unchanged. To keep the total length of the red driving cable constant, l_3 needs to be reduced by $\Delta\theta \cdot r$. Correspondingly, the decoupling fixed pulley must move downward by a distance of Δl, whose value can be calculated using Eq. (3.1),

$$\Delta l/2 + \Delta l/(2 + \cos\varphi) = \Delta\theta \cdot r \tag{3.1}$$

It is crucial to highlight that the green decoupling cable shares an identical route with the red driving cable. Additionally, the decoupling pulley and the driving pulley are coaxial and have the same diameter. Therefore, when joint i rotates counterclockwise by $\Delta\theta$, the length of the arc where the decoupling cable wraps around the fixed decoupling pulley at joint i increases by $\Delta\theta \cdot r$. This causes the movable decoupling pulley to be pulled downward by a distance of Δl, thereby achieving the objective of keeping the total length of the red driving cable unchanged and ensuring motion decoupling.

3.2.1.2 Motion Decoupling Based on Rotational Compensation

Figure 3.5 demonstrates the principle of motion decoupling via rotational compensation. As shown in Figure 3.5a, the axes of fixed pulleys 1 and 2 are concentric and parallel to those of movable pulleys 1 and 2, with all axes perpendicular to the plane. Each pulley rotates around its respective concentric axis. Fixed pulley 1 acts as the driving pulley, connected to the rear joint, while fixed pulley 2 serves as the primary decoupling pulley, driven by the follower decoupling cable. Movable pulleys 1 and 2 function as secondary decoupling pulleys, rotating alongside fixed pulley 2 while also spinning on their own axes. For directional clarity, the direction perpendicular to the paper's surface and pointing inward is defined as the front end (closer to the base), and the outward direction as the rear end (farther from the base). Taking the left-side driving cable as an example: it enters at point A_{roa}, wraps around fixed pulley 2 along the $B_{\text{roa}}C_{\text{roa}}$ segment, then loops around movable pulley 1 along the $C_{\text{roa}}D_{\text{roa}}$ segment, continues to wind around fixed pulley 1 along the $D_{\text{roa}}E_{\text{roa}}$ segment, exits at point F_{roa}, and finally connects to the rear joint. When fixed pulley 1 rotates, fixed pulley 2 as well as movable pulleys 1 and 2 rotate in the same direction but at a lower angular velocity.

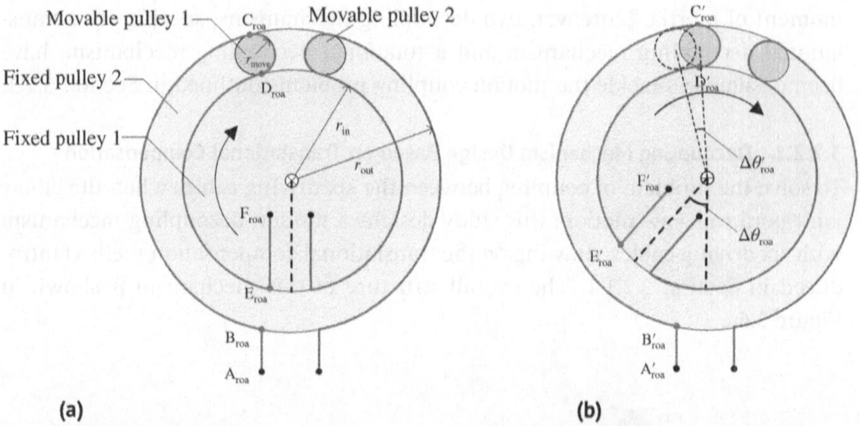

Figure 3.5 The principle of motion decoupling based on rotational compensation. (a) Before the rotation of the driving pulley. (b) After the rotation of the driving pulley.

As illustrated in Figure 3.4, fixed pulley 1 rotates clockwise by $\Delta\theta_{\mathrm{roa}}$ shifting from the state shown in Figure 3.5a to that in Figure 3.4b. Suppose fixed pulley 2 and movable pulleys 1 and 2 stay motionless; in this case, the length of segment $D'_{\mathrm{roa}}F'_{\mathrm{roa}}$ will shorten by $\Delta\theta_{\mathrm{roa}} \cdot r_{\mathrm{in}}$. Meanwhile, the lengths of segments $A_{\mathrm{roa}}B_{\mathrm{roa}}$, $C_{\mathrm{roa}}D_{\mathrm{roa}}$, and $E_{\mathrm{roa}}F_{\mathrm{roa}}$ remain unchanged. Thus, to ensure that the equation $|A'_{\mathrm{roa}}F'_{\mathrm{roa}}| = |A_{\mathrm{roa}}F_{\mathrm{roa}}|$ holds after rotation, movable pulley 1 needs to rotate clockwise by $\Delta\theta'_{\mathrm{roa}}$ around the axis of fixed pulleys 1 and 2, which must satisfy Eq. (3.2),

$$\Delta\theta_{\mathrm{roa}} \cdot r_{\mathrm{in}} = \Delta\theta'_{\mathrm{roa}} \cdot r_{\mathrm{in}} + \Delta\theta'_{\mathrm{roa}} \cdot r_{\mathrm{out}} \tag{3.2}$$

When fixed pulley 1 rotates, the reduction in the length of segment $D'_{\mathrm{roa}}E'_{\mathrm{roa}}$ is exactly offset by the combined increase in the lengths of segments $B'_{\mathrm{roa}}C'_{\mathrm{roa}}$ and $D'_{\mathrm{roa}}E'_{\mathrm{roa}}$ caused by the rotation of movable pulley 1, thereby achieving motion decoupling. From Eq. (3.3), it can be derived that $\Delta\theta'_{\mathrm{roa}} = \Delta\theta_{\mathrm{roa}} \cdot r_{\mathrm{in}}/(r_{\mathrm{in}} + r_{\mathrm{out}})$.

3.2.2 Mechanical Design of 7-DOF RFCHR

As depicted in Figure 3.6, a 7-DOF cable-driven serial spray-painting robot integrated with motion-decoupling mechanisms has been developed [14]. Its adjacent joints are arranged perpendicularly to one another, and the robot features a typical nonbiased spherical–roll–spherical configuration. Seven groups of servo actuators are centrally mounted on the base, with driving forces transmitted via multiple cables to the robot's seven rotational joints. This design reduces the weight of the robot body (excluding the driving mechanism) and lowers its moment of inertia. Moreover, two decoupling mechanisms, specifically a translational decoupling mechanism and a rotational decoupling mechanism, have been designed to tackle the motion coupling problems outlined in Section 3.2.1.

3.2.2.1 Decoupling Mechanism Design Based on Translational Compensation

To solve the problem of coupling between the six driving cables when the elbow joint performs yaw motion, this study designs a motion-decoupling mechanism with six driving cables, drawing on the translational compensation method introduced in Section 3.2.1.1. The overall structure of this mechanism is shown in Figure 3.6a.

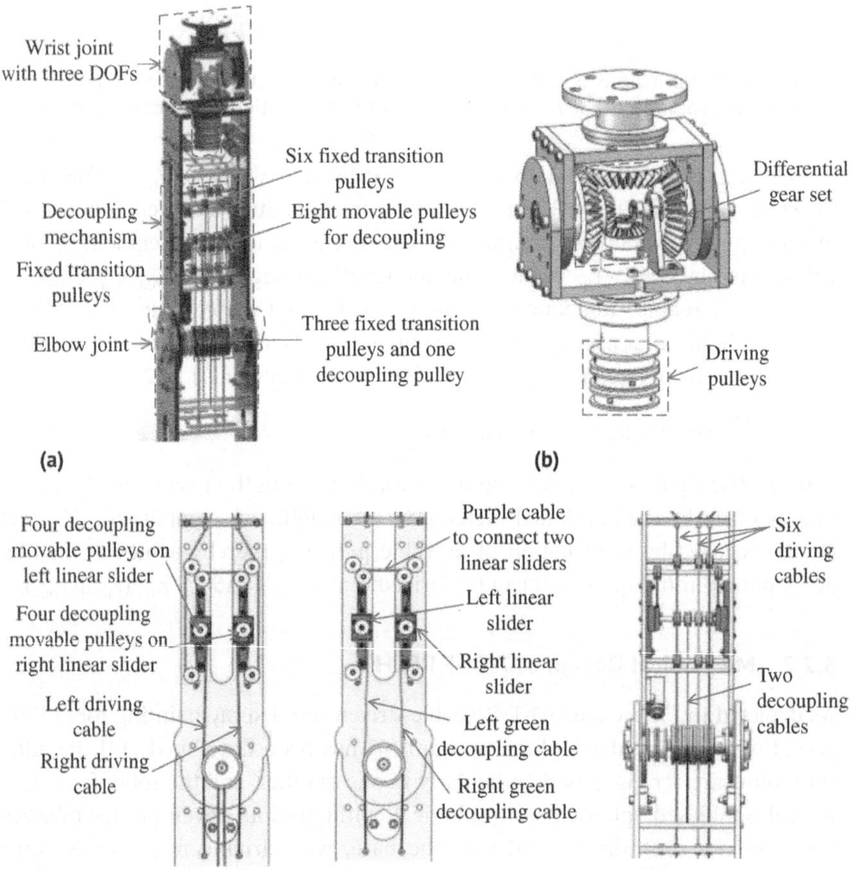

Figure 3.6 (a) Design of six-cable decoupling mechanism base on translational compensation; (b) wrist joints; (c) side view of driving cable winding; (d) side view of decoupling cable winding; (e) front view. *Source*: [14]/with permission of Elsevier.

Finally, a wrist joint with three DOFs is integrated, characterized by three rotational axes that adhere to the principles elaborated in this section; specifically, the axes are mutually perpendicular and intersect at a single point. The primary function of the robot's wrist joints is to adjust the posture. As shown in Figures 3.5b and 3.6a, the wrist joint utilizes a cable-driven mechanism to transmit driving forces from the base to the terminal actuated pulley. Motion is then transferred to the differential gear via shafts and bushings, enabling the 3-DOF motion of the wrist joint.

In the proposed mechanism, eight decoupling movable pulleys are coaxially mounted between linear slides, and motion coupling compensation is mainly achieved through the synchronous linear movement of these eight pulley sets. The translational compensation-based decoupling mechanism, as shown in Figure 3.4, operates on a similar principle. The cable routing method and decoupling principle of this proposed mechanism are illustrated in Figure 3.6c–e. When the elbow joint rotates counterclockwise, the length of the green decoupling cable wrapped around the elbow joint's decoupling pulley increases by $\Delta\theta \cdot r$, causing the right-side decoupling movable pulley to move downward by Δl, as depicted in Figures 3.4c,d and 3.6d. At the same time, the purple cable fixed to the upper end of the right linear slide pulls the left linear slide upward by an equal distance Δl, and the length of the green decoupling cable on the elbow joint's decoupling pulley decreases by $\Delta\theta \cdot r$ accordingly. As shown in Figure 3.6c,e, the decoupling movable pulleys on the linear slide are coaxial and have the same diameter. When the elbow joint rotates counterclockwise as described, the right-side green decoupling cable pulls the linear slide downward by Δl. According to Eq. (3.1), the increase in length of the right red driving cable at the elbow joint is exactly offset by the decrease in length of the red driving cable after the slide moves downward. This ensures that the total length of the red driving cable remains unchanged, thereby achieving motion decoupling of the driving cable. The decoupling principles for the other four driving cables of the two wrist joints follow the same logic.

3.2.2.2 Decoupling Mechanism Design Based on Rotational Compensation

Building on the rotational compensation-based decoupling method detailed in Section 3.2.1.2, two rotational compensation decoupling mechanisms, featuring 8 cables and 16 cables, respectively, are proposed, as shown in Figure 3.7. These mechanisms are specifically designed to achieve motion decoupling for joint 1 and joint 3 during rolling rotation. Each decoupling mechanism comprises driving rotary drums and decoupling rotary drums: the driving rotary drums are linked to the rear linkages, while the decoupling rotary drums are connected to the front linkages.

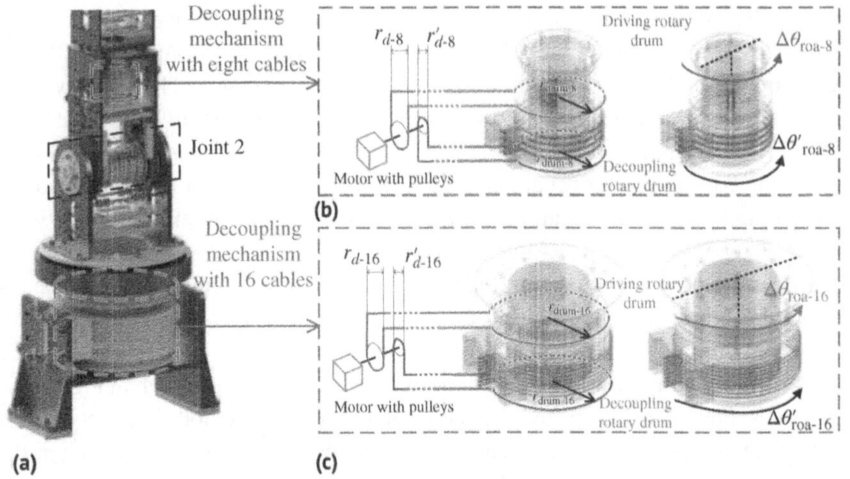

Figure 3.7 (a) Design of eight-cable and 16-cable decoupling mechanism base on rotational compensation. (b) Decoupling principle of eight-cable mechanism. (c) Decoupling principle of 16-cable mechanism. *Source*: [14]/with permission of Elsevier.

Taking the eight-cable decoupling mechanism as an example, and with reference to Figure 3.5, the outer diameter of the driving rotary drum is $r_{\text{in-8}}$, while that of the decoupling rotary drum is $r_{\text{out-8}}$. The corotational movement of these two drums around the central axis enables motion decoupling during the rolling rotation of the joint. The rotational angles must satisfy Eq. (3.2), namely, $\Delta\theta'_{\text{roa}} = \Delta\theta_{\text{roa}} \cdot r_{\text{in-8}}/(r_{\text{in-8}} + r_{\text{out-8}})$. As a result, the radius of the rotary drum for the driving decoupling mechanism must comply with Eq. (3.3),

$$\begin{cases} r_{d\text{-}8} \cdot \Delta\theta_d = r_{\text{drum-8}} \cdot \Delta\theta_{\text{roa-8}} \\ r'_{d\text{-}8} \cdot \Delta\theta_d = r_{\text{drum-8}} \cdot \Delta\theta'_{\text{roa-8}} \end{cases} \tag{3.3}$$

Here, $r_{d\text{-}8}$ and $r'_{d\text{-}8}$ refer to the radii of the coaxial driving pulleys on the motors, corresponding respectively to the driving rotary drum and the decoupling rotary drum. $\Delta\theta_d$ stands for the rotational angle of the driving motor, and $r_{\text{drum-8}}$ is the radius shared by both the driving and decoupling drums. By substituting $\Delta\theta'_{\text{roa}} = \Delta\theta_{\text{roa}} \cdot r_{\text{in-8}}/(r_{\text{in-8}} + r_{\text{out-8}})$ into Eq. (3.3), we derive $r'_{d\text{-}8} = r_{d\text{-}8} \cdot r_{\text{in-8}}/(r_{\text{in-8}} + r_{\text{out-8}})$. In the same way, for the 16-cable decoupling mechanism based on rotational compensation, the relationship $r'_{d\text{-}16} = r_{d\text{-}16} \cdot r_{\text{in-16}}/(r_{\text{in-16}} + r_{\text{out-16}})$.

3.2.2.3 Tension-amplification Mechanism Design for Joints 2 and 4

A tension amplifier utilizing movable pulleys is capable of not only boosting cable tension but also improving the stiffness of cable-driven joints. Both single-DOF and multi-DOF tension amplification systems have been applied to rolling

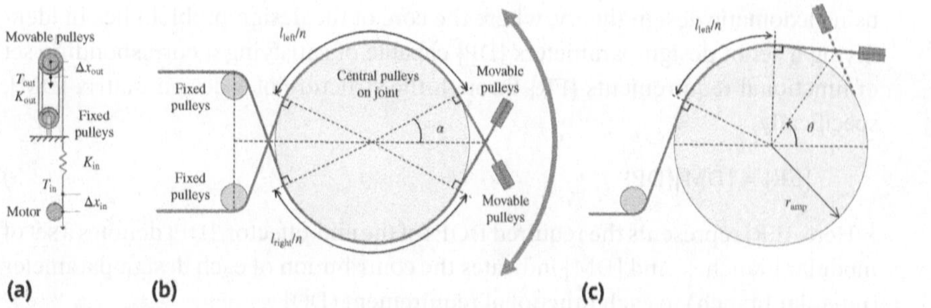

Figure 3.8 (a) Principle diagram of the tension amplification mechanism using movable pulleys. (b) Tension amplification mechanism for rotary joints. (c) Relationship between cable length and joint angle.

joints, enabling the design of high-stiffness structures in humanoid robots [15, 16]. Figure 3.8a demonstrates the fundamental principle of tension amplification based on a block and tackle pulley system. The tension (T_{in}) generated by the driving motor is transmitted through a pulley system equipped with n (=4) driving cables. The output tension exerted on the movable pulley can be computed using the formula $T_{out} = nT_{in}$. Meanwhile, the input displacement (Δx_{in}) of the driving motor is n times the output displacement (Δx_{out}) of the movable pulley. As a result, the stiffness of the pulley system can be calculated using Eq. (3.4),

$$K_{out} = \frac{T_{out}}{\Delta x_{out}} = \frac{nT_{in}}{\Delta x_{in}/n} = n^2 \frac{T_{in}}{\Delta x_{in}} = n^2 K_{in} \tag{3.4}$$

Here, K_{in} represents the stiffness of the input driving cable, and K_{out} denotes the output stiffness of the movable pulley mechanism. Figure 3.8b,c shows the motion diagrams of the designed tension amplification mechanism for rotary joints. The movable pulley rotates around its axis within the range of $[-\pi + a, \pi - a]$, which matches the motion range of the joint. Taking a group of movable pulleys as an example, the variation in the length of the driving cable can be expressed in terms of the joint angle as $l_{left}/2 = \theta r_{amp}$.

3.3 Modular and Reconfigurable Mechanism Design of RFCHRs

3.3.1 Design Methodology

The design concept for the mechanism design of rigid–flexible coupling hoisting robot (MDRFCHR) hinges on the utilization of modular branches as fundamental building blocks. The design methodology employed draws on the design for reconfigurability approach. The design process of the MDRFCHR can be described using axiomatic design theory, where the core of the design problem lies in identifying a set of design parameters {DP} capable of satisfying a corresponding set of functional requirements {FR} through the structure of a design matrix [DM], specifically,

$$\{\mathrm{FR}\} = [\mathrm{DM}]\{\mathrm{DP}\} \tag{3.5}$$

Here, {FR} represents the required DOFs of the end-effector, {DP} denotes a set of modular branches, and [DM] indicates the contribution of each design parameter (modular branch) to each functional requirement (DOF).

Thus, the design objective for the MDRFCHR is defined as maximizing the number of functional requirements (FR) while minimizing the number of modular branches. (DP). This can be formulated as a bi-objective design optimization problem expressed as

$$\underset{k}{\mathrm{Max}}\underset{n}{\mathrm{Min}}\ (\{\mathrm{FR}(k), \mathrm{DP}(n)\}) \tag{3.6}$$

where n represents the number of modular branches, and k stands for the DOFs of the end-effector.

3.3.2 Mechanical Description

The three-dimensional model and physical prototype of the MDRFCHR are presented in Figures 3.9 and 3.10, respectively. The main components of the MDRFCHR include a circular orbit, 6 sets of telescopic and rotatable modular branches, 6 cables, 6 tension sensors, 12 rope displacement sensors, an end-effector, a hydraulic station, a control cabinet, and an industrial personal computer, among others. Each modular branch is primarily composed of a mobile

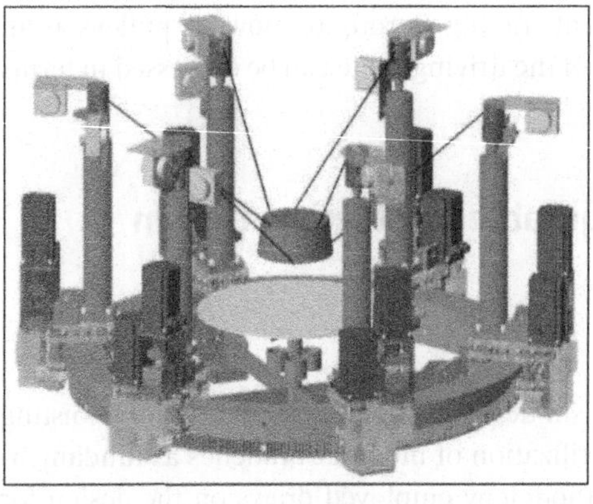

Figure 3.9 Three-dimensional model of the MDRFCHR.

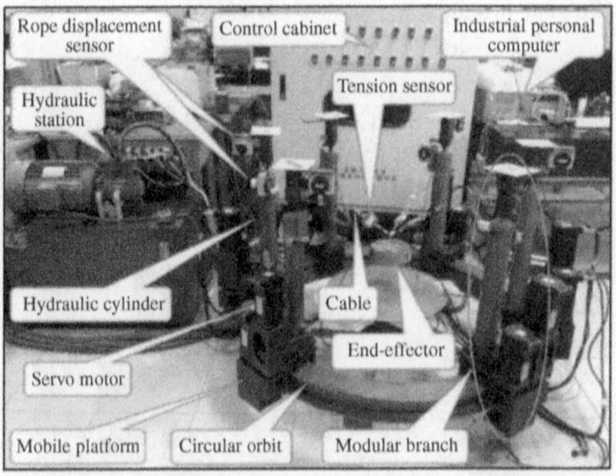

Figure 3.10 Physical prototype of the MDRFCHR. *Source*: [17]/Sage Publications/CC BY 4.0.

platform, 2 servo motors, 2 motor seats, a hydraulic cylinder, 2 rope displacement sensors, a fixed pulley, a fixed pulley bracket, 4 bevel gears, a cylindrical gear, 4 shafts, a gear shaft, a drum, 4 guide wheels, 17 bearings, 2 couplings, etc. The assembly and disassembly diagram of the modular branch is illustrated in Figure 3.11.

Each modular branch is capable of rotating along the circular orbit via gear transmission, and its height can be adjusted by a hydraulic cylinder. The mobile platform and drum of each modular branch are driven by separate servo motors. Four guide wheels, mounted inside the mobile platform of each modular branch, enhance the stability and smoothness of the branch's rotation along the circular orbit. For each cable, one end is connected to the corresponding attachment point on the end-effector, while the other end passes through the fixed pulley at the top of the modular branch and is wound around a drum fixed to the mobile platform. It is important to note that when the end-effector moves near the boundaries of the workspace, significant cable entanglement may occur relative to the fixed pulley stents. To address this, each fixed pulley at the top of the modular branch is installed on an omnidirectional fixed pulley stent, allowing it to adapt to changes in the end-effector's position. This arrangement ensures the cable remains perpendicular to the drum axis, enabling reliable coiling and uncoiling. A tension sensor is connected in series with each cable near the end-effector to measure cable tension. Additionally, two rope displacement sensors are mounted on the top of each modular branch, which accurately measure the cable length and the height of the hydraulic cylinder, respectively.

Based on diverse task requirements, the spatial topology of the MDRFCHR can be reconfigured by manually adding or removing modular branches and adjusting the connection points on the end-effector. Unlike traditional cable-driven parallel

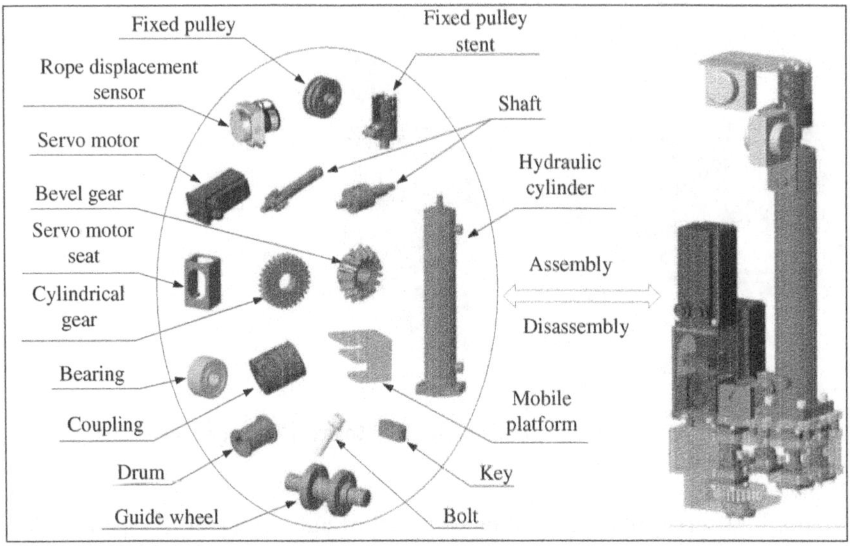

Figure 3.11 Assembly and disassembly diagram of modular branch.

robots (CPRs), the proposed MDRFCHR features reconfigurability, enabling it to take on multiple different configurations using six identical modular branches. Specifically, by detaching one, two, three, or four modular branches respectively, the MDRFCHR can be reconfigured from a 6-DOF system to a 4-DOF, 3-DOF, or 2-DOF system.

3.3.3 Typical Configuration

A variety of typical configurations for the MDRFCHR can be achieved through reconfiguration, which involves adjusting the quantity of modular branches and modifying the connection points on the end-effector, tasks carried out with manual assistance, as illustrated in Figure 3.12.

Typically, CPRs with n DOFs driven by m cables can be broadly categorized into three main types: incompletely restrained CPRs (IRCPRs) when $n + 1 > m$, completely restrained CPRs (CRCPRs) when $n + 1 = m$, and redundantly restrained CPRs (RRCPRs) when $n + 1 < m$. As shown in Figure 3.12, configuration I comprises two modular branches and two cables, enabling two translational motions within a 2D plane. For the 2-DOF CPRs, $\{FR\}$ can be specifically defined as $\{x, z\}^T$. Configuration II consists of three modular branches and three cables, allowing three translational motions in 3D space. Configuration III includes four modular branches and four cables, also capable of realizing three translational motions in 3D space. For these two 3-DOF CPRs, $\{FR\}$ can be specifically defined as $\{x, y, z\}^T$. Configuration IV is made up of five modular branches and five cables, which can achieve three translational motions and one rotational motion in 3D space; for

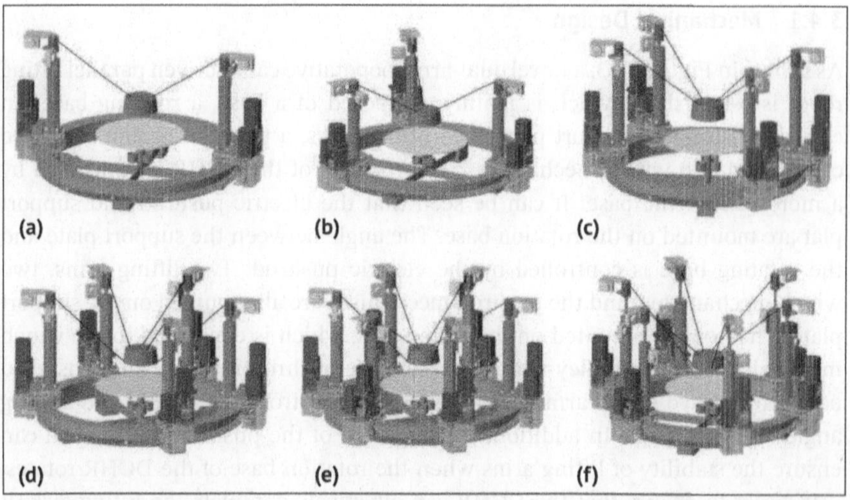

Figure 3.12 Several typical configurations of the MDRFCHR. (a) Configuration I, (b) configuration II, (c) configuration III, (d) configuration IV, (e) configuration V, and (f) configuration VI.

this 4-DOF, {FR} can be specifically defined as $\{x, y, z, \beta\}^T$. Configurations V and VI consist of six modular branches and six cables, enabling three translational motions and three rotational motions in 3D space. For the 6-DOF CDPR, {FR} can be specifically defined as $\{x, y, z, \alpha, \beta, \gamma\}^T$. In all configurations, x, y, z denote the three translational motions of the end-effector along the X-, Y-, and Z-axes of the global coordinate frame, respectively, while α, β, γ represent the three rotational motions of the end-effector which rotate about Z-, Y-, and X-axes in global coordinate frame, respectively. Additionally, the heights and/or angles of the modular branches in each configuration can be adjusted when the hydraulic cylinders and/or mobile platforms of the modular branches are not locked, thereby increasing the number of motions compared to their original configurations. Meanwhile, the MDRFCHR exhibits the characteristics of IRCPRs, CRCPRs, and RRCPRs. In other words, the parameters of the MDRFCHR's various configurations can be adjusted according to specific task requirements, which enhances flexibility and reduces costs in optimization processes.

3.4 Integrated Mechanism Design of Dual-machine Collaborative RFCHRs

This section presents a mechanical prototype of the dual-machine collaborative rigid–flexible coupling hoisting robot (DCHR), along with the construction of its kinematic and dynamic models.

3.4.1 Mechanical Design

As shown in Figure 3.13, a novel dual-arm cooperative cable-driven parallel lifting robot is established, which is mainly composed of a base, a rotating base, an electric pushrod, a support plate, two lifting arms, a pushrod mechanism, two cables, and two winch mechanisms. The rotation of the DCHR is controlled by a motor under the base. It can be seen that the electric pushrod and support plat are mounted on the rotation base. The angle between the support plate and the rotating base is controlled by the electric pushrod. Two lifting arms, two winch mechanisms, and the pushrod mechanism are all mounted on the support plate. The hook is mounted on the lifting arm, which is connected to the winch mechanism through pulley and cable reel. The pushrod mechanism is installed between the two lifting arms, which is used to control the opening and closing angle of lifting arms. In addition, the existence of the pushrod mechanism can ensure the stability of lifting arms when the rotation base of the DCHR rotates.

In order to ensure that the DCHR can effectively complete the lifting task, it is designed as a mechanism with five DOFs. Taking one of the lifting arms as an example, it consists of a pulley, a sliding sleeve, a ball screw, a coupling, a motor, and a connecting sleeve. When the push rod mechanism controls the opening and closing angle of lifting arms, the rotating shaft in the connecting sleeve rotates accordingly. In addition, the lifting arm is equipped with retractable function.

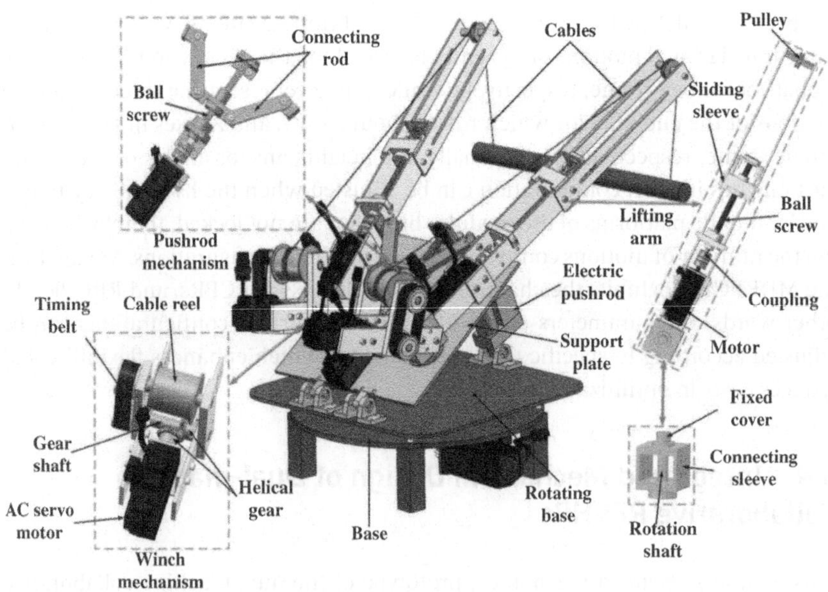

Figure 3.13 Mechanical prototype of the DCHR.

The extension length of the lifting arm is controlled by the motor, effectively increasing the working space of the DCHR in practical engineering applications. In addition, the pose of the payload is controlled by two cables and two lifting arms. Meanwhile, it also can change with the size of the rotation angle of the DCHR and the speed of change.

3.4.2 Kinematic Modeling

The structural schematic of the DCHR is shown in Figure 3.14. A global coordinate system $\{B\}: O\text{–}XYZ$ is located at the midpoint of the hinged connection line between the support plate and rotating base. A moving coordinate system $\{P\}: O_P\text{–}X_PY_PZ_P$ is located at the center of the payload. A_1 and A_2 are located at the hinge points of the lifting arm and support plate, respectively; B_1 and B_2 are located at the connection points of the lifting arms and the cables, respectively; C_1 and C_2 are located at both ends of the payload, respectively; l is the length between points C_1 and C_2; L_i is the length of the ith lifting arm; S_i is the length of the ith cable; D is the distance between the origin of global coordinate system O and the midpoint of A_1A_2; L_s is the length of the square support plate; L_r is the length of the winch mechanism; d is the length between points A_1 and A_2; γ is the amplitude angle; α is the angle of the lifting arm rotation on the plane of the support plate; and θ is the angle of the moving coordinate system around the Y-axis.

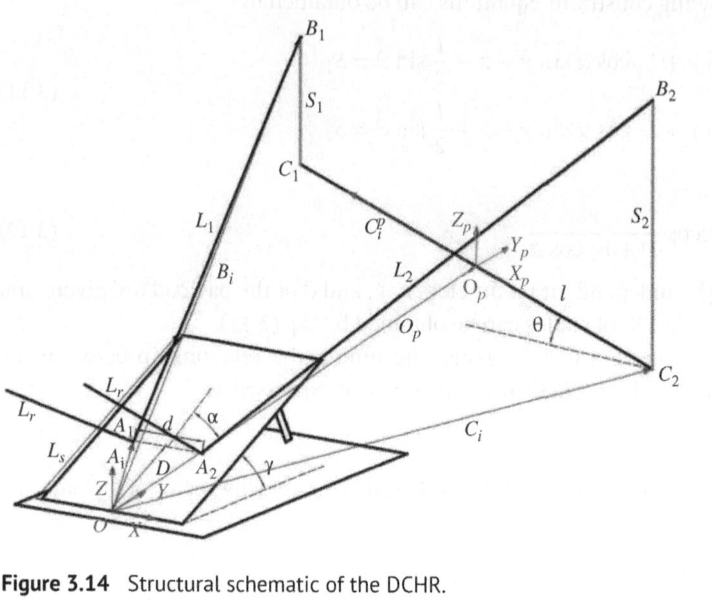

Figure 3.14 Structural schematic of the DCHR.

The attitude rotation matrix $\boldsymbol{R}$ of the moving coordinate system $\{P\}$ relative to the global coordinate system $\{B\}$ can be expressed as

$$\boldsymbol{R} = \begin{bmatrix} \cos\theta & 0 & \sin\theta \\ 0 & 1 & 0 \\ -\sin\theta & 0 & \cos\theta \end{bmatrix} \tag{3.7}$$

The position vector of the moving coordinate system $\{P\}$ relative to the global coordinate system $\{B\}$ is $\boldsymbol{O_p} = \begin{pmatrix} x & y & z \end{pmatrix}^T$.

The rotation angle α of the lifting arm changes with the increase of the angle of the moving coordinate system around the Y-axis θ, and

$$\alpha = \arcsin\frac{l\cos\theta - d}{2L_1} \tag{3.8}$$

In addition, the distances between points B_i and C_i in the directions of X-axis and Y-axis are the same, and

$$y = D\cos\gamma + L_1\cos\alpha\cos\gamma \tag{3.9}$$

Based on the closed vector principle, the length of the ith cable S_i can be calculated by

$$S_i = \|\boldsymbol{C}_i - \boldsymbol{B}_i\|,\ i = 1, 2 \tag{3.10}$$

and the following constraint equations can be obtained by

$$\begin{cases} D\sin\gamma + L_1\cos\alpha\sin\gamma - z - \dfrac{l}{2}\sin\theta = S_1 \\ D\sin\gamma + L_2\cos\alpha\sin\gamma - z + \dfrac{l}{2}\sin\theta = S_2 \end{cases} \tag{3.11}$$

where

$$\gamma = \arccos\frac{y}{D + L_1\cos\alpha} \tag{3.12}$$

Note that the independent parameters z, y, and θ of the payload are given, and the lengths S_1 and S_2 of cables can be obtained by Eq. (3.11).

By differentiating Eq. (3.11) against the time t, the relationship between the velocity vectors of the payload and cables can be expressed as

$$\dot{\boldsymbol{S}} = \boldsymbol{J}\dot{\boldsymbol{\alpha}} \tag{3.13}$$

where $\boldsymbol{J}$ is the Jacobian matrix of the DCHR; $\dot{\boldsymbol{\alpha}}$ is the velocity vector of the payload; and

$$\dot{\boldsymbol{S}} = \begin{bmatrix} \dot{S_1} \\ \dot{S_2} \end{bmatrix} \tag{3.14}$$

$$\dot{\boldsymbol{a}} = \begin{bmatrix} \dot{z} & \dot{y} & \dot{\theta} \end{bmatrix}^T \tag{3.15}$$

$$\boldsymbol{J} = \begin{pmatrix} \frac{\partial S}{\partial z} & \frac{\partial S}{\partial y} & \frac{\partial S}{\partial \theta} \end{pmatrix} \tag{3.16}$$

3.4.3 Dynamic Modeling

The robot is assumed to be well lubricated, allowing the frictional forces within the system to be neglected. The Lagrangian method is adopted to establish dynamic modeling in this subsection. The payload is considered as a uniform rigid rod with mass m and length l. Thus, the moment of inertia of the payload around point O_p can be calculated by

$$I_m = \frac{1}{12} m l^2 \tag{3.17}$$

Based on the kinematics of DCHR, the velocity matrix of the payload can be expressed as

$$\begin{bmatrix} v_y \\ v_z \\ w \end{bmatrix} = \begin{bmatrix} \frac{\partial y}{\partial \alpha}\dot{\alpha} + \frac{\partial y}{\partial \gamma}\dot{\gamma} + \frac{\partial y}{\partial S_1}\dot{S}_1 + \frac{\partial y}{\partial S_2}\dot{S}_2 \\ \frac{\partial z}{\partial \alpha}\dot{\alpha} + \frac{\partial z}{\partial \gamma}\dot{\gamma} + \frac{\partial z}{\partial S_1}\dot{S}_1 + \frac{\partial z}{\partial S_2}\dot{S}_2 \\ \frac{\partial \theta}{\partial \alpha}\dot{\alpha} + \frac{\partial \theta}{\partial \gamma}\dot{\gamma} + \frac{\partial \theta}{\partial S_1}\dot{S}_1 + \frac{\partial \theta}{\partial S_2}\dot{S}_2 \end{bmatrix} \tag{3.18}$$

where

$$z = (D + L_1 \cos \alpha) \sin \gamma - \frac{S_2 + S_1}{2} \tag{3.19}$$

$$\theta = \arcsin\left(\frac{S_2 - S_1}{l}\right) \tag{3.20}$$

in which the linear velocity vector of the payload in the global coordinate system $\{B\}$ can be expressed as

$$v = \begin{bmatrix} 0 & v_y & v_z \end{bmatrix}^T \tag{3.21}$$

The total kinetic energy K_m of the payload is defined as the sum of its translational and rotational kinetic energies. Accordingly, the total kinetic energy of the payload can be expressed as

$$K_m = \frac{1}{2} m v^T v + \frac{1}{2} I_m w^2 \tag{3.22}$$

The potential energy of the payload consists of gravitational potential energy, which is given by

$$U_m = mgz \tag{3.23}$$

where g is the acceleration of gravity.

It can be seen from the Figure 3.14 that the single lifting arm has two rotation angles. One is rotation angle α, and the other is rotation angle γ. Thus, the calculation process of the moment of inertia of the single lifting arm also has two parts. The mass of the lifting arm is m_L, which is evenly distributed and $L_1 = L_2 = L$. The moment of inertia of the lifting arm L_1 around the hinge between the support plate and the lifting arm L_1 can be expressed as

$$I_{L_1}^1 = \frac{1}{3} m_L L^2 \tag{3.24}$$

The moment of inertia of the lifting arm L_1 around the hinge between the support plate and rotating base can be expressed as

$$I_{L_1}^2 = \frac{m_L}{3L} \left((D + L \cos \alpha)^3 - D^3 \right) \tag{3.25}$$

The total kinetic energy of the lifting arm can be calculated by

$$K_L = \frac{1}{2} I_{L_1}^1 \dot{\alpha}^2 + \frac{1}{2} I_{L_1}^2 \dot{\gamma}^2 \tag{3.26}$$

The center of gravity of the lifting arm coincides with its geometric center. Therefore, the center of gravity of the lifting arm along the Z-axis of the global reference coordinate system $\{B\}$ can be expressed as

$$Z_L = D \sin \gamma + \frac{1}{2} L \cos \alpha \sin \gamma \tag{3.27}$$

The potential energy of the lifting arms can be calculated by

$$U_L = m_L g Z_L = m_L g \left(D \sin \gamma + \frac{1}{2} L \cos \alpha \sin \gamma \right) \tag{3.28}$$

Similarly, the kinetic energy K_r and potential energy U_r of the single winch mechanism can be respectively calculated by

$$I_{r_1}^1 = m_r \left(L_r \cos \left(\frac{\pi}{4} \right) \right)^2 \tag{3.29}$$

$$I_{r_1}^2 = m_r \left(\left(L_r \cos \left(\frac{\pi}{4} \right) \cos \alpha - D \right)^2 + \left(L_r \sin \left(\frac{\pi}{4} \right) \right)^2 \right) \tag{3.30}$$

$$K_r = \frac{1}{2} I_{r_1}^1 \dot{\alpha}^2 + \frac{1}{2} I_{r_1}^2 \dot{\gamma}^2 \tag{3.31}$$

$$U_r = m_r g \left(L_r \sin\left(\frac{\pi}{4}\right) \cos(\gamma) + \left(D - L_r \cos\left(\frac{\pi}{4}\right)\right) \sin\gamma \right) \tag{3.32}$$

The support plate is a square sheet with uniform mass distribution. The mass of the support plate is m_0; the length of the support plate is L_s; and the center of gravity is $\left(0, \frac{1}{2}L_s \cos\gamma, \frac{1}{2}L_s \sin\gamma\right)$. Thus, the moment of inertia of the support plate can be expressed as

$$I_0 = \frac{1}{3} m_0 L_s^2 \tag{3.33}$$

The kinetic energy of the support plate can be calculated by

$$K_0 = \frac{1}{2} I_0 \dot{\gamma}^2 \tag{3.34}$$

The gravitational potential energy of the support plate can be calculated by

$$U_0 = \frac{1}{2} m_0 L_s g \sin\gamma \tag{3.35}$$

Based on Eqs. (3.22)–(3.34), the total Lagrangian for the robot according to the Lagrangian dynamic formulation can be expressed as

$$L = K - U \tag{3.36}$$

where

$$K = K_m + 2K_L + 2K_r + K_0 \tag{3.37}$$

$$U = U_m + 2U_L + 2U_r + U_0 \tag{3.38}$$

The vector of generalized coordinates $\boldsymbol{q}$ of the robot can be expressed as

$$\boldsymbol{q} = (\alpha, \gamma, S_1, S_2) \tag{3.39}$$

By taking derivatives of Eq. (3.30), the dynamic equation for the robot can be expressed as

$$\tau = \frac{d}{dt}\frac{\partial L}{\partial \dot{\boldsymbol{q}}} - \frac{\partial L}{\partial \boldsymbol{q}} = \boldsymbol{M}(\boldsymbol{q})\,\ddot{\boldsymbol{q}} + \boldsymbol{C}(\boldsymbol{q}, \dot{\boldsymbol{q}}) + \boldsymbol{G}(\boldsymbol{q}) \tag{3.40}$$

where τ are the external forces and torques, and $\tau = \begin{bmatrix} T_1 & T_2 & T_3 & T_4 \end{bmatrix}^T$; $\dot{\boldsymbol{q}}$ and $\ddot{\boldsymbol{q}}$ are the velocity and acceleration vectors, respectively; $\boldsymbol{M}(\boldsymbol{q})$ is the inertia matrix; $\boldsymbol{C}(\dot{\boldsymbol{q}}, \boldsymbol{q})$ is the centrifugal and Coriolis forces; and $\boldsymbol{G}(\boldsymbol{q})$ is the gravity matrix.

3.5 Conclusion

This chapter outlines the design and realization of a 7-DOF rigid–flexible coupling hoisting robot integrated with motion-decoupling mechanisms. The research accomplished independent motion decoupling for the robot's joints, presenting a feasible resolution to the problem of coupling in cable-driven joint movements. Moreover, a nonsingular inverse kinematics method was put forward, which incorporates dual-arm angles and configuration control parameters. This method enables both the control of the robot's configuration and the avoidance of singularities in inverse kinematics calculations. Initially, an analysis was conducted on the motion coupling phenomenon in cable-driven serial robots. On the basis of this analysis, translational and rotational compensation approaches were introduced. Afterward, translational and rotational decoupling mechanisms were developed according to these approaches, designed specifically to decouple the rolling motion of joint 1 and the pitching motion of joint 4 in cable-driven robots. Subsequently, an investigation into the robot's redundancy and singularities was performed, which led to the proposition of the aforementioned nonsingular inverse kinematics method that combines dual-arm angles and configuration control parameters.

The decoupling mechanisms presented herein can function as modular solutions, suitable for adaptation in other rigid–flexible coupling robots. However, in practical hoisting operations involving large-scale equipment, certain factors are unavoidable, including elastic deformation of cables, joint compliance, and system friction, all of which impact the overall performance of the robotic system. As a result, our future research efforts will focus mainly on scaling up the physical prototype. This will entail further exploring methods to alleviate the aforementioned adverse effects, such as optimizing the structural design, strategically increasing the number of driving cables, and conducting analyses of the robot's rigid–flexible coupling dynamics.

References

1 Jiang S R, Wang Y Y, Li B B, et al. Design and verification of a novel motion-decoupled cable-driven manipulator. Proceedings of the Institution of Mechanical Engineers, Part I: Journal of Systems and Control Engineering, 2020, 234(6): 690–700.

2 Quigley M, Asbeck A, Ng A. A low-cost compliant 7-DOF robotic manipulator. 2011 IEEE International Conference on Robotics and Automation (ICRA), 2011: 6051–6058.

3 Liang Y L, Du Z Z, Wang W D, et al. An improved scheme for eliminating the coupled motion of surgical instruments used in laparoscopic surgical robots. Robotics and Autonomous Systems, 2019, 112: 49–59.

4 Lee J K, Choi C H, Yoon K H, et al. Design and evaluation of cable-driven manipulator with motion-decoupled joints. 2008 International Conference on Smart Manufacturing Application (ICSMA), 2008: 575–580.
5 Lee J K, Choi C H, Yoon K H, et al. Design of a servo manipulator with tendon transmission. 2008 International Conference on Control, Automation and Systems (ICCAS), 2008: 1653–1656.
6 Lee J K, Lee H J, Park B S, et al. Bridge-transported bilateral master–slave servo manipulator system for remote manipulation in spent nuclear fuel processing plant. Journal of Field Robotics, 2012, 29(1): 138–160.
7 Yuan J J, Zhang W J, Song Y T, et al. Structural analysis of steel-cable-driven multi-joint robot manipulator. 2011 IEEE International Conference on Mechatronics and Automation (ICMA), 2011: 58–63.
8 Jiang S R, Hua D R, Wang Y Y, et al. Design and modeling of motion-decoupling mechanism for cable-driven joints. Advances in Mechanical Engineering, 2018, 10(5): 1–10.
9 Zhao B, Nelson C A. Decoupled cable-driven grasper design based on planetary gear theory. Journal of Medical Devices, 2013, 7(2): 020918.
10 Shah D, Parmiggiani A, Kim Y J. Constant length tendon routing mechanism through axial joint. 2020 IEEE/ASME International Conference on Advanced Intelligent Mechatronics (AIM), 2020: 753–758.
11 Chen Q Z, Chen W H, Liu R, et al. Mechanism design and tension analysis of a cable-driven humanoid-arm manipulator with joint angle feedback. Journal of Mechanical Engineering, 2010, 46(13): 83–90.
12 Liu F, Xu W F, Huang H L, et al. Design and analysis of a high-payload manipulator based on a cable-driven serial-parallel mechanism. Journal of Mechanisms and Robotics, 2019, 11(5): 051006.
13 Niku S B. Introduction to Robotics: Analysis, Systems, Applications. Prentice Hall, 2001.
14 Xu F, Zi B, Yu Z, et al. Design and implementation of a 7-DOF cable-driven serial spray-painting robot with motion-decoupling mechanisms. Mechanism and Machine Theory, 2024, 192: 105549.
15 Kim Y J. Anthropomorphic low-inertia high-stiffness manipulator for high-speed safe interaction. IEEE Transactions on Robotics, 2017, 33(6): 1358–1374.
16 Pang S X, Shang W W, Zhang F, et al. Design and stiffness analysis of a novel 7-DOF cable-driven manipulator. IEEE Robotics and Automation Letters, 2022, 7(2): 2811–2818.
17 Zhao T, Zi B, Qian S, et al. Typical configuration analysis of a modular reconfigurable cable-driven parallel robot. International Journal of Advanced Robotic Systems, 2019, 16(2): 1729881419834756.

4

Optimization Design of Rigid–Flexible Coupling Hoisting Robots

4.1 Preamble

In practical engineering applications, the performance of robots directly affects the efficiency, accuracy, environmental adaptability, and other aspects of robotic flexible cable robots (RFCRs). Zhou et al. [1] put forward a new optimization algorithm to optimize the objective function that integrates workspace and dexterity indices, thus realizing the performance improvement of RFCRs. Song et al. [2] established a continuous tension distribution method for the configuration optimization of RFCRs and proposed a real-time cable tension distribution algorithm. This method is applied to the optimization of cable tension distribution in RFCRs, achieving the optimal size layout and structural optimization of RFCRs. Yu et al. [3] constructed a stable feasible workspace based on the stability of RFCRs and optimized the size of the end-effector with the goal of maximizing the stable feasible workspace. Hamida et al. [4] proposed a multi-objective optimization method that minimizes cable tension and footprint, realizing the performance optimization of RFCRs. Sun et al. [5] proposed a design method for RFCRs that can passively adjust the cable attachment points on the end-effector, which can effectively expand the workspace of RFCRs without increasing kinematic redundancy, thus improving the working performance of the robot. However, the above studies did not take into account the impact of uncertainty sources on the performance optimization of RFCRs.

To summarize, although existing research has explored dynamic modeling, uncertainty, and performance optimization of the rigid–flexible coupling hoisting robot (RFCHR), there is a notable gap: few performance optimization methods integrate uncertainty quantification analysis with both the uncertainty sources and dynamics of the RFCHR. Addressing this, a multi-objective optimization design method is proposed that accounts for uncertainty sources and integrates reliability, workspace, and static stiffness indices. The method realizes dynamic uncertainty quantification of the RFCHR by introducing uncertain parameters into the dynamic model and applying the random interval perturbation method (RIPM) [6]. Based on the quantification results, a reliability index is established. Subsequently, the objective function of the multi-objective optimization model—incorporating reliability, workspace, and static stiffness indices—is optimized using an appropriate optimization algorithm. Ultimately, this enables the configuration optimization design of the RFCHR. It is expected to provide valuable references for the performance optimization of RFCRs in practical engineering applications, where uncertainty factors often play a significant role [7].

4.2 Multi-objective Optimization Design for Workspace and Dexterity

In this section, the kinematic and static models of RFCHR are derived, and based on this model, performance indicators including workspace and dexterity are derived, ultimately leading to the establishment of a multi-objective optimization model. By introducing an adaptive adjustment inertia weight to the particle swarm optimization (PSO) algorithm, the adaptive adjustment inertia weight particle swarm optimization (AAIWPSO) algorithm is proposed.

4.2.1 Kinematic Modeling and Static Modeling of RFCHR

As shown in Figure 4.1, the structural schematic diagram was simplified based on the structural characteristics. The purpose of this is to facilitate theoretical analysis. Owing to the distance of a certain cable from the reel of driving unit to the

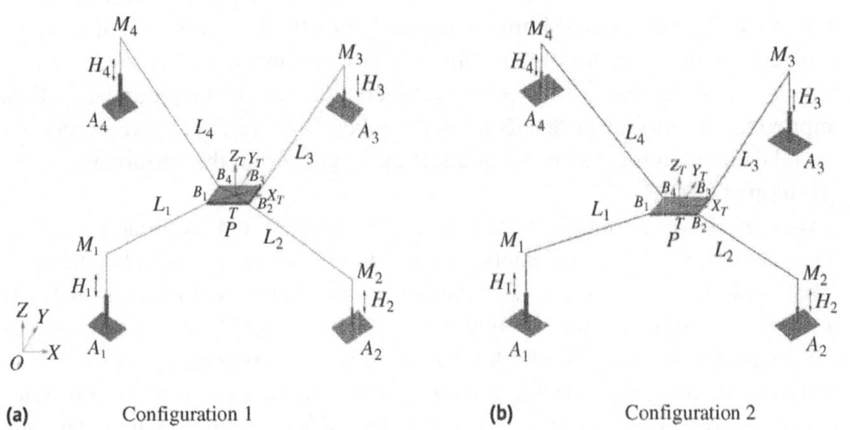

Figure 4.1 Structural schematic of the RFCHR.

corresponding pulley block on the workbench is equal to the distance of the cable to the fixed end of the driving unit after bypassing the pulley block, the amplitude and direction of the force on the two ropes are the same in the meantime. Thus, a group of parallel cables is regarded as a cable. Noteworthily, the length of elongation or shortening of a cable is half the actual length of it. As shown in Figure 4.1, two configurations of CDPR are given in this section. As shown in Figure 4.1a, characteristics of configuration 1 are $H_1 = H_4, H_2 = H_3, H_1\ (H_4) > H_2\ (H_3)$; as shown in Figure 4.1b, characteristics of configuration 2 are $H_1 = H_3, H_2 = H_4, H_1\ (H_3) > H_2\ (H_4)$.

The mechanical structure of the mechanical gripper is shown in Figure 4.2. The mechanical gripper is composed of a palm, cables, two cable reels, springs, a coupling, a servo motor, two driving motors, pulleys, knuckles, and fingers. Every knuckle is connected to a driving motor through a cable and the driving motor is fixed to the palm. Every cable connects with a driving motor to realize the change

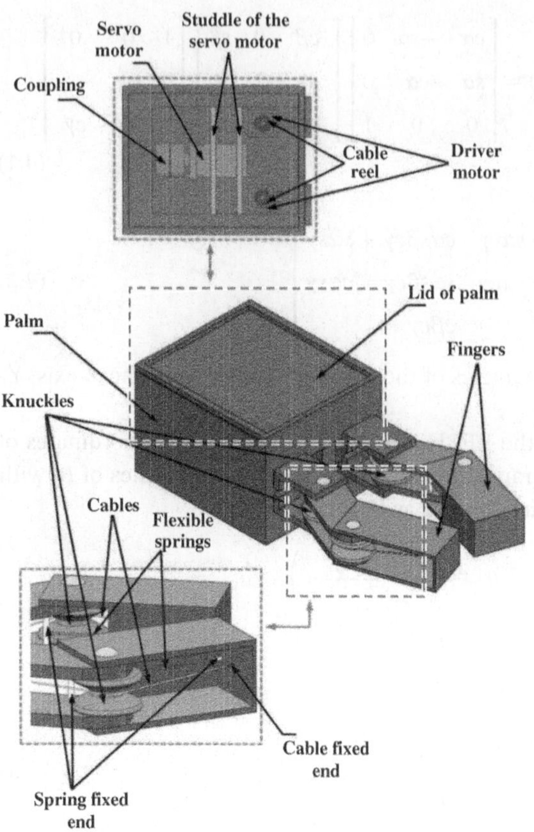

Figure 4.2 Mechanical structure of the mechanical gripper.

of motion states of knuckles and fingers through pulleys. To keep the curved status of fingers in some positions, a flexible spring is located between the first and second knuckles of a finger. As shown in Figure 4.2, when the end-effector reaches a designated position, the mechanical gripper is driven by an electric screw in the X-axis direction to arrive at the vicinity of the grasped object. Subsequently, fingers are driven by pulling cables through driving motors inside the palm to grasp the object.

As shown in Figure 4.1, A_i are intersection points between columns and the ground, M_i denote the attachment points of the cables on the driving units, while B_i represent the attachment points on the end-effector. A fixed Cartesian frame O–XYZ is established on the base plane, with its origin O at the center of the plane formed by A_1, A_2, A_3, and A_4, and the X-axis aligned parallel to A_1A_2. A moving Cartesian frame T–$X_TY_TZ_T$ is attached to the end-effector P with its origin T coinciding with the center of it, and the Y-axis is parallel to B_1B_4. Thus, the pose matrix of the moving frame relative to the fixed frame is expressed as

$$ {}_T^O\boldsymbol{R} = \boldsymbol{R}_Z(\alpha)\boldsymbol{R}_Y(\beta)\boldsymbol{R}_X(\gamma) = \begin{bmatrix} c\alpha & -s\alpha & 0 \\ s\alpha & c\alpha & 0 \\ 0 & 0 & 1 \end{bmatrix} \begin{bmatrix} c\beta & 0 & s\beta \\ 0 & 1 & 0 \\ -s\beta & 0 & c\beta \end{bmatrix} \begin{bmatrix} 1 & 0 & 0 \\ 0 & c\gamma & -s\gamma \\ 0 & s\gamma & c\gamma \end{bmatrix} \tag{4.1} $$

$$ {}_T^O\boldsymbol{R} = \begin{bmatrix} c\alpha c\beta & c\alpha s\beta s\gamma - s\alpha c\gamma & c\alpha s\beta c\gamma + s\alpha s\gamma \\ s\alpha c\beta & s\alpha s\beta s\gamma + c\alpha c\gamma & s\alpha s\beta c\gamma - c\alpha s\gamma \\ -s\beta & c\beta s\gamma & c\beta c\gamma \end{bmatrix} \tag{4.2} $$

where α, β, and γ are rotation angles of the moving frame around the Z-axis, Y-axis, and X-axis, respectively.

The kinematic diagram of the RFCHR is shown in Figure 4.3. Coordinates of M_i with respect to the fixed frame are denoted as $\boldsymbol{a}_i$, and coordinates of B_i with respect to the moving frame are denoted as $\boldsymbol{b}_i$, and

$$ \begin{aligned} a_1 &= \left(-\frac{f_1}{2} \quad -\frac{f_2}{2} \quad h\right)^T, \quad a_2 = \left(\frac{f_1}{2} \quad -\frac{f_2}{2} \quad e\right), \\ a_3 &= \left(\frac{f_1}{2} \quad \frac{f_2}{2} \quad e\right)^T, \quad a_4 = \left(-\frac{f_1}{2} \quad \frac{f_2}{2} \quad h\right)^T \end{aligned} \tag{4.3} $$

$$ \begin{aligned} b_1 &= \left(-\frac{s_1}{2} \quad -\frac{s_2}{2} \quad \frac{s_3}{2}\right)^T, \quad b_2 = \left(\frac{s_1}{2} \quad -\frac{s_2}{2} \quad -\frac{s_3}{2}\right)^T, \\ b_3 &= \left(\frac{s_1}{2} \quad \frac{s_2}{2} \quad -\frac{s_3}{2}\right)^T, \quad b_4 = \left(-\frac{s_1}{2} \quad \frac{s_2}{2} \quad \frac{s_3}{2}\right)^T \end{aligned} \tag{4.4} $$

where f_2 are lengths of A_1A_4 and A_2A_3; f_1 are lengths of A_1A_2 and A_3A_4; h are heights of A_1M_1 and A_4M_4; e are heights of A_2M_2 and A_3M_3; s_2 are lengths

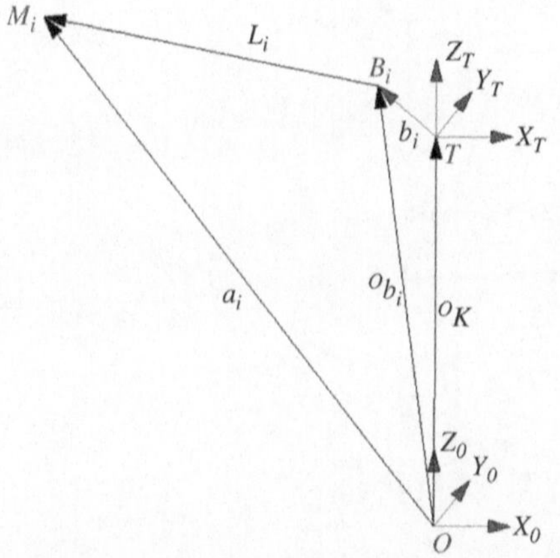

Figure 4.3 Kinematic diagram.

of B_1B_4 and B_2B_3; s_1 are lengths of B_1B_2 and B_3B_4; and s_3 is the thickness of the end-effector.

The position vector $^O\boldsymbol{K}$ is set from O to T, and it can be expressed as $^O\boldsymbol{K} = (x\ y\ z)^T$. $\boldsymbol{a}_i$ and $\boldsymbol{b}_i$ are denoted as position vectors of $\boldsymbol{OM}_i$ and $\boldsymbol{TB}_i$, respectively. $^O\boldsymbol{b}_i$ is denoted as the position vector of $\boldsymbol{OB}_i$. Thus, the ith cable vector $\boldsymbol{L}_i$ can be obtained by

$$\boldsymbol{L}_i = \boldsymbol{a}_i - {}^O\boldsymbol{b}_i = \boldsymbol{a}_i - {}^O\boldsymbol{K} - {}^O_T\boldsymbol{R}\boldsymbol{b}_i,\ i = 1,\ 2, \ldots,\ 4 \tag{4.5}$$

The length of the ith cable l_i can be calculated by

$$l_i = \|\boldsymbol{L}_i\| = \sqrt{\boldsymbol{L}_i^T\boldsymbol{L}_i} \tag{4.6}$$

Taking the derivative of Eq. (4.6) with respect to time yields

$$\dot{\boldsymbol{L}} = \boldsymbol{J}\dot{\boldsymbol{q}} \tag{4.7}$$

where $\boldsymbol{J}$ is the Jacobian matrix of the RFCHR; $\dot{\boldsymbol{q}}$ is the velocity vector of the end-effector; and

$$\dot{\boldsymbol{L}} = \begin{pmatrix} \dot{l}_1 & \ldots & \dot{l}_4 \end{pmatrix}^T \tag{4.8}$$

$$\dot{\boldsymbol{q}} = \begin{pmatrix} \dot{x} & \dot{y} & \dot{z} \end{pmatrix}^T \tag{4.9}$$

$$\boldsymbol{J} = -\begin{pmatrix} \dfrac{\partial \boldsymbol{L}}{\partial x} & \dfrac{\partial \boldsymbol{L}}{\partial y} & \dfrac{\partial \boldsymbol{L}}{\partial z} \end{pmatrix} \tag{4.10}$$

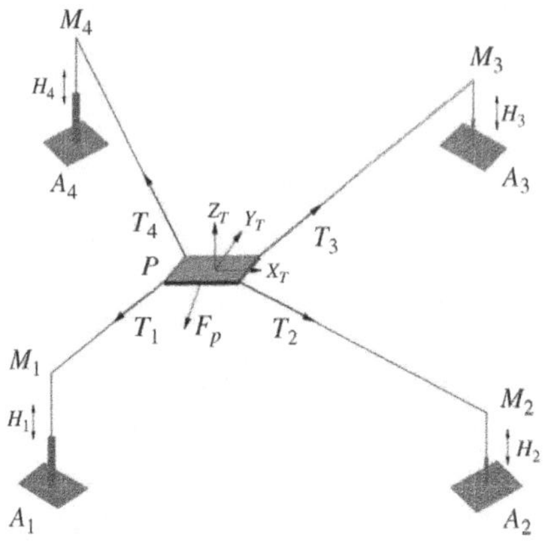

Figure 4.4 Force analysis diagram of the RFCHR.

The force analysis diagram of the CDPR is shown in Figure 4.4. P is the center of the central mobile platform of the CDPR. For simplicity, the cables are assumed to remain in tension and are considered as ideal straight lines without weight or elasticity. Based on the principle of moment balance, the static equation of the CDPR is formulated as

$$\begin{cases} \sum_{i=1}^{4} \boldsymbol{T}_i + \boldsymbol{F}_P = 0 \\ \sum_{i=1}^{4} \boldsymbol{b}_i \times \boldsymbol{T}_i + \boldsymbol{M}_Q = 0 \end{cases}, \quad i = 1,\ 2, \ldots, 4 \tag{4.11}$$

where $\boldsymbol{T}_i = T_i \boldsymbol{u}_i$ is the vector of tension of the ith cable; T_i is the scalar of tension of the ith cable; $\boldsymbol{u}_i = \frac{\boldsymbol{L}_i}{l_i}$ is the unit vector of the ith cable; $\boldsymbol{F}_P$ is the net force on P; $\boldsymbol{b}_i$ is vector of B_i in the moving Cartesian frame; $\boldsymbol{M}_Q$ is the external torque on P; $\boldsymbol{W} = \{\boldsymbol{F}_P^T, \boldsymbol{M}_Q^T\}^T$ is the wrench on P. Thus, based on Eq. (4.11), one has

$$\begin{Bmatrix} \boldsymbol{u}_1 & \cdots & \boldsymbol{u}_n \\ \boldsymbol{b}_1 \times \boldsymbol{u}_1 & \cdots & \boldsymbol{b}_n \times \boldsymbol{u}_n \end{Bmatrix} \begin{Bmatrix} T_1 \\ \vdots \\ T_n \end{Bmatrix} = -\boldsymbol{W},\ n = 4 \tag{4.12}$$

where $\boldsymbol{S}_i = \begin{Bmatrix} \boldsymbol{u}_i \\ \boldsymbol{b}_i \times \boldsymbol{u}_i \end{Bmatrix}$ is the linear vector of the ith cable. Thus, based on Eq. (4.12), one has

$$\boldsymbol{S} \cdot \boldsymbol{T} = -\boldsymbol{W} \tag{4.13}$$

where

$$\begin{cases} \boldsymbol{T} = \left\{ \boldsymbol{T}_1 \ \dots \ \boldsymbol{T}_n \right\}^T \\ \boldsymbol{S} = \left\{ \boldsymbol{S}_1 \ \dots \ \boldsymbol{S}_n \right\}^T \end{cases}, n = 4 \tag{4.14}$$

where $\boldsymbol{T}$ and $\boldsymbol{S}$ are vectors of tensions of cables and the structure matrix of the static model, respectively.

Based on Eqs. (4.12)–(4.14), using the principle of virtual work yields

$$\boldsymbol{T}^T \cdot \Delta \boldsymbol{l} + \boldsymbol{W} \cdot \Delta \boldsymbol{x} = 0 \tag{4.15}$$

where

$$\begin{cases} \Delta \boldsymbol{l} = \begin{pmatrix} \boldsymbol{l}_1 & \dots & \boldsymbol{l}_n \end{pmatrix}^T \\ \Delta \boldsymbol{x} = \begin{pmatrix} \dot{\boldsymbol{x}} & \dot{\boldsymbol{y}} & \dot{\boldsymbol{z}} \end{pmatrix}^T \end{cases}, \ n = 4 \tag{4.16}$$

where $\Delta \boldsymbol{l}$ and $\Delta \boldsymbol{x}$ are displacement increments of cables and the virtual displacement of the center P, respectively. Based on Eqs. (4.10), (4.13), and (4.15), the static model of the CDPR can be expressed as

$$\boldsymbol{J}^T \cdot \boldsymbol{T} = -\boldsymbol{W} \tag{4.17}$$

4.2.2 Performance Indices of RFCHR

In this section, based on kinematic and static models in Section 4.2.1, the performance indices including workspace and dexterity are derived.

4.2.2.1 Workspace

The workspace of the RFCHR depends on the configuration of the mechanism, including the number of cables, the connection points between each cable and the static platform, as well as those between each cable and the moving platform [8]. The workspace of the RFCHR is classified into five categories: the static equilibrium workspace, the wrench closure workspace, the wrench feasible workspace, the dynamic workspace, and the collision-free workspace. This subsection focuses on the static equilibrium workspace of the RFCHR, which is defined as the set of positions and orientations where the robot can remain stationary when only the gravity of the end-effector is considered [9]. Based on the above factors and kinematic modeling in Chapter 2, the workspace index for this RFCHR can be expressed as

$$W_G = \frac{\int dV_Q}{V} = \frac{\iiint dxdydz}{V} \tag{4.18}$$

where W_G is the workspace index for the RFCHR; V_Q is the actual workspace of the RFCHR; V is the equivalent workspace (i.e., the circumscribed cuboid of the actual workspace). Thus, the efficiency of the workspace is directly proportional to W_G. Note that ranges of W_G change from 0 to 1, and W_G is a standardized index. Furthermore, the closer the workspace index W_G gets to 1, the larger the actual workspace is. If W_G is equal to 1, which means that the RFCHR can move to any position in the task space.

4.2.2.2 Dexterity

For the central mobile platform of the RFCHR, it can reach a certain working position under the pull of cables. Thus, the better the coordinated response performance of the end-effector at this working position, the more dexterous the RFCHR is. Generally, dexterity is evaluated by the dexterous solid angle (the ratio of the volume of a reachable area to the volume of a sphere encircling the reachable area). However, this operation has to calculate redundant solutions for every working position. Thus, a quantitative analysis method is proposed for calculating dexterity [10]. For the RFCHR, the reciprocal of the condition number of the Jacobian matrix ranges between 0 and 1, it is considered an important index for evaluating dexterity.

The dexterity of the RFCHR refers to the transitivity of velocity and coordination of cables and end-effector. The Jacobian matrix can relate the velocity of the end-effector to velocities of cables. Thus, the Jacobian equation is expressed in Eq. (4.7). Based on the structure description of the RFCHR in Figure 4.1, and Eqs. (4.7) and (4.8), the length vector of cables can be denoted as $\boldsymbol{L} = \begin{pmatrix} l_1 & \dots & l_n \end{pmatrix}^T$, $n = 4$. Thus, the Jacobian matrix $\boldsymbol{J}$ can be expressed as

$$\boldsymbol{J} = -\left(\boldsymbol{u}_i^T ({}^O_T\boldsymbol{R}\boldsymbol{b}_i \times \boldsymbol{u}_i)^T\right) \tag{4.19}$$

Note that $\boldsymbol{J}$ is related to lengths of cables determined by the position of the end-effector, and the Jacobian matrix $\boldsymbol{J}$ is a 4×6 matrix. Thus, the pseudo-inverse matrix $\boldsymbol{J}^{-1}$ is regarded as the inverse matrix of $\boldsymbol{J}$. Subsequently, the unit velocity vector of the end-effector satisfies

$$\dot{\boldsymbol{q}}^T \cdot \dot{\boldsymbol{q}} = (\boldsymbol{J}^{-1}\dot{\boldsymbol{L}})^T \cdot (\boldsymbol{J}^{-1}\dot{\boldsymbol{L}}) = \dot{\boldsymbol{L}}^T (\boldsymbol{J}\boldsymbol{J}^T)^{-1} \dot{\boldsymbol{L}} = \boldsymbol{1} \tag{4.20}$$

If $\boldsymbol{D}$ is denoted as $\boldsymbol{J}\boldsymbol{J}^T$, Eq. (4.20) can be equivalently transformed to

$$\dot{\boldsymbol{L}}^T \boldsymbol{D}^{-1} \dot{\boldsymbol{L}} = \boldsymbol{1} \tag{4.21}$$

Note that the matrices $\boldsymbol{D}$ and $\boldsymbol{D}^{-1}$ are symmetric positive-definite matrices. For any $\boldsymbol{D}^{-1}$, the velocity vector of cables $\dot{\boldsymbol{L}}$ satisfies the condition in Eq. (4.21). This proves that the matrix $\boldsymbol{D}$ can reflect the dexterity of the end-effector under the pull of cables. Thus, the condition number of the matrix $\boldsymbol{D}$ is qualified to measure the dexterity of the RFCHR, and

$$m(\boldsymbol{D}) = \|\boldsymbol{D}\|_2 \cdot \left\|\boldsymbol{D}^{-1}\right\|_2 = \frac{\sigma_{\max}(\boldsymbol{D})}{\sigma_{\min}(\boldsymbol{D})} \geq 1 \tag{4.22}$$

where $\|\boldsymbol{D}\|_2$ represents the 2-norm of the matrix $\boldsymbol{D}$; $\sigma_{\max}(\boldsymbol{D})$ and $\sigma_{\min}(\boldsymbol{D})$ are the maximal and minimal eigenvalues of the matrix $\boldsymbol{D}$. c is defined as the reciprocal of condition number of $\boldsymbol{J}$, and

$$c = \frac{1}{m(\boldsymbol{D})} \tag{4.23}$$

The index c can measure the ability to move along every axis in a certain configuration state. Owing to the Jacobian matrix is determined by the configuration of the RFCHR, c is a local index for the dexterity of the RFCHR. Subsequently, it is necessary to evaluate the global dexterity index for the RFCHR. The global dexterity index is defined as the integral of local dexterity index over the reachable workspace [10]. Thus, the global dexterity index is expressed as

$$\eta = \frac{\int_{E_0}^{E_n} c\, dE}{\int_{E_0}^{E_n} dE} \tag{4.24}$$

where E_0 and E_n denote the first and last points of the whole workspace, respectively.

As a whole, η, between 0 and 1, is a global dexterity index of the RFCHR. Note that the closer the index η to 1, the better dexterous of the RFCHR is, which means that the RFCHR is fully isotropic. Obviously, when the index η approaches 0, the poorer dexterous of the RFCHR is, which means that the RFCHR is located in a singular state.

4.2.3 Multi-objective Optimal Design

In this section, a multi-objective optimization model is developed. By integrating an adaptive adjustment of inertia weight into the PSO algorithm, the AAIWPSO algorithm is proposed.

4.2.3.1 Multi-objective Optimization Model

The workspace and dexterity indices are two important performance characteristics for the multi-objective optimal design of the RFCHR. Generally, there are common methods to transform multi-objective optimization problem to single-objective optimization problem. The multi-objective multiplication and division method [11] of the optimal theory is adopted. Thus, a multi-objective optimization model of the RFCHR is established by the product of W_G and η , which can be defined as follows:

$$\begin{cases} \max K = W_G\,(d) \cdot \eta\,(d) \\ \text{s.t. } d_{i_{\min}} < d_i < d_{i_{\max}},\ i = 1,\ \ldots,\ n \end{cases} \tag{4.25}$$

where $d = (d_1, \ldots, d_i, \ldots, d_n)$ is the set of constraint functions; n is the total number of constraint functions; W_G is the workspace index for the RFCHR given in Eq. (4.18); and η is the global dexterity index given in Eq. (4.24). In this optimization process, the set of constraint functions includes m and H_i, which represent the mass of the end-effector and heights of driving units, respectively, as shown in Figure 4.2. The value range of m is 6 kg $< m <$ 20 kg, and the value range of H_i $(i = 1, 2, 3, 4)$ is 2 m $< H_i <$ 5 m.

4.2.3.2 PSO Algorithm

The basic idea of the PSO algorithm is to find the optimal solution based on the cooperation and information sharing between individuals in a group. The flowchart of the PSO algorithm is shown in Figure 4.5, and particles need to update positions and velocities after determining individual and global extremes, and the iterative equation is given as follows

$$\begin{cases} v_{jD}^{M+1} = w v_{jD}^{M} + c_1 r_1 \left(p_{jD}^{M} - t_{jD}^{M}\right) + c_2 r_2 \left(b_{jD}^{M} - t_{jD}^{M}\right), \\ t_{jD}^{M+1} = t_{jD}^{M+1} + t_{jD}^{M+1} \end{cases} \tag{4.26}$$

where M is the number of iterations; c_1 and c_2 are learning factors of the algorithm; r_1 and r_2 are random numbers in the range of [0, 1]; w is the inertia weight; D is the dimension of the problem; v_j is the set of velocities of the jth particle, and $v_j = v_{j1}, \ldots, v_{jD}$, $j = 1, 2, \ldots, D$; j is the subscript of the jth particle; p_j and b_j are sets of individual and global extremes of the optimal position for current iteration number, respectively, and $p_j = p_{j1}, \ldots, p_{jD}, b_j = b_{j1}, \ldots, b_{jD}$, $j = 1, 2, \ldots, D$; and t_j is the set of positions of the jth particle, $t_j = t_{j1}, \ldots, t_{jD}$, $j = 1, 2, \ldots, D$.

To locate the optimal values of W_G and η for the RFCHR, the PSO algorithm is used to deal with the optimization problem. As shown in Figure 4.5, based on the flowchart of the PSO algorithm, the optimization process is summarized as follows:

1. Implement the PSO algorithm in MATLAB based on the defined objective function and constraints.

2. Set initial parameters for the PSO algorithm: fitness function (the fitness function equals to the objective function) K, position coordinates of the end-effector x, $y \in (-2500, 2500)$ and $z \in (0, 5000)$, particle population size $N = 10$, the maximum number of iterations $M = 300$, inertia weight $w = 0.8$, learning factors $c_1 = c_2 = 2$.
3. Run the program. Optimal results for the RFCHR with configurations 1 and 2 are listed in Table 4.1. From Table 4.1, one finds that, for configuration 1 with masses of the end-effector being 6 and 20 kg, the optimization results by the PSO algorithms can achieve $W_G = 0.6241$, $\eta = 0.5065$, $K = 0.3162$ and $W_G = 0.2125$, $\eta = 0.2042$, $K = 0.0434$, respectively; for configuration 2, the optimization results by the PSO algorithms can achieve $W_G = 0.7049$, $\eta = 0.5457$, $K = 0.3847$ and $W_G = 0.3369$, $\eta = 0.2119$, $K = 0.0714$, respectively.

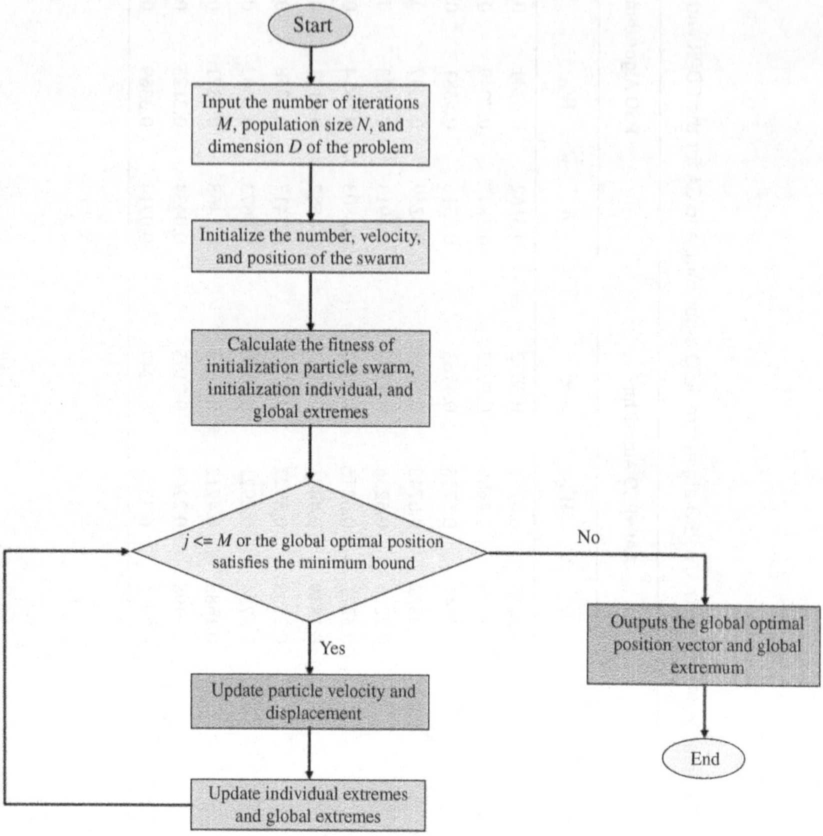

Figure 4.5 Flowchart of the PSO algorithm.

Table 4.1 Optimal results of the AAIWPSO algorithm, PSO algorithm, and GA for the CDPR with configurations 1 and 2.

		AAIWPSO Algorithm			PSO Algorithm			GA		
Configurations	m	K	W_G	η	K	W_G	η	K	W_G	η
1	6	0.3818	0.6731	0.5672	0.3162	0.6241	0.5065	0.3015	0.6268	0.4810
2		0.4558	0.7689	0.5927	0.3847	0.7049	0.5457	0.3793	0.6784	0.5591
1	10	0.2741	0.5755	0.4763	0.2316	0.5381	0.4304	0.2068	0.5172	0.3998
2		0.3822	0.6742	0.5669	0.3269	0.6347	0.5150	0.2915	0.6031	0.4833
1	12	0.2246	0.5236	0.4290	0.1913	0.4823	0.3966	0.1702	0.4369	0.3896
2		0.3271	0.6275	0.5213	0.2803	0.5814	0.4821	0.2642	0.5521	0.4785
1	14	0.1849	0.4485	0.4123	0.1582	0.4131	0.3829	0.1163	0.3562	0.3302
2		0.2799	0.5922	0.4726	0.2417	0.5358	0.4511	0.2142	0.5117	0.4186
1	18	0.0784	0.2921	0.2620	0.0673	0.2682	0.2509	0.0589	0.2447	0.2407
2		0.1685	0.4713	0.3575	0.1468	0.4371	0.3358	0.1175	0.3899	0.3013
1	20	0.0498	0.2269	0.2195	0.0434	0.2125	0.2042	0.0397	0.2013	0.1972
2		0.0815	0.3632	0.2244	0.0714	0.3369	0.2119	0.0581	0.2764	0.2102

4.2.3.3 AAIWPSO Algorithm

The PSO algorithm can avoid crossover and mutation operations and has been applied in the performance index optimization with a fast search speed. Furthermore, owing to the simple iterative process and few tuning parameters, the PSO algorithm can be directly applied to engineering implementation. However, due to the lack of dynamic adjustment of search speed, the PSO algorithm is prone to the state of local optimal. Therefore, the convergence accuracy of the PSO algorithm is usually low. Note that an inertia weight w can be described as the effect of previous velocity of a particle on current velocity. The capability of local and global optimization of the PSO algorithm is decided by the size of the inertia weight w. Generally, the range of w is 0–1. As w is close to 0, the capability of local optimization of the PSO algorithm is stronger; and as w is close to 1, the capability of global optimization of the PSO algorithm is stronger.

To gain better optimization results of performance indices of the RFCHR, an AAIWPSO algorithm is proposed based on the PSO algorithm and the multi-objective optimization model, and the inertia equation of the inertia weight w is expressed as

$$w = \begin{cases} w_{\mathrm{MAX}} - (w_{\mathrm{MAX}} - w_{\mathrm{MIN}}) \cdot \dfrac{f\left(x_{jn}^{N}\right) - f_{\mathrm{MIN}}^{N}}{f_{\mathrm{av}}^{N} - f_{\mathrm{MIN}}^{N}}, & f\left(x_{jn}^{N}\right) \le f_{\mathrm{av}}^{N}, \\ w_{\mathrm{MIN}}, & f\left(x_{jn}^{N}\right) > f_{\mathrm{av}}^{N} \end{cases} \tag{4.27}$$

where w_{MIN} and w_{MAX} are the maximum and minimum coefficients of inertia weight, respectively; f_{MIN}^{N} is the minimum fitness of all particles with Nth iteration, and $f_{\mathrm{MIN}}^{N} = \min\left\{f\left(x_{j1}^{N}\right), f\left(x_{j2}^{N}\right), f\left(x_{j3}^{N}\right), \ldots, f\left(x_{jn}^{N}\right)\right\}$; f_{av}^{N} is the average fitness of all particles with Nth iteration; $f\left(x_{jn}^{N}\right)$ is the fitness value of current particle with Nth iteration. Similarly, the equation of the position and velocity of the particles is expressed in Eq. (4.26). As shown in Eq. (4.27), the updating mode of w is nonlinear. The value of w varies with change of the fitness value of the particle. When $f\left(x_{jn}^{N}\right)$ approaches the local optimal value, the value of w increases; otherwise, it decreases. When $f\left(x_{jn}^{N}\right) > f_{\mathrm{av}}^{N}$, the value of $w = w_{\mathrm{MIN}}$ is small, which protects the current particle; when $f\left(x_{jn}^{N}\right) \le f_{\mathrm{av}}^{N}$, the value of w calculated based on Eq. (4.27) is large, which allows the particle to move toward a better search area.

The flowchart of the AAIWPSO algorithm is shown in Figure 4.6, and the optimization process is summarized as follows:

1. Execute the PSO algorithm in MATLAB in accordance with the specified objective function and constraints.
2. Set initial parameters for the AAIWPSO algorithm: fitness function K, position coordinates of the end-effector x, $y \in (-2500, 2500)$ and $z \in (0, 5000)$,

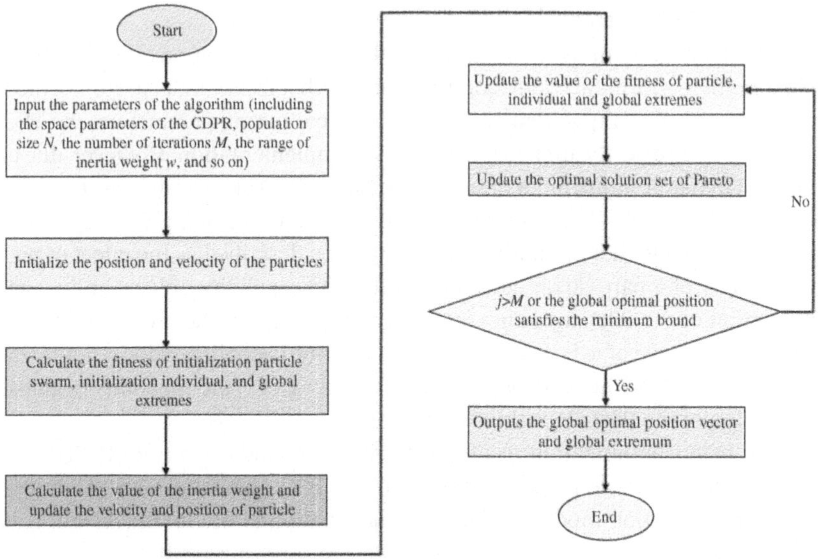

Figure 4.6 Flowchart of the AAIWPSO algorithm.

particle population size $N = 10$, the maximum number of iterations $M = 300$, inertia weight $w \in (0.4, 0.9)$, learning factors $c_1 = c_2 = 2$.

3. Run the program. Optimal results for the RFCHR with configurations 1 and 2 are listed in Table 4.1. From Table 4.1, one finds that, for configuration 1 with masses of the end-effector being 6 and 20 kg, the optimization results by the AAIWPSO algorithm can achieve $W_G = 0.6731$, $\eta = 0.5672$, $K = 0.3818$ and $W_G = 0.2269$, $\eta = 0.2195$, $K = 0.0498$, respectively; for configuration 2, the optimization results by the AAIWPSO algorithm can achieve $W_G = 0.7689$, $\eta = 0.5927$, $K = 0.4558$ and $W_G = 0.3632$, $\eta = 0.2244$, $K = 0.0815$, respectively.

For a solution of the optimization problem, if no other solutions can dominate the solution, then the solution is called a Pareto optimal solution [12]. The collection of all optimal solutions is referred to as the Pareto optimal solution set. In this study, the dense distance method is applied to update the Pareto optimal solution set. Meanwhile, the updated formula is shown in Eq. (4.28). The update method is as follows:

1. Judge whether the number of solutions exceeds the upper limit, the upper limit of set is 300 in the optimal solution set of Pareto. If not, the current optimal solution is added to the set; otherwise, one proceeds to the next step.

2. Calculate the dense distance between the two closest solutions to the current optimal solution in the optimal solution set of Pareto based on Eq. (4.28).
3. Compare two dense distances calculated in step (2); the solution with smaller dense distance is eliminated. Meanwhile, the current optimal solution is added to the optimal set.

The density distance is calculated as follows

$$I(x_i) = \frac{|W_G(x_j) - W_G(x_k)|}{W_{G\max}} + \frac{|\eta(x_j) - \eta(x_k)|}{\eta_{\max}} \tag{4.28}$$

where x_j and x_k are the two closest particles to x_i; $W_G(x_j)$ is the value of W_G for particle x_j; $W_G(x_k)$ is the value of W_G for particle x_k; $\eta(x_j)$ is the value of η for particle x_j; $\eta(x_k)$ is the value of η for particle x_k; and $W_{G\max}$ and $\eta_{\max}$ are the maximums for W_G and η, respectively.

4.3 Multi-objective Optimization Design for Reliability, Workspace, and Stiffness

4.3.1 Performance Indices of RFCHR

This section focuses on quantifying the dynamic uncertainty of the RFCHR, and on the basis of the quantification outcomes, it analyzes and establishes the reliability index. Moreover, two other performance metrics, the workspace and static stiffness of the RFCHR, are also analyzed and defined herein. For the structural and dynamic models of the RFCHR, refer to Section 2.2.

4.3.1.1 Uncertainty Quantification

Uncertainty quantification entails the quantitative depiction and handling of uncertainty across diverse application scenarios. Its purpose is to assess the uncertainty inherent in parameters and gauge the corresponding levels of confidence, serving as an essential preliminary step for the analysis, design, and optimization of structures characterized by uncertainty.

The RFCHR's design parameters, structural parameters, motion parameters, and loading parameters are bound to have uncertainties, which stem from a variety of factors such as design flaws, processing and assembly errors, friction, and wear. As a result, it is necessary to further characterize these uncertain parameters induced by the aforementioned factors. In this study, the input random variables and interval variables of the RFCHMS are defined as $\vec{\boldsymbol{M}}_R$ and $\vec{\boldsymbol{N}}_I$, respectively.

The equivalent form of the dynamic model that incorporates all random variables and interval variables can be expressed as follows:

$$\boldsymbol{F}_s = \boldsymbol{K}\left[\boldsymbol{J}(\vec{\boldsymbol{M}}_R, \vec{\boldsymbol{N}}_I)\right] \tag{4.29}$$

where $\boldsymbol{J}(\vec{\boldsymbol{M}}_R, \vec{\boldsymbol{N}}_I)$ denotes the matrix of dynamic parameters that includes both random variables and interval variables. To streamline the analysis, we assume that random variables and interval variables are mutually independent, allowing the dynamic equation containing these variables to be expressed as $f(\vec{\boldsymbol{M}}_R, \vec{\boldsymbol{N}}_I)$. By applying the RIPM, combined with first-order Taylor series expansion and perturbation theory, $f(\vec{\boldsymbol{M}}_R, \vec{\boldsymbol{N}}_I)$ can be expressed as

$$
\begin{aligned}
f(\vec{\boldsymbol{M}}_R, \vec{\boldsymbol{N}}_I) = {} & f\left(E\left(\vec{\boldsymbol{M}}_R\right), \vec{\boldsymbol{N}}_I^C\right) + \sum_{j=1}^{k} \left\{ \left.\frac{\partial f(\vec{\boldsymbol{M}}_R, \vec{\boldsymbol{N}}_I)}{\partial n_j^I}\right|_{E(\vec{\boldsymbol{M}}_R), \vec{\boldsymbol{N}}_I^C} \right\} \Delta n_j^I \\
& + \sum_{j=1}^{k} \left\{ \left.\frac{\partial f(\vec{\boldsymbol{M}}_R, \vec{\boldsymbol{N}}_I)}{\partial m_i^R}\right|_{E(\vec{\boldsymbol{M}}_R), \vec{\boldsymbol{N}}_I^C} \right. \\
& \left. + \sum_{j=1}^{k} \left\{ \left.\frac{\partial^2 f(\vec{\boldsymbol{M}}_R, \vec{\boldsymbol{N}}_I)}{\partial m_i^R \partial n_J^I}\right|_{E(\vec{\boldsymbol{M}}_R), \vec{\boldsymbol{N}}_I^C} \right\} \Delta n_j^I \right\} \cdot (m_i^R - E(m_i^R))
\end{aligned}
\tag{4.30}
$$

In this context, h and k correspond to the quantities of random variables and interval variables, respectively. $E\left(m_i^R\right)$ and $E\left(\vec{\boldsymbol{M}}_R\right)$ represent the expected values of the random variable m_i^R and the random vector $\vec{\boldsymbol{M}}_R$, respectively; $\vec{\boldsymbol{N}}_I^C$ denotes the midpoint (average) of the interval vector $\vec{\boldsymbol{N}}_I$, and Δn_j^I stands for the interval deviation of the jth interval variable n_j^I.

Using the random interval moment method, and by neglecting higher-order terms and random variables, the expectation and variance of the dynamic response $f(\vec{\boldsymbol{M}}_R, \vec{\boldsymbol{N}}_I)$ can be expressed as

$$
E(f(\vec{\boldsymbol{M}}_R, \vec{\boldsymbol{N}}_I)) = f\left(E\left(\vec{\boldsymbol{M}}_R\right), \vec{\boldsymbol{N}}_I^C\right) + \sum_{j=1}^{k} \left\{ \left.\frac{\partial f(\vec{\boldsymbol{M}}_R, \vec{\boldsymbol{N}}_I)}{\partial n_j^I}\right|_{E(\vec{\boldsymbol{M}}_R), \vec{\boldsymbol{N}}_I^C} \right\} \Delta n_j^I
\tag{4.31}
$$

$$
\begin{aligned}
\operatorname{Var}(f(\vec{\boldsymbol{M}}_R, \vec{\boldsymbol{N}}_I)) = {} & \sum_{i=1}^{h} \left\{ \left.\frac{\partial f(\vec{\boldsymbol{M}}_R, \vec{\boldsymbol{N}}_I)}{\partial m_i^R}\right|_{E(\vec{\boldsymbol{M}}_R), \vec{\boldsymbol{N}}_I^C} \right. \\
& \left. + \sum_{j=1}^{k} \left\{ \left.\frac{\partial^2 f(\vec{\boldsymbol{M}}_R, \vec{\boldsymbol{N}}_I)}{\partial m_i^R \partial n_J^I}\right|_{E(\vec{\boldsymbol{M}}_R), \vec{\boldsymbol{N}}_I^C} \right\} \Delta n_j^I \right\}^2 \cdot \operatorname{Var}(m_i^R)
\end{aligned}
\tag{4.32}
$$

Here, $\operatorname{Var}(m_i^R)$ refers to the variance of the random variable m_i^R.

Owing to the structural coupling effect—such as varying cable tension states and interstructural coupling vibrations—when the RFCHMS drives the lifting arm, the nonlinear deformation of the cables during the lifting process is significantly more pronounced than their linear deformation. For this particular robot, calculations based on the elastic modulus indicate that the elastic deformation of the cables induced by the driving force can be disregarded. However, traditional analytical methods struggle to accurately characterize the precise nonlinear deformation of the cables. Moreover, there exists cognitive uncertainty throughout the analysis process. These factors collectively contribute to uncertainties in the dynamic system parameters of the robot. Additionally, the random loads acting on the hoisted target introduce uncertainties in the input variables, which in turn may compromise the actuation precision of the robot.

For the RFCHR, cable deformation is treated as an interval parameter, and its impact on the motion parameters of the lifting arm can be examined through kinematic analysis. Furthermore, in line with the descriptions in relevant references, the cable deformation proposed is a comprehensive variable that incorporates uncertainties in cable length arising from factors such as cable tension, pose errors caused by pulley manufacturing and installation gaps, and discrepancies between the actual and modeled positions of the flexible cable. As such, this assumption is applicable to all RFCHRs. In practical engineering applications, however, this assumption should be adjusted appropriately based on the specific mechanism design. Additionally, if the payload mass falls within the allowable lifting range of the RFCHR, slight cable deformations are already accounted for within the scope of uncertainty, thereby ensuring the validity of this assumption in practical engineering scenarios. At this point, the cable length l_{Oi} can be expressed as

$$l_{Oi} = l_i + \Delta l_i, i = 1, 2, 3 \tag{4.33}$$

Here, Δl_i represents the interval value of the cable length. Drawing on the research in [13], let $\boldsymbol{J}_u = \left(\boldsymbol{u}_i^T({}_B^O\boldsymbol{R}\boldsymbol{b}_i \times \boldsymbol{u}_i)^T\right)$; the error transfer formula for the cable length can then be expressed as

$$\Delta\boldsymbol{e} = -\boldsymbol{J}_u^{-1}\Delta\boldsymbol{l} = -\left(\boldsymbol{u}_i^T({}_B^O\boldsymbol{R}\boldsymbol{b}_i \times \boldsymbol{u}_i)^T\right)^{-1}\Delta\boldsymbol{l} \tag{4.34}$$

Here, $\Delta\boldsymbol{e}$ denotes the errors in the position and attitude of the lifting arm caused by cable length variations; $\Delta\boldsymbol{l}$ is a vector consisting of evidence variables. Owing to the presence of the luffing rigid rod, cable deformation primarily impacts the attitude angle β in the position and attitude of the lifting arm. Based on Eq. (4.35), this can be calculated as

$$\Delta\beta = \boldsymbol{J}_{\beta u}\Delta\boldsymbol{l} \tag{4.35}$$

Here, $\boldsymbol{J}_{\beta u}$ represents the row vector corresponding to the matrix $-\boldsymbol{J}_u^{-1}$ in Eq. (4.35) when calculating $\Delta\beta$. Taking into account the monotonicity of error propagation, the angle interval variable resulting from cable length errors can be expressed as

$$\beta_I = \beta + \Delta\beta = \beta + \boldsymbol{J}_{\beta u}\Delta\boldsymbol{l} = \theta_2 + (\theta_3 + \Delta\theta_3) \tag{4.36}$$

where $\Delta\theta_3$ denotes the uncertain parameter value of θ_3. Additionally, the random loading acting on the hoisted target is primarily induced by natural factors like wind loads. Taking into account its influencing factors, which include the mass and surface area of the hoisted target, the random loading can be expressed as

$$f_R = m_{\text{ou}}g + \frac{1}{\sqrt[2]{2\pi\sigma}}e^{-\frac{(x-\mu)^2}{2\sigma^2}} \tag{4.37}$$

$$\sigma = km_{\text{ou}} + js_{\text{ou}} \tag{4.38}$$

$$\boldsymbol{f}_{\text{Rou}} = [0, 0, f_R]^T \tag{4.39}$$

Here, m_{ou} refers to the mass of the hoisted target, s_{ou} denotes the surface area of the hoisted target, while j and k represent the influence coefficients of these two variables on the standard deviation, respectively.

To effectively assess how uncertainty sources impact the dynamic response of the RFCHR, the aforementioned random variables and interval variables are substituted into the dynamic model to derive the dynamic response. Consequently, the dynamic model incorporating uncertain parameters can be expressed as

$$\tau_{\text{RI}} = \boldsymbol{M}(\theta + \Delta\theta)\,\ddot{\theta} + \boldsymbol{C}(\theta + \Delta\theta, \dot{\theta}) + G(\theta + \Delta\theta) + \tau_o(\boldsymbol{f}_{\text{Rou}}) \tag{4.40}$$

Let the mean value and expectation of the input be denoted as θ_{3e} and f_{Re}, respectively. Based on the dynamic response of the robot, the actual joint torque under the combined effect of random variables and interval variables can be calculated as

$$\begin{aligned} \tau_{\text{RI}} &= \tau_{\text{RI}}(\theta_{3e}, f_{\text{Re}}) + \left.\frac{\partial\tau_{\text{RI}}}{\partial\theta_3}\right|_{\theta_{3e}, f_{\text{Re}}} \cdot \Delta\theta_3 \\ &+ \left\{ \left.\frac{\partial\tau_{\text{RI}}}{\partial f_R}\right|_{\theta_{3e}, f_{\text{Re}}} + \left.\frac{\partial^2\partial\tau_{\text{RI}}}{\partial f_R \partial\theta_3}\right|_{\theta_{3e}, f_{\text{Re}}} \cdot \Delta\theta_3 \right\} \cdot (f_R - f_{\text{Re}}) \end{aligned} \tag{4.41}$$

The expectation and variance of τ_{RI} are

$$E\left(\tau_{\text{RI}}(\theta_{3e}, f_{\text{Re}})\right) = \tau_{\text{RI}}(\theta_{3e}, f_{\text{Re}}) + \left.\frac{\partial\tau_{\text{RI}}}{\partial\theta_3}\right|_{\theta_{3e}, f_{\text{Re}}} \cdot \Delta\theta_3 \tag{4.42}$$

$$\mathrm{Var}(\tau_{\mathrm{RI}}(\theta_{3e}, f_{\mathrm{Re}})) = \left\{ \left.\frac{\partial \tau_{\mathrm{RI}}}{\partial f_R}\right|_{\theta_{3e}, f_{\mathrm{Re}}} + \left.\frac{\partial^2 \partial \tau_{\mathrm{RI}}}{\partial f_R \partial \theta_3}\right|_{\theta_{3e}, f_{\mathrm{Re}}} \cdot \Delta\theta_3 \right\}^2 \cdot \mathrm{Var}(f_R) \quad (4.43)$$

Based on Eqs. (4.39) and (4.40), the respective expectations of the upper and lower bounds, as well as the variance of the dynamic response, can be calculated as follows:

$$\begin{cases} E\left(\tau_{\mathrm{RI}}^U\right) = \tau_{\mathrm{RI}}(\theta_{3e}, f_{\mathrm{Re}}) + \left.\frac{\partial \tau_{\mathrm{RI}}}{\partial \theta_3}\right|_{\theta_{3e}, f_{\mathrm{Re}}} \cdot \Delta\theta_U, \\ E\left(\tau_{\mathrm{RI}}^L\right) = \tau_{\mathrm{RI}}(\theta_{3e}, f_{\mathrm{Re}}) + \left.\frac{\partial \tau_{\mathrm{RI}}}{\partial \theta_3}\right|_{\theta_{3e}, f_{\mathrm{Re}}} \cdot \Delta\theta_L \end{cases} \quad (4.44)$$

$$\begin{cases} \mathrm{Var}\left(\tau_{\mathrm{RI}}^U\right) = \left\{ \left.\frac{\partial \tau_{\mathrm{RI}}}{\partial f_R}\right|_{\theta_{3e}, f_{\mathrm{Re}}} + \left.\frac{\partial^2 \partial \tau_{\mathrm{RI}}}{\partial f_R \partial \theta_3}\right|_{\theta_{3e}, f_{\mathrm{Re}}} \cdot \Delta\theta_U \right\}^2 \mathrm{Var}(f_R), \\ \mathrm{Var}\left(\tau_{\mathrm{RI}}^U\right) = \left\{ \left.\frac{\partial \tau_{\mathrm{RI}}}{\partial f_R}\right|_{\theta_{3e}, f_{\mathrm{Re}}} + \left.\frac{\partial^2 \partial \tau_{\mathrm{RI}}}{\partial f_R \partial \theta_3}\right|_{\theta_{3e}, f_{\mathrm{Re}}} \cdot \Delta\theta_L \right\}^2 \mathrm{Var}(f_R) \end{cases} \quad (4.45)$$

Here, τ_{RI}^U and τ_{RI}^L represent upper and lower bounds of τ_{RI}, respectively. Similarly, $\Delta\theta_U$ and $\Delta\theta_L$ stand for upper and lower bounds of $\Delta\theta$, respectively.

Building on the aforementioned analysis and the derivations presented in Section 2.3, the distribution of the dynamic response $\boldsymbol{F}_S(\Delta\boldsymbol{l}, \boldsymbol{f}_{\mathrm{Rou}})$ of the RFCHMS, which incorporates both interval and random variables, can be further computed.

4.3.1.2 Reliability

During the lifting operation of the RFCHR, it is impacted by factors like random loads and deformation, which can readily reduce the probability of it fulfilling the motion function requirements.

For the inevitable uncertainty sources inherent in the robot, the impact of uncertain parameters on the robot's dynamic response can be mitigated by adjusting design variables, thereby further enhancing the robot's reliability. The reliability of the robot is assessed based on the distribution of dynamic responses under these uncertain parameters. As elaborated in Section 4.3.1.1, both random variables and interval variables exert an influence on the robot's dynamic response, meaning that the dynamic response manifests as an interval with random characteristics. To effectively evaluate the reliability under the limit state, the failure range of the dynamic response is defined as

$$\begin{cases} \boldsymbol{F}_S\left(\Delta \boldsymbol{l}, \boldsymbol{f}_{\text{Rou}}\right) > E\left(\boldsymbol{F}_S^U\right) + \text{Sd}\left(\boldsymbol{F}_S^U\right), \\ \boldsymbol{F}_S\left(\Delta \boldsymbol{l}, \boldsymbol{f}_{\text{Rou}}\right) < E\left(\boldsymbol{F}_S^L\right) + \text{Sd}\left(\boldsymbol{F}_S^L\right) \end{cases} \tag{4.46}$$

where $E(\boldsymbol{F}_S^U)$ and $E(\boldsymbol{F}_S^L)$ refer to the expected values of the upper and lower bounds of the specified dynamic response of the robot, respectively; $\text{Sd}(\boldsymbol{F}_S^U)$ and $\text{Sd}(\boldsymbol{F}_S^L)$ represent the standard deviations of the upper and lower bounds of the robot's dynamic response, respectively. Moreover, the failure probability can be computed as

$$\begin{cases} p_f^{\max} = \Pr\left\{E\left(\boldsymbol{F}_S^U\right) + \text{Sd}\left(F_S^U\right) - \boldsymbol{F}_S\left(\Delta \boldsymbol{l}, \boldsymbol{f}_{\text{Rou}}\right) < 0\right\}, \\ p_f^{\min} = \Pr\left\{\boldsymbol{F}_S\left(\Delta \boldsymbol{l}, \boldsymbol{f}_{\text{Rou}}\right) - E\left(\boldsymbol{F}_S^L\right) - \text{Sd}\left(F_S^L\right) < 0\right\} \end{cases} \tag{4.47}$$

The reliability of the RFCHR is expressed as

$$p_r = 1 - \frac{p_f^{\min} + p_f^{\max}}{2} \tag{4.48}$$

Furthermore, in order to evaluate how uncertain parameters affect reliability and to better improve system performance, a sensitivity analysis of uncertain parameters, encompassing cable deformation and random loading, is carried out.

In sensitivity analysis, the extent to which input parameters affect calculation results is mainly characterized by the sensitivity index. Here, the sensitivity index $S(y_i, x_i)$ is constructed using a non-regularized method, which is expressed as

$$S\left(y_i, x_j\right) = \frac{\partial y_i}{\partial_{x_j}} \approx \frac{\Delta y_i}{\Delta x_j} \tag{4.49}$$

At the same time, to make it easier to compare the results of sensitivity analysis across different parameters, each parameter undergoes normalization. The normalization formula employed is as follows

$$\beth_{\text{normalized}} = \frac{\beth - \beth_{\min}}{\beth_{\max} - \beth_{\min}} \tag{4.50}$$

where $\beth_{\text{normalized}}$ denotes the normalized value; $\beth$ is the value of the parameter; $\beth_{\max}$ represents the maximum value of the parameter; and $\beth_{\min}$ stands for the minimum value of the parameter. From the above analysis and the reliability index, it can be determined that the uncertainty parameters associated with reliability include Δl_1, Δl_2, Δl_3, and f_R. The baseline values and value ranges of these uncertain parameters are presented in Table 4.2. To analyze the sensitivity of each parameter, the trend of reliability is observed as the parameter changes (with the values of other parameters fixed at their baseline values). As described in Section 4.3.2, the value ranges of Δl_1, Δl_2, and Δl_3 are equivalent ranges that take into account various error factors.

Table 4.2 The baseline values and intervals for uncertain parameters.

Parameter	Baseline Value	Value Range
Δl_1	1 mm	[−8,10]
Δl_2	1 mm	[−8,10]
Δl_3	0.5 mm	[−9,9]
f_R	11 N	[10,12]

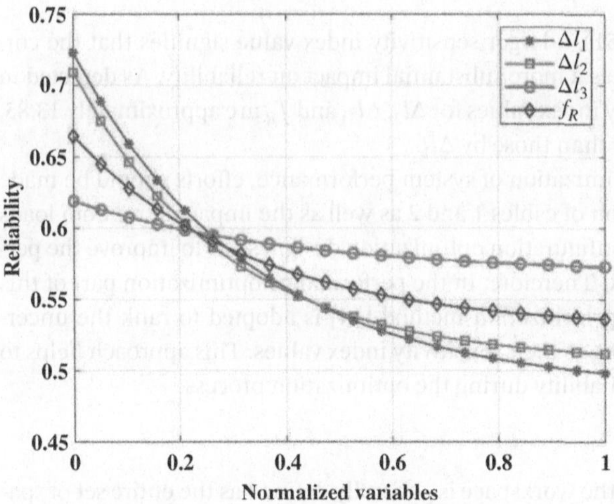

Figure 4.7 The trend of reliability with a single variable changing at the baseline value.

To eliminate discrepancies in magnitude among different parameters, each parameter is normalized using Eq. (4.50). Following normalization, the value range of every parameter becomes [0,1], and it exhibits a monotonically increasing trend. The variation pattern of reliability as a single variable changes is illustrated in Figure 4.7.

As indicated in Figure 4.7, by examining the amplitude of the curves, the order of the influence of each parameter on reliability can be determined as follows:

$$\Delta l_1 \approx \Delta l_2 > f_R > \Delta l_3 \tag{4.51}$$

It can be observed that the uncertain parameters Δl_1 and Δl_2 exert the most significant impact on reliability, followed by f_R. Δl_3 has the smallest influence on reliability. Based on Eq. (4.51), the calculated sensitivity indices for these uncertain parameters are presented in Table 4.3.

Table 4.3 The calculation results of sensitivity indices for uncertain parameters.

Parameter	Sensitivity Index
Δl_1	0.2368
Δl_2	0.204
Δl_3	0.0504
f_R	0.1352

As defined by Eq. (4.51), a larger sensitivity index value signifies that the corresponding parameter has a more substantial impact on reliability. As depicted in Figure 4.3, the sensitivity index values for Δl_2, Δl_3, and f_R are approximately 13.85, 78.72, and 42.91% lower than those by Δl_1.

In the subsequent optimization of system performance, efforts should be made to reduce the deformation of cables 1 and 2 as well as the impact of random loads on reliability through configuration optimization design, so as to improve the performance of the RFCHR. Therefore, in the performance optimization part of this section, the parameter prioritization method [14] is adopted to rank the uncertain parameters according to their sensitivity index values. This approach helps to maximize the system reliability during the optimization process.

4.3.1.3 Workspace

In the robotics domain, the workspace is generally defined as the entire set of spatial points accessible to the end-effector. In this study, based on the robot's structural characteristics, it is readily apparent that the end-effector (lifting arm) of the robot possesses three degrees of freedom. Unlike the RFCR, where all supporting chains are composed of cables, the cable structure of the robot designed herein is less susceptible to interference. It is important to highlight that the structure itself constitutes a key factor restricting the robot's workspace during actual movement.

The structural constraints of the robot primarily involve the operational length of the electric telescopic rod and the angular deviation of the luffing rigid rod relative to the lifting arm. Additionally, the height of the robot's lifting arm must not be lower than the plane of the rotating base. Consequently, the constraint conditions can be expressed as

$$\begin{cases} 0 \le k \le k^U, \\ \theta_3^L \le \theta_3 \le \theta_3^U, \\ l_{AB} \cos \theta_2 + l_{BB_4} \cos (\theta_2 + \theta_3) \ge h^L \end{cases} \tag{4.52}$$

Here, k represents the extension length of the electric telescopic rod; k^U denotes the maximum extension length of the electric telescopic rod; θ_3^U and θ_3^L stand for the upper and lower limits of θ_3, respectively.

Additionally, as analyzed in Section 2.2, $O_1B_4 = \left(\sqrt{x^2+y^2}, 0, z\right)^T$ serves as the position vector of the lifting arm's end. Using the cosine law, θ_2 and θ_3 can be expressed as

$$\begin{cases} \theta_2 = \delta + \arccos\left(\left(l_{BB_4}^2 - l_{AB}^2 - d^2 - z^2\right) / \left(2l_{AB}\sqrt{d^2+z^2}\right)\right), \\ \theta_3 = \arccos\left(\left(l_{AB}^2 + l_{BB_4}^2 - d^2 - z^2\right) / 2l_{AB}l_{BB_4}\right) \end{cases} \tag{4.53}$$

Based on Eqs. (4.52) and (4.53), the value ranges of θ_2 and θ_3 can be computed. Accordingly, the initial structural parameters of the robot are presented in Table 4.4.

Using the basic structural parameters of the RFCHR and the Monte Carlo method (MCM), the workspace can be computed. In this study, the workspace index can be considered as

$$W_{\text{ind}} = V_{\text{act}} \tag{4.54}$$

where W_{ind} denotes the workspace index of the RFCHR; and V_{act} represents the actual workspace of the RFCHR. Moreover, the workflow for calculating the workspace using the MCM is summarized as follows:

1. Using the MCM, the kinematically corresponding reachable positions of the RFCHR's end-effector are solved. In this process, the number of random sampling points is set to 20,000.
2. Create a point cloud graph to illustrate the workspace of the RFCHR.
3. Compute the area of the workspace, which is also referred to as the workspace index and denoted by V_{act} (W_{ind}).

Table 4.4 Initial structure parameters of the RFCHR.

Parameter	Values	Parameter	Values
θ_1	$[0, 2\pi]$	$\boldsymbol{a}_1$	(389, 398, −1.5) mm
θ_2	$[-0.5\pi, -0.236\pi]$	$\boldsymbol{a}_2$	(389, −398, −1.5) mm
θ_3	$[-0.3\pi, 0.3\pi]$	$\boldsymbol{a}_3$	(220, 0, 8) mm
l_{O_1A}	100 mm	$\boldsymbol{b}_1$	(0, −298, 125) mm
l_{AB}	700 mm	$\boldsymbol{b}_2$	(0, −298, −125) mm
l_{BB_4}	686 mm	$\boldsymbol{b}_3$	(0, 100, 0) mm

4.3.1.4 Static Stiffness

Stiffness characterizes a robot's capacity to withstand deformation under external loads, making it one of the key metrics for evaluating the performance of the RFCR. Furthermore, prior to conducting in-depth analysis, the following assumptions are put forward in light of the unique properties of cables:

1. The gravitational influence on the cables can be disregarded.
2. The elastic deformation of the cables is considered ideal.
3. The weight of each individual cable is negligible.
4. The connections between each cable and both the frame and the end-effector are treated as ideal hinge joints.

As illustrated in Figure 4.8, on the assumption that the end-effector (lifting arm) has infinite stiffness, it can be treated as a rigid body. Moreover, the pose screw vector of any point on the end-effector relative to point B is expressed as $\mathbf{S} = \left(\boldsymbol{d}^T \quad \boldsymbol{\Phi}^T\right)^T$, where $\boldsymbol{d} = (x, y, z)^T$ represents the coordinate vector of the point in the local coordinate system $B\text{–}X_B Y_B Z_B$; $\boldsymbol{\Phi} = \left(\Phi_x, \ \Phi_y, \ \Phi_z\right)^T$ stands for the attitude angle vector of the end-effector.

Based on the structure of the RFCHR, it can be seen that when the robot's end-effector is in an ideal working state, it is affected by cable tension, the supporting force from the rigid rod, and its own gravity. (In the subsequent derivation of static stiffness, when the end-effector is subjected to generalized external forces, both displacement and changes in attitude angles are considered. In reality, any point on the end-effector only experiences variations in attitude angles relative to the local coordinate system $B\text{–}X_B Y_B Z_B$)

Therefore, in accordance with the definition of static stiffness, the infinitesimal external force $\boldsymbol{\delta}\mathbf{exf}$ exerted on the end-effector maintains a linear relationship

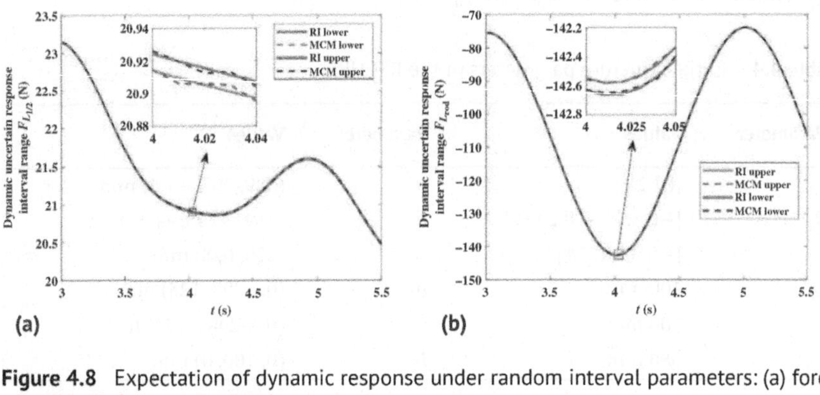

Figure 4.8 Expectation of dynamic response under random interval parameters: (a) force of cable; (b) force of rigid rod.

with the infinitesimal pose screw $\delta\boldsymbol{S}$ induced by this external force, and this relationship can be expressed as

$$\delta\mathbf{exf} = \boldsymbol{K} \cdot \delta\boldsymbol{S} \tag{4.55}$$

where $\boldsymbol{K}$ represents the static stiffness matrix.

As illustrated in Figure 4.8, the end-effector of the RFCHR is equipped with four parallel supporting chains, which consist of three cables and a rigid rod. Accordingly, by analyzing the static equilibrium state of the end-effector under the generalized external force $\mathbf{exf}$, together with the cable tensions and the supporting force from the rigid rod $\boldsymbol{F} = \begin{pmatrix} F_{L_1} & F_{L_2} & F_{L_3} & F_{L_{\text{rod}}} \end{pmatrix}$, we can derive

$$\boldsymbol{A}^T\boldsymbol{F} + \mathbf{exf} = \mathbf{0} \tag{4.56}$$

where $\boldsymbol{A}^T = \begin{bmatrix} \boldsymbol{u}_1 & \boldsymbol{u}_2 & \boldsymbol{u}_3 & \boldsymbol{u}_{\text{rod}} \\ \boldsymbol{r}_1 \times \boldsymbol{u}_1 & \boldsymbol{r}_2 \times \boldsymbol{u}_2 & \boldsymbol{r}_3 \times \boldsymbol{u}_3 & \boldsymbol{r}_{\text{rod}} \times \boldsymbol{u}_{\text{rod}} \end{bmatrix}$ represents the system's structural matrix; $\boldsymbol{u}_1, \boldsymbol{u}_2, \boldsymbol{u}_3$, and $\boldsymbol{u}_{\text{rod}}$ are the unit vectors along cables A_1B_1, A_2B_2, A_3B_3, and rigid rod AB, respectively; $\boldsymbol{r}_1, \boldsymbol{r}_2, \boldsymbol{r}_3$, and $\boldsymbol{r}_{\text{rod}}$ are the position vectors of the force application points of cables A_1B_1, A_2B_2, A_3B_3, and rigid rod AB on the end-effector. Additionally, F_{L_1}, F_{L_2}, and F_{L_3} are the tensions in cables A_1B_1, A_2B_2, and A_3B_3, and $F_{L_{\text{rod}}}$ is the support force of rigid rod AB. When the system is in static equilibrium, differentiating Eq. (4.57) gives

$$\delta\mathbf{exf} = -\delta\boldsymbol{A}^T\boldsymbol{F} - \boldsymbol{A}^T\delta\boldsymbol{F} \tag{4.57}$$

where $\delta\boldsymbol{A}^T$ represents the derivative of the structural matrix with respect to $\boldsymbol{S}$, $\delta\boldsymbol{F}$ denotes the infinitesimal force variation. By substituting Eq. (4.58) into Eq. (4.56), we obtain

$$\boldsymbol{K} \cdot \delta\boldsymbol{S} = -\delta\boldsymbol{A}^T\boldsymbol{F} - \boldsymbol{A}^T\delta\boldsymbol{F} \tag{4.58}$$

Equation (4.58) can be expressed as

$$\boldsymbol{K} = -\frac{\delta\boldsymbol{A}^T}{\delta\boldsymbol{S}}\boldsymbol{F} - \boldsymbol{A}^T\frac{\delta\boldsymbol{F}}{\delta\boldsymbol{S}} = \boldsymbol{K}_1 + \boldsymbol{K}_2 \tag{4.59}$$

Equation (4.59) reveals that static stiffness consists of two components. The first, termed active stiffness $\boldsymbol{K}_1$, is largely linked to the position and orientation of the end-effector. The second, known as passive stiffness $\boldsymbol{K}_2$, is mainly associated with the deformation and deflection arising from pose changes of the cables and rigid rod. An analysis of the RFCHR's structure shows that it is a redundantly constrained positioning mechanism, whose static stiffness is primarily determined by the passive stiffness $\boldsymbol{K}_2$. For the RFCHR, the active stiffness $\boldsymbol{K}_1$ is of a significantly smaller order of magnitude compared to $\boldsymbol{K}_2$. Therefore, in this study, the passive stiffness $\boldsymbol{K}_2$ is taken as the static stiffness $\boldsymbol{K}$. Additionally, based on the RFCHR model, the following is derived:

$$\delta \boldsymbol{F} = -\boldsymbol{K}_I \delta \boldsymbol{l} = -\text{diag}\,(k_1, k_2, k_3, k_{\text{rod}})\delta \boldsymbol{l} \tag{4.60}$$

where $\boldsymbol{K}_I$ is a diagonal matrix made up of the stiffness coefficients of the cables and rigid rod, and $\delta \boldsymbol{l}$ represents a small deformation. It should be emphasized that when the displacement increment of the end-effector is in the same direction as the force exerted by the cable or rigid rod, the force increment takes a negative value. Moreover, the stiffness coefficients of the cables and rigid rod can be expressed as

$$k_i = \frac{E_i A_i}{l_i},\ i = 1{,}2{,}3, \text{rod} \tag{4.61}$$

where E_i and A_i refer to the elastic modulus and cross-sectional area of the cable or rigid rod, respectively. According to the mapping relationship between the end-effector's pose and the driving joint, $\delta \boldsymbol{l}$ can be described as

$$\delta \boldsymbol{l} = \boldsymbol{A} \delta \boldsymbol{S} \tag{4.62}$$

By substituting Eqs. (4.61) and (4.62) into Eq. (4.60), it becomes

$$\delta \boldsymbol{F} = -\boldsymbol{K}_I \delta \boldsymbol{l} = -\text{diag}\left(\frac{EA}{l_1}, \frac{EA}{l_2}, \frac{EA}{l_3}, \frac{E_{\text{rod}} A_{\text{rod}}}{l_{\text{rod}}}\right) \boldsymbol{A} \delta \boldsymbol{S} \tag{4.63}$$

$E_1 = E_2 = E_3 = E$, $A_1 = A_2 = A_3 = A$. Thus, the $\boldsymbol{K}_2$ can be expressed as

$$\boldsymbol{K}_2 = \boldsymbol{A}^T \cdot \text{diag}\left(\frac{EA}{l_1}, \frac{EA}{l_2}, \frac{EA}{l_3}, \frac{E_{\text{rod}} A_{\text{rod}}}{l_{\text{rod}}}\right) \cdot \boldsymbol{A} \tag{4.64}$$

Based on Eqs. (4.59) and (4.64), the static stiffness $\boldsymbol{K}$ is written as

$$\begin{aligned} \boldsymbol{K} = \boldsymbol{K}_2 = & \begin{bmatrix} \boldsymbol{u}_1 & \boldsymbol{u}_2 & \boldsymbol{u}_3 & \boldsymbol{u}_{\text{rod}} \\ \boldsymbol{r}_1 \times \boldsymbol{u}_1 & \boldsymbol{r}_2 \times \boldsymbol{u}_2 & \boldsymbol{r}_3 \times \boldsymbol{u}_3 & \boldsymbol{r}_{\text{rod}} \times \boldsymbol{u}_{\text{rod}} \end{bmatrix} \\ & \cdot \text{diag}\left(\frac{EA}{l_1}, \frac{EA}{l_2}, \frac{EA}{l_3}, \frac{E_{\text{rod}} A_{\text{rod}}}{l_{\text{rod}}}\right) \\ & \cdot \begin{bmatrix} \boldsymbol{u}_1 & \boldsymbol{u}_2 & \boldsymbol{u}_3 & \boldsymbol{u}_{\text{rod}} \\ \boldsymbol{r}_1 \times \boldsymbol{u}_1 & \boldsymbol{r}_2 \times \boldsymbol{u}_2 & \boldsymbol{r}_3 \times \boldsymbol{u}_3 & \boldsymbol{r}_{\text{rod}} \times \boldsymbol{u}_{\text{rod}} \end{bmatrix}^T \end{aligned} \tag{4.65}$$

where $\boldsymbol{K}$ is the 6×6 symmetrical matrices. $\boldsymbol{K}$ is taken as an indicator to assess the static stiffness of the RFCHR's end-effector across various positions.

4.3.2 Multi-objective Optimization Design

In this section, a multi-objective optimization design approach is put forward, which takes into account uncertainty sources (such as cable deformation and random loads) and integrates indices related to reliability, workspace, and static stiffness. At the same time, a multi-objective optimization model for the RFCHR is constructed. Subsequently, the structural optimization of the RFCHR is realized on the basis of the three indices established in Chapter 3.

4.3.2.1 Multi-objective Optimization Model

To enhance the reliability of the RFCHR under uncertain parameters, while ensuring the workspace is maximized and the static stiffness remains sufficiently high, consideration is given to adjusting two key design variables of the RFCHR. These variables are the center of gravity position $\boldsymbol{l}_{m3}$ of the lifting arm and the positions $\boldsymbol{a}_i$ $(i = 1,2)$ of the cable exit points. Altering the center of gravity position of the lifting arm will inevitably lead to a change in the arm's mass m_3, which in turn modifies the total mass of the RFCHR, as well as the forces acting on the cables and the rigid rod. Similarly, changing the positions of the cable exit points will impact the forces exerted on the cables and the rigid rod, along with the workspace of the RFCHR.

The design parameter variables can be denoted as $\mathbf{arg} = \left[\boldsymbol{l}_{m3} \quad \boldsymbol{a}_i\right]$. The constraint range for the structural parameters is expressed as $\Omega = \{\mathbf{arg} \in \boldsymbol{R} : \mathbf{arg}^L \leq \mathbf{arg} \leq \mathbf{arg}^U\}$.

In the configuration optimization design of a robot, numerous optimization objectives can be taken into account, and these are typically determined based on the robot's application functions and structure. Taking the robot discussed as an example, the objective functions considered are to achieve higher reliability, a larger workspace of the RFCHR, and higher static stiffness. Hence, the multi-objective optimization model of the RFCHR is established by integrating p_r, W_{ind}, and K, which can be defined as follows:

$$\begin{cases} \max \quad \text{opt} = c_1 p_r\,(\text{arg}) + c_2 W_{\text{ind}}\,(\text{arg}) + c_3 K\,(\text{arg}) \\ \text{s.t.} \quad \text{arg}_i^L < \text{arg}_i < \text{arg}_i^U, \quad i = 1, \dots, n \end{cases} \tag{4.66}$$

where opt denotes the objective function for multi-objective optimization; c_i $(i = 1,2,3)$ are the weight coefficients corresponding to each parameter; p_r represents the reliability index of the RFCHR as given in Eq. (4.49); W_{ind} is the workspace index specified in Eq. (4.55); K stands for the static stiffness index defined in Eq. (4.65); arg is the scalar form of the vector $\mathbf{arg}$; arg_i^L is the lower bound of arg_i; $\text{arg}_{i\max}$ is the upper bound of arg_i; n is the total number of variables in arg. Furthermore, the reliability can be expressed in a form that incorporates uncertain parameters

$$\eta\,(\Delta \boldsymbol{l},\ \boldsymbol{f}_{\mathrm{Rou}}) = p_r \tag{4.67}$$

4.3.2.2 Analysis of Dynamic Response and Optimization Design Method

An analysis and numerical simulations of the dynamic response of the RFCHR under uncertain parameters will be carried out. To obtain valid simulation results, a relatively complex motion trajectory within the main working range of the RFCHR is selected, as it can effectively demonstrate the robot's motion under relatively extreme operating conditions. The trajectory of the RFCHR's end-effector is specified as follows:

$$\begin{cases} x = 886 + 100\cos(\pi t)\,, \\ y = 100\sin(\pi t)\,, \\ z = 10\pi t + 700 \end{cases} \tag{4.68}$$

The initial structural parameters of the RFCHR are presented in Table 4.2. In the meantime, the interval parameters and random parameters associated with the RFCHR are specified as follows:

$$\begin{cases} \Delta l_1 = \Delta l_2 = [-8{,}10]\ \text{mm}, \\ \Delta l_3 = [-9{,}9]\ \text{mm}, \\ m_{\mathrm{ou}} = 1\ \text{kg}, \\ k = 0.0204, \\ j = 0 \end{cases} \tag{4.69}$$

Based on the trajectory specified in Eq. (4.68), the dynamic response results obtained using the RIPM and MCM are illustrated in Figures 4.8 and 4.9, covering the robot's movement period from 3 to 5.5 sec.

As depicted in Figures 4.8 and 4.9, it is readily observable that the forces of the cables and rigid rod computed via the RIPM are comparable to those obtained using the MCM. With variations in the trajectory, the expected upper and lower bounds as well as the variance of the dynamic response also undergo changes. As shown in Figure 4.8, the relative error in the expectation between the upper and lower bounds derived from the RIPM and MCM is within 1%. In Figure 4.9, the relative error in the variance between the upper and lower bounds of the RIPM and MCM is within 2%. Furthermore, considering the structure of the RFCHR, both the center of mass of the end-effector and the suspended load are positioned in front of joint 3. The torque at joint 3 is sufficient to meet the torque demands for

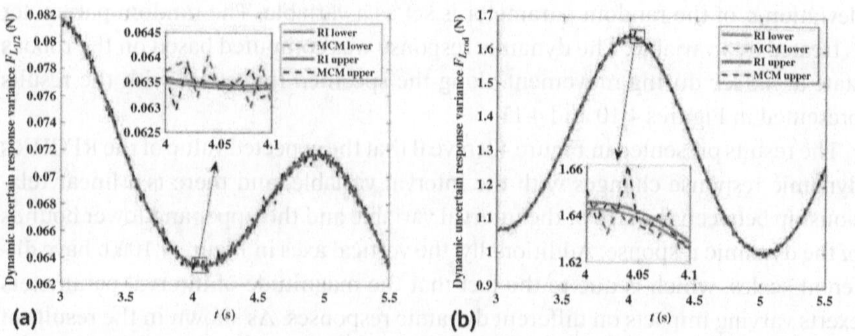

Figure 4.9 Variance of dynamic response under random interval parameters: (a) force of cable; (b) force of rigid rod.

Table 4.5 Calculation efficiency and relative error of different methods.

			Expectation		Standard Deviation	
			Upper Bound	**Lower Bound**	**Upper Bound**	**Lower Bound**
MCM	F_{L_1}	Calculation time (s)	30.155	29.997	145.303	141.960
	$F_{L_{\text{rod}}}$		29.669	31.986	153.250	148.011
RIPM	F_{L_1}	Calculation time (s)	0.239	0.222	0.203	0.182
		Error	1.531e−03N	1.506e−03N	2.388e−04	2.509e−04
	$F_{L_{\text{rod}}}$	Calculation time (s)	0.233	0.219	0.211	0.209
		Error	6.370e−03N	6.105e−03N	4.479e−03	4.718e−03

the rapid descent of the lifting arm. Consequently, the tension in cable 3 remains zero throughout the dynamic response analysis.

Owing to the massive data simulation feature of the MCM, the curves at each node lack sufficient smoothness. The relevant results are presented in Table 4.5, which indicates that the calculation time of the RIPM has been significantly improved in comparison with the MCM, and the calculation error is small. Thus, the RIPM boasts good efficiency in quantifying the uncertainty of dynamic responses. For this reason, it is both effective and reasonable to employ the RIPM for further exploration of the RFCHR's reliability.

To further examine how the dynamic response of the RFCHR varies with interval variables and random variables, calculations were performed on the dynamic response under varying interval parameters and varying standard deviations of random parameters. The coefficient of variation for interval variables is set as ε, so the variable intervals are defined as $\{[1-\varepsilon, 1+\varepsilon]$ mm, $[1-\varepsilon, 1+\varepsilon]$ mm, $[-\varepsilon, \varepsilon]$ mm$\}$. Meanwhile, the standard deviation σ of the random parameter is set as a variable. The random parameter is treated as a variable. The dynamic response was computed based on the robot's state at 4.5 sec during movement along the specified trajectory, with the results presented in Figures 4.10 and 4.11.

The results presented in Figure 4.9 reveal that the expected value of the RFCHR's dynamic response changes with the interval variable, and there is a linear relationship between the size of the interval variable and the upper and lower bounds of the dynamic response. Additionally, the vertical axes in Figure 4.10a,b have different scales, which is due to the fact that the magnitude of interval parameters exerts varying impacts on different dynamic responses. As shown in the results of Figure 4.11, as the standard deviation of the random parameter increases, the variance of the dynamic response grows with the square of this increase. This indicates that the standard deviations of the upper and lower bounds of the dynamic

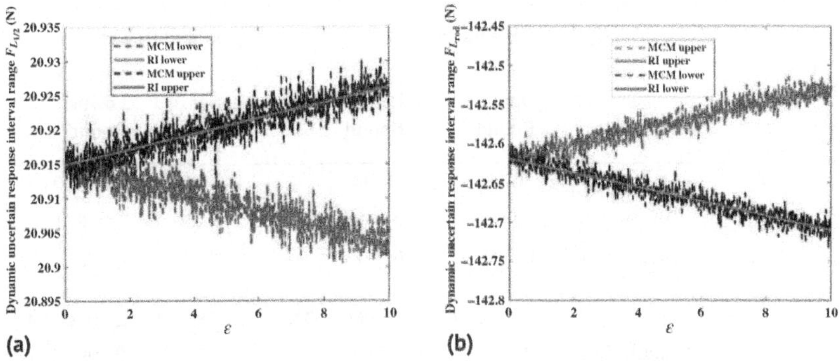

Figure 4.10 Expectation of dynamic response with the changing of interval size: (a) force of cable; (b) force of rigid rod.

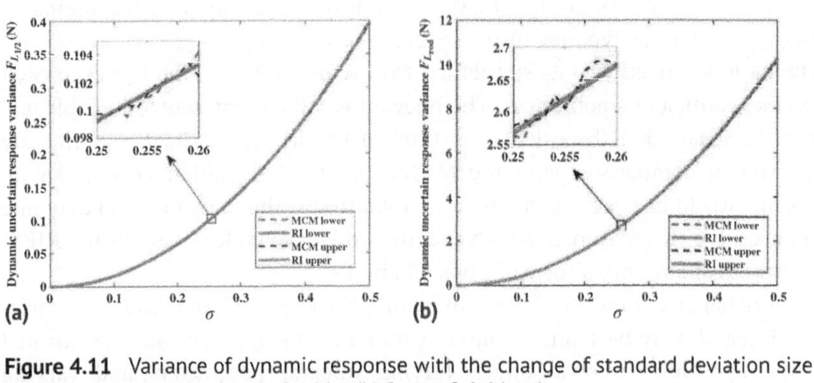

Figure 4.11 Variance of dynamic response with the change of standard deviation size of random parameter: (a) force of cable; (b) force of rigid rod.

response have a linear relationship with the standard deviation of the random variables. In conclusion, the accuracy of the RIPM is further verified by comparing its results with those of the MCM.

4.3.2.3 Optimization Design Method

The PSO algorithm is adopted to implement the multi-objective optimization design. Moreover, to guarantee the validity of the optimized parameters, the genetic algorithm (GA) is chosen as a control group for comparison.

As stated in Section 4.1, the design parameter variable of the RFCHR is expressed as $\mathbf{arg} = \begin{bmatrix} \boldsymbol{l}_{m3} & \boldsymbol{a}_i \end{bmatrix}$. Considering the structure of the RFCHR, let

$$\begin{cases} \boldsymbol{a}_1 = [x_a, y_a, -1.5] \text{ mm}, \\ \boldsymbol{a}_2 = [x_a, -y_a, -1.5] \text{ mm}, \\ \boldsymbol{l}_{m3} = [0, l_{m3y}, 0] \text{ mm} \end{cases} \tag{4.70}$$

the constraint range is expressed as

$$\Omega = \{x_a, y_a, l_{m3y} \in \boldsymbol{R}: 320 \text{ mm} \leq x_a \leq 400 \text{ mm}, 200 \text{ mm} \leq y_a \leq 400 \text{ mm}, 50 \text{ mm} \leq l_{m3y} \leq 400 \text{ mm}\} \tag{4.71}$$

Regarding the reliability index, in accordance with the distribution characteristics of random parameters, the upper and lower bounds of the reliability constraints for the RFCHR are set as follows:

$$\begin{cases} \text{Sd}\left(\boldsymbol{F}_S^U\right) = 0.1 * E\left(\boldsymbol{F}_S^U\right), \\ \text{Sd}\left(\boldsymbol{F}_S^L\right) = 0.15 * E\left(\boldsymbol{F}_S^L\right) \end{cases} \tag{4.72}$$

Therefore, the optimization process for the RFCHR can be summarized as follows:

1. Implement the PSO algorithm within the MATLAB environment, taking into account the objective function (fitness function), the design parameters, and the relevant constraints.
2. Set the parameters for PSO algorithm: particle population size $N = 20$, the maximum number of iterations $M = 300$, inertia weight $w = 0.8$, learning factors $r_1 = r_2 = 0.5$, and $c_1 = 10$, $c_2 = 1$, $c_3 = 0.01$.
3. Run the program.

It is important to highlight that the selected values for the weight coefficients are intended to maximize the alignment of the orders of magnitude among the three indicators. This ensures that the optimization algorithm can simultaneously attain optimal levels of reliability, static stiffness, and workspace. To secure superior optimization outcomes and strengthen the robustness of the results, a sensitivity analysis of the weight coefficients is performed. The baseline values and interval ranges for these weight coefficients are presented in Table 4.6.

Likewise, each parameter undergoes normalization in accordance with Eq. (4.50). Figure 4.12 illustrates how the reliability changes as a single variable is adjusted under the baseline value, with all other parameters of the RFCHR remaining at their initial settings.

As depicted in Figure 4.12, drawing on the sensitivity analysis findings and normalization principles, the weight coefficients are determined to be 10.26, 1.008, and 0.01, respectively. For the convenience of engineering applications, these

Table 4.6 The baseline values and intervals for weight coefficients.[a]

Parameter	Baseline Value	Value Range
c_1	10	[8,11]
c_2	1	[0,1.4]
c_3	0.05	[0,0.1]

a The value ranges for c_1, c_2, and c_3 can be adjusted according to the actual configuration of the RFCHR.

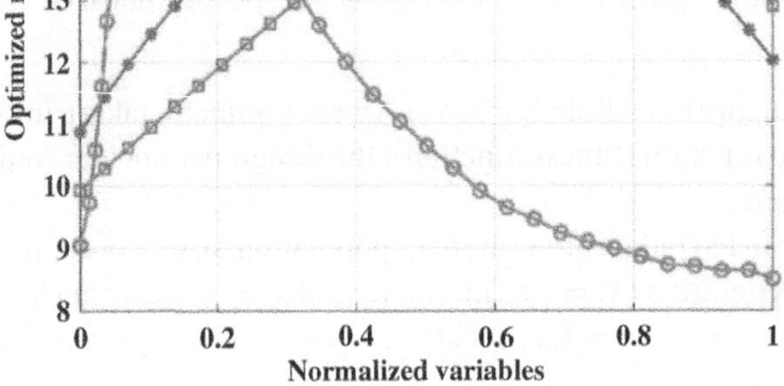

Figure 4.12 The trend of optimized result with a single variable changing at the baseline value.

weight coefficients are adjusted to 10, 1, and 0.01. It should be noted that the influence of the differences between these two sets of weight coefficient values on the optimization results can be regarded as negligible.

Furthermore, based on the analysis in Section 4.3.1, the active stiffness $\boldsymbol{K}_1$ is excluded from consideration. Therefore, in the optimization process, $\boldsymbol{K}$ is taken as $\boldsymbol{K}_2$. The outcomes of the multi-objective optimization design for the RFCHR are presented in Table 4.7.

As indicated in Table 4.7, the design parameters that allow the RFCHR to attain high performance are $(x_a, y_a, l_{m3y}) = (400, 200, 169.58)$. After optimization via the PSO algorithm, all three performance indices of the RFCHR, reliability, workspace, and static stiffness, have been enhanced. Specifically, the reliability index p_r, workspace index W_{ind}, and static stiffness index K have risen by 10.7, 13.4, and 8.02%, respectively, compared to their pre-optimization levels. Additionally, Figure 4.13 displays a comparison of the RFCHR's workspace before and after optimization, from which it is evident that the reachable range within the optimized workspace has increased significantly.

Meanwhile, as shown in Table 4.7, the final optimization results obtained by the PSO algorithm and the GA are roughly comparable. Nevertheless, in the course of the optimization process, the PSO algorithm requires fewer iterations than the GA to arrive at the results. The primary factors contributing to the performance disparities between the GA and the PSO algorithm are as follows [15]:

1. Because the GA requires a series of operations like crossover and mutation during its iterative process, its computational complexity is notably higher than that of the PSO algorithm. Consequently, it consumes more computational resources.

Table 4.7 The results of the multi-objective optimization design for RFCHR.

	$\lVert F_{L_{\text{rod}}} \rVert$ (N)	$\lVert F_{L_i} \rVert$ (N)	Design parameter variable (mm)	p_r	W_{ind} (m^3)	$K(\text{N} \cdot \text{m}^{-1})$
Original parameter	21.0878	107.7511	$(x_a, y_a, l_{m3y}) =$ (389, 398, 240.5)	0.5905	5.1371	398
PSO	18.7092	104.8446	$(x_a, y_a, l_{m3y}) =$ (400, 200, 169.58)	0.6537	5.8253	429.9
GA	18.703	104.7235	$(x_a, y_a, l_{m3y}) =$ (398.83, 200, 169.74)	0.6537	5.7364	426.53

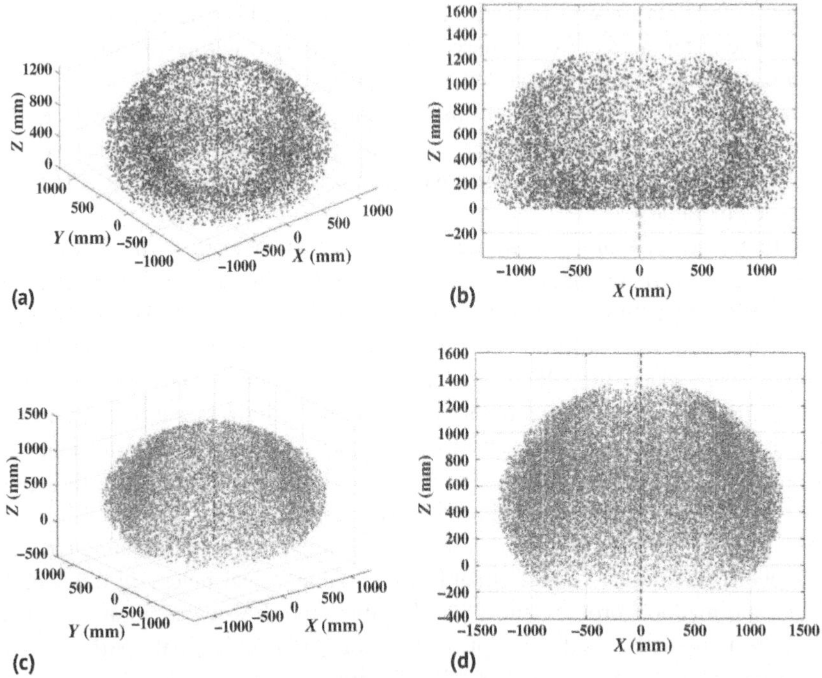

Figure 4.13 Comparison of the workspace of the RFCHR before and after optimization: (a, b) workspace before optimization; (c, d) workspace after optimization.

2. The optimization outcomes of the GA are affected by a variety of parameters, such as crossover rate, mutation rate, and population size. In comparison, the PSO algorithm has relatively fewer parameters that need tuning. Therefore, when dealing with complex computational problems, the convergence speed of the GA is relatively slow.
3. The crossover and mutation operations in the GA, when combined, might lead to the search space failing to fully cover regions where global optima are located. This makes the GA more prone to getting stuck in local optima than the PSO algorithm.

In conclusion, the PSO algorithm holds distinct advantages over the GA in performance optimization. When tackling complex optimization problems in practical engineering scenarios, the PSO algorithm is more likely to find a global optimal solution and operates with higher efficiency. For these reasons, the PSO algorithm is chosen as the fundamental optimization method for carrying out the multi-objective optimization of the RFCHR.

4.4 Experiment and Verification

In this section, the multi-objective optimization design of workspace and stiffness is taken as an example. The experiment primarily verifies the reliability of the robot following multi-objective optimization. As illustrated in Figure 4.14, a physical prototype of the RFCHR is constructed. To better validate the optimization results, the robot configuration is adjusted according to the optimized design. Furthermore, the motion of the RFCHR is tracked using the Nokov multi-camera visual capture system. In addition, force sensors are installed on the cables and the rigid rod to accurately measure the dynamic response of the RFCHR during motion.

The configuration of the RFCHR is then adjusted according to the results presented in Table 4.7, resulting in four configurations labeled as configurations 1, 2, 3, and 4. The original configuration is $(x_{a1}, y_{a1}, l_{m3y}) = (389, 389, 240.5)$. After modifying the center of gravity of the lifting arm, the configuration becomes $(x_{a1}, y_{a1}, l_{m3y}) = (389, 398, 160)$. The optimized configuration, serving as the control group based on the reliability assessment of the RFCHR, is $(x_{a1}, y_{a1}, l_{m3y}) = (400, 200, 240.5)$, while the worse configuration is $(x_{a1}, y_{a1}, l_{m3y}) = (345, 200, 240.5)$, as illustrated in Figure 4.15. The optimized and worse configurations are derived by adjusting the positions of the cable exit points from the original configuration. The inclusion of the worse configuration facilitates a more effective comparison to validate the optimization design. Furthermore, as illustrated in Figure 4.15b, the center of mass was modified by adding weight to the lifting arm.

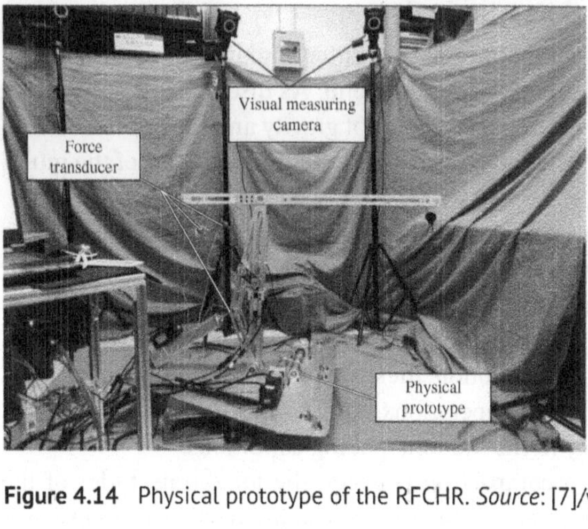

Figure 4.14 Physical prototype of the RFCHR. *Source*: [7]/with permission of Elsevier.

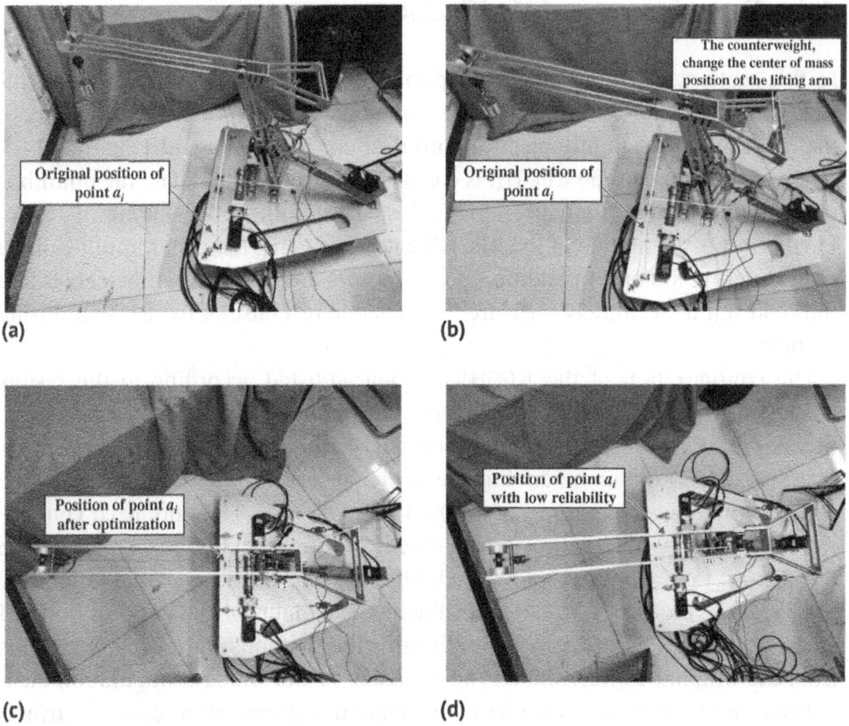

Figure 4.15 Different configurations of the RFCHR: (a) the original configuration; (b) the configuration after changing the center of gravity of the lifting arm; (c) the optimized configuration; (d) the worse configuration. *Source*: [7]/with permission of Elsevier.

Starting from the original configuration, the center of gravity of the lifting arm is modified by attaching a counterweight at its end. As discussed in Section 4.3, both the configuration with the adjusted center of gravity and the optimized configuration exhibit improved reliability, consistent with the outcomes of the multi-objective optimization design.

Given that parameter variations caused by the uncertainty sources of the RFCHR are relatively minor, while friction and other factors also exist, the experiment focuses solely on the uncertainty introduced by random loading, as it is more practical to replicate. A 1 kg weight is suspended at the end of the lifting arm, which oscillates irregularly during motion, thereby simulating significant random loading encountered in real-world operations. Moreover, the same trajectory is adopted to ensure consistency in the random loading conditions.

Figures 4.16–4.19 present a comparison between the force trajectories of the RFCHR under different configurations and the corresponding ideal trajectories. It is also observed that the forces in cables 1 and 2 are not identical, which results

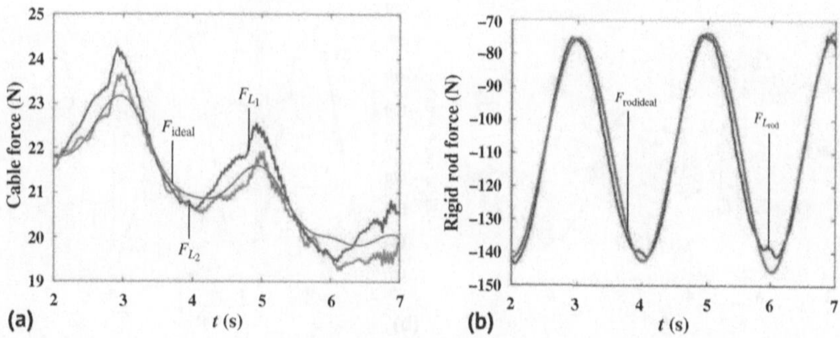

Figure 4.16 Actual dynamic response in the original configuration: (a) force of cable; (b) force of rigid rod.

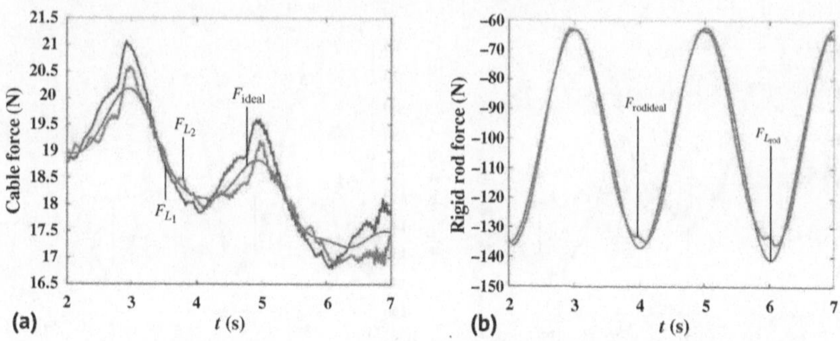

Figure 4.17 Actual dynamic response in the configuration after changing the center of gravity of the lifting arm: (a) force of cable; (b) force of rigid rod.

from the inability of the winding wheel to exert perfectly synchronized control over both cables during the operation of the RFCHR.

It is evident that, relative to the other three configurations, the RFCHR under the optimized configuration exhibits a dynamic response trajectory that more closely aligns with the ideal value. Moreover, the curve patterns in Figures 4.16 and 4.17 show strong similarity, as do those in Figures 4.18 and 4.19. This similarity arises from modifications to the cable exit point positions of the RFCHR, which differ between the first two configurations and the latter two.

To more precisely validate the enhancement of configuration reliability after optimization, the relative error between the actual force $\boldsymbol{F}_S$ trajectory and the ideal force trajectory is computed. The error is defined as follows:

$$\boldsymbol{e}(\boldsymbol{F}_S) = \frac{\sum \left| \dfrac{\boldsymbol{F}_S - \boldsymbol{F}_{L_{i\text{ideal}}}}{\boldsymbol{F}_{L_{i\text{ideal}}}} \right|}{N}, i = 1,2,\text{rod} \tag{4.73}$$

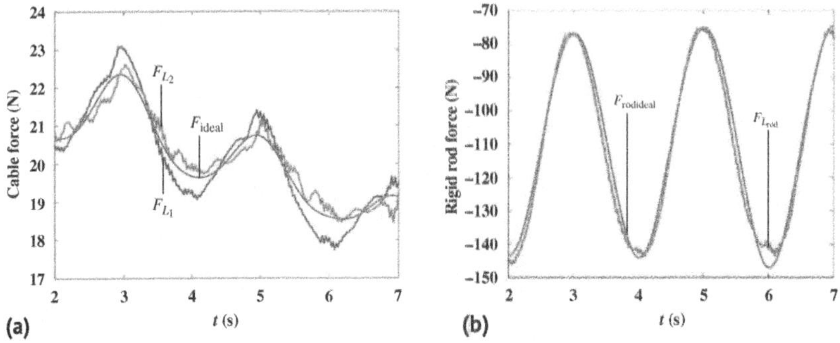

Figure 4.18 Actual dynamic response in the optimized configuration: (a) force of cable; (b) force of rigid rod.

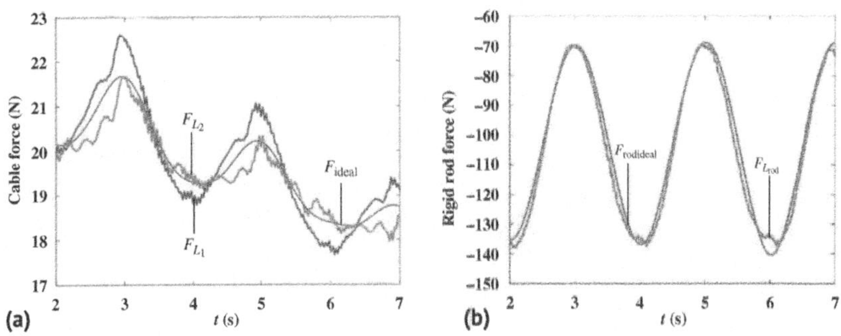

Figure 4.19 Actual dynamic response in the worse configuration: (a) force of cable; (b) force of rigid rod.

where N is the number of data collected by the force sensor. The relevant calculation results are shown in Figure 4.20 and Table 4.8.

As illustrated in Figure 4.20 and summarized in Table 4.8 (where configurations 1, 2, 3, and 4 correspond to the original configuration, the configuration after adjusting the center of gravity of the lifting arm, the optimized configuration, and the worse configuration, respectively), the force errors relative to the ideal values are as follows: for the original configuration, 1.63, 1.11, and 2.52% for the two cables and the rigid rod; for the configuration with the adjusted center of gravity, 1.51, 0.9, and 2.33%; for the optimized configuration, 1.63, 0.9, and 2.32%; and for the worse configuration, 1.75, 1.18, and 2.73%.

The results indicate that the dynamic relative errors of both the optimized configuration and the configuration with the adjusted center of gravity of the lifting arm are smaller than those of the original and worse configurations. Specifically,

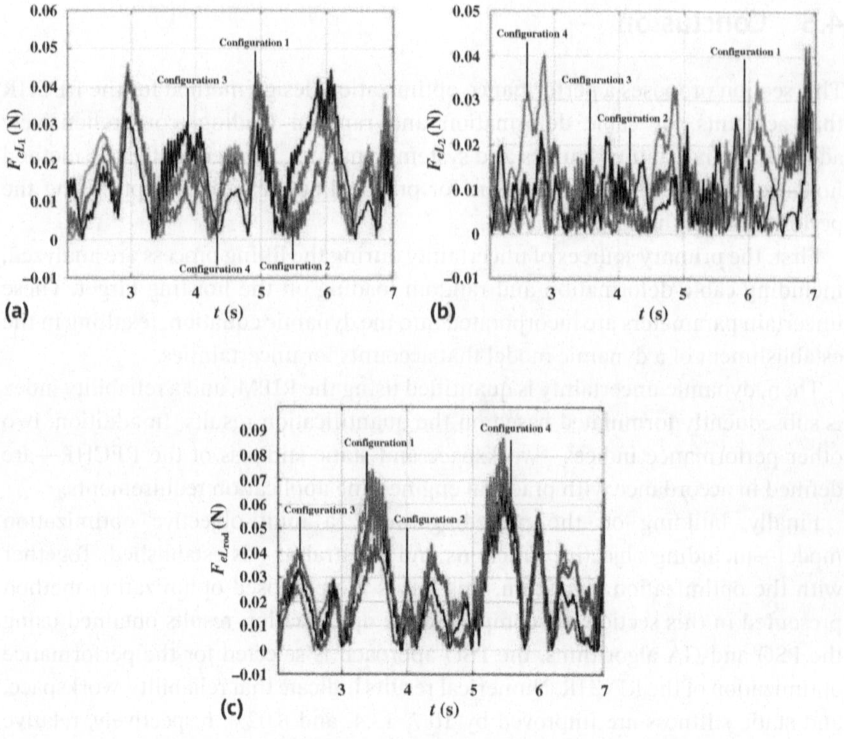

Figure 4.20 Relative error between the actual force and the ideal force: (a) force of cable 1; (b) force of cable 2; (c) force of rigid rod.

Table 4.8 Relative errors of actual and theoretical forces in different configurations.

Different Configuration	$e(F_{L_1})$ (%)	$e(F_{L_2})$ (%)	$e(F_{L_{rod}})$ (%)
Configuration 1	1.63	1.11	2.52
Configuration 2	1.51	0.9	2.33
Configuration 3	1.63	0.9	2.32
Configuration 4	1.75	1.18	2.73

the average values of $e(F_{L_1})$, $e(F_{L_2})$, and $e(F_{L_{rod}})$ for configuration 2 are approximately 9.89 and 14.31% lower than those of configurations 1 and 4, respectively. For configuration 3, the corresponding reductions are about 7.79 and 16.25%. These findings experimentally confirm the enhanced reliability of the RFCHR following multi-objective optimization.

4.5 Conclusion

This section proposes a performance optimization design method for the RFCHR that accounts for cable deformation and random loading, comprehensively addressing uncertainty sources and system dynamics. Consequently, the method holds substantial engineering value for practical applications in optimizing the performance of the RFCHR.

First, the primary sources of uncertainty during the lifting process are analyzed, including cable deformation and random loading on the hoisting target. These uncertain parameters are incorporated into the dynamic equation, resulting in the establishment of a dynamic model that accounts for uncertainties.

Then, dynamic uncertainty is quantified using the RIPM, and a reliability index is subsequently formulated based on the quantification results. In addition, two other performance indices—workspace and static stiffness of the RFCHR—are defined in accordance with practical engineering application requirements.

Finally, building on the preceding work, a multi-objective optimization model—including objective functions and constraints—is established. Together with the optimization algorithm, this forms the proposed optimization method presented in this section. By comparing the optimization results obtained using the PSO and GA algorithms, the PSO approach is selected for the performance optimization of the RFCHR. Numerical results indicate that reliability, workspace, and static stiffness are improved by 10.7, 13.4, and 8.02%, respectively, relative to their pre-optimization values. The optimal configuration of the RFCHR is also determined. Subsequently, experimental validation confirms the accuracy and effectiveness of the proposed optimization design method in enhancing the RFCHR's reliability.

In future work, we plan to further refine the robot's structure and driving mode, enhance its precision, and minimize the impact of structural interference on experimental results.

References

1 Zhou B, Li S P, Zi B, et al. Multi-objective optimal design of a cable-driven parallel robot based on an adaptive adjustment inertia weight particle swarm optimization algorithm. Journal of Mechanical Design, 2023, 145(8): 083301.

2 Song D, Zhang L X, Xue F. Configuration optimization and a tension distribution algorithm for cable-driven parallel robots. IEEE Access, 2018, 6: 33928–33940.

3 Yu J S, Tao J G, Wang G X, et al. Stability analysis and optimal design of a cable-driven parallel robot. Robotics and Autonomous Systems, 2024, 173: 104611.

4 Hamida I B, Laribi M A, Mlika A, et al. Multi-objective optimal design of a cable driven parallel robot for rehabilitation tasks. Mechanism and Machine Theory, 2021, 156: 104141.

5 Sun C, Gao H B, Liu Z, et al. Design and optimization of three-degree-of-freedom planar adaptive cable-driven parallel robots using the cable wrapping phenomenon. Mechanism and Machine Theory, 2021, 166: 104475.

6 Fu C, Xu Y D, Yang Y F, et al. Response analysis of an accelerating unbalanced rotating system with both random and interval variables. Journal of Sound and Vibration, 2020, 466: 115047.

7 Zhou B, Li S, Zi B, et al. Performance analysis and optimization design of a rigid-flexible coupling hoisting robot considering cable deformation and random loading. Mechanical Systems and Signal Processing, 2025, 223: 111903.

8 Hu B. Kinematically identical manipulators for the Exechon parallel manipulator and their comparison study. Mechanism and Machine Theory, 2016, 103: 117–137.

9 Martin-Parra A, Juarez-Perez S, Gonzalez-Rodriguez A, et al. A novel design for fully constrained planar cable-driven parallel robots to increase their wrench-feasible workspace. Mechanism and Machine Theory, 2023, 180: 105159.

10 Wang J, Lau H Y K. Dexterity analysis based on Jacobian and performance optimization for multi-segment continuum robots. Journal of Mechanisms and Robotics, 2021, 13(6): 061012.

11 Brauers W K M, Zavadskas E K. Robustness of MULTIMOORA: a method for multi-objective optimization. Informatica, 2012, 23(1): 1–25.

12 Xiong H, Cao H H, Zeng W F, et al. Real-time reconfiguration planning for the dynamic control of reconfigurable cable-driven parallel robots. Journal of Mechanisms and Robotics, 2022, 14(6): 060913.

13 Gao J, Zhou B, Zi B, et al. Kinematic uncertainty analysis of a cable-driven parallel robot based on an error transfer model. Journal of Mechanisms and Robotics, 2022, 14(5): 051008.

14 Xiao Q G, Li C B, Tang Y, et al. Energy efficiency modeling for configuration-dependent machining via machine learning: a comparative study. IEEE Transactions on Automation Science and Engineering, 2020, 18(2): 717–730.

15 Roberge V, Tarbouchi M, Labonté G. Comparison of parallel genetic algorithm and particle swarm optimization for real-time UAV path planning. IEEE Transactions on Industrial Informatics, 2012, 9(1): 132–141.

5 Kinematic Analysis of Rigid–Flexible Coupling Hoisting Robots with Uncertainty

5.1 Preamble

For complex mechanical systems, the dynamic responses of flexible multibody systems under various conditions have been extensively studied [1, 2]. For instance, using a combination of finite element and analytical methods, Zrnić et al. derived the dynamic response equations to determine transverse and longitudinal vibrations of a gantry crane system subjected to an elastically suspended moving body [3]. Beyond single-crane models, dual or multiple automobile cranes are increasingly employed to handle more demanding operations in modern construction projects [4, 5]. For example, Leban et al. applied the Newton–Euler equations to analyze the dynamic model of a dual shipboard crane system and proposed an inverse kinematics control strategy to address the underdetermined kinematic problem [6]. Furthermore, based on the Lagrange equation and kinematic analysis, the complete dynamic model of cooperative cable parallel manipulators for multiple mobile cranes (CPMMC) was established [7].

However, the kinematic response of dual automobile crane systems (DACS) has not yet been systematically investigated. In most traditional analyses of single automobile crane models or CPMMC, external excitations and system parameters are treated as deterministic [1, 3–7]. In reality, uncertainties in system parameters for single or dual crane systems may arise from both internal and external factors [8, 9], such as mechanical tolerances (e.g., design, manufacturing, or assembly errors), unpredictable external excitations (e.g., vibrations, sea waves), and complex environmental conditions (e.g., temperature variations, wind loads) [10–13]. In general, deterministic methods can only provide approximate solutions due to the inherent randomness of system parameters. Therefore, it is essential to analyze the stochastic response field problem of DACS.

The interval algorithm is a most convenient approach to deal with uncertainty problems when the inner probability information of an uncertain parameter is completely missing. Up to now, there are two main approaches for interval UQ problems. The first type is small interval uncertainty analysis methods, where the interval model is constructed based on interval variables with small uncertain levels. Recently, Sofi et al. [14] incorporated the interval algorithm with the interval field for uncertainty propagation analysis of structures. To resolve the dependency problem, Luo et al. [15] presented a nonstochastic framework for calculating bounds of the structural performance with a certain spatial correlation. The second type is the subinterval theory-based approach, where the interval model consists of interval variables with large uncertain levels. Xia et al. [16] introduced the subinterval theory to the modified interval method for analyzing an uncertain structure with the large bounded uncertainty. Wang et al. [17] combined the subinterval theory and fuzzy theory for an uncertain heat problem with the large uncertainty.

Due to the practical existence of uncertain factors including cable deformation, mechanical errors, and cable slipping, it is difficult to accurately predict kinematic performances of cable-driven parallel robots (CDPRs) without considering the effects of uncertain parameters. Thus, the uncertainty in CDPRs is inevitable and the research of it is of great significance. Uncertainty quantification strategies have started to be taken into consideration in CDPRs recently, and uncertainty plays an important role in the kinematic performances of CDPRs [18]. In general, the uncertainty is categorized into the aleatory uncertainty and epistemic uncertainty [19]. The aleatory uncertainty relates to random behavior, and epistemic uncertainty relates to the lack of information [20]. Up to now, there are many approaches to describe the uncertainty, such as the interval algorithm [8, 21], the fuzzy sets [22], and evidence theory [23, 24]. Evidence theory has a strong ability to deal with the epistemic uncertainty [25, 26]. Evidence theory not only has a more flexible framework than other modeling techniques, but also can deal with conflicting information from experts [27]. Thus, evidence theory has been applied to the analysis of uncertainty in many engineering fields. By introducing evidence theory, Yang et al. [28] proposed a dual-rotor misalignment fault quantitative identification method. Liang et al. [29] proposed a new structural damage identification approach based on the Dempster–Shafer theory. Jian et al. [30] studied the uncertainty quantification of the flutter analysis of the rudder using evidence theory. Besides, the evidence method has also been used in the fault diagnosis [31], network security [32], and so on [33].

In this section, a hybrid perturbation compound function-based moment method is proposed to predict the LAR field of the DACS with both random and interval parameters, extending the approach in [8] to three-dimensional problems. Building on our previous work in [9], the hybrid uncertain parameter model (HUPM) is established, and the hybrid uncertain LAR equilibrium equation of the DACS with random and interval parameters is subsequently derived. Leveraging the random interval perturbation method and the random interval moment method from [34], two novel approaches are introduced by integrating the compound function technique: the random interval perturbation compound function-based method and the random interval compound function-based moment method. In the former, the LAR expression of the DACS is formulated using a first-order Neumann series expansion, while in the latter, the expectations and variances of the LAR bounds are determined via monotonic analysis. Numerical simulations demonstrate that the proposed methods achieve higher efficiency and lower computational cost compared with HMCM and IPM.

The remainder of this chapter is organized as follows. Section 5.2 introduces the kinematic uncertainty analysis with random parameters. Section 5.3 introduces the kinematic uncertainty analysis with interval variables. Section 5.4 briefly introduces evidence theory, and the kinematic uncertainty analysis based on evidence theory. Section 5.5 introduces the kinematic uncertainty analysis with hybrid random and interval parameters. Finally, some conclusions are summarized in Section 5.6.

5.2 Kinematic Uncertainty Analysis with Random Parameters

5.2.1 Mechanism Description

Figure 5.1 shows the prototype of the DACS, which consists of two automobile cranes and a payload. Both the cranes and the payload are constrained within a 2D plane. As described by Leban et al. [6], the system allows the payload to translate along the Y- and Z-axes and rotate about a single axis perpendicular to the O–YZ plane. The kinematics of the DACS are analyzed in detail.

Figure 5.1 Prototype of the DACS. *Source*: https://news.sina.com.cn/c/2006-08-19/03299789078s.shtml, last accessed July 25, 2025.

As illustrated in Figure 5.2, each automobile crane consists of a lifting arm, a luffing cylinder, a rotating table, a hoisting rope, and a suspended weight. In this study, the luffing motions of both mobile cranes are considered simultaneously. The luffing motion is generated by the luffing cylinder rotating about the hinge point fixed to the rotating table.

5.2.2 Inverse Kinematics

The kinematic model of a 3-DOF DACS is depicted in Figure 5.3. The base frame $\{B\}$: O–YZ is fixed at the midpoint of the line connecting points A_1 and A_2, which correspond to the hinge points of the lifting arm and rotating table of each automobile crane. The moving frame $\{P\}$: O_p–Y_pZ_p is attached to the centroid of the payload defined by points C_1C_2. Let D and d represent the lengths of A_1A_2 and C_1C_2, respectively.

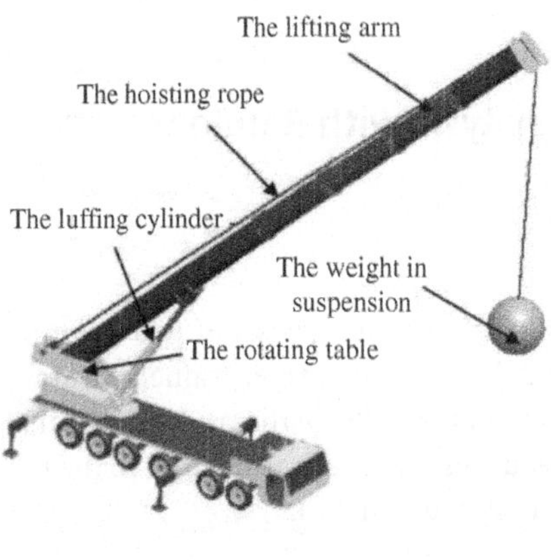

Figure 5.2 Three-dimensional model of each automobile crane.

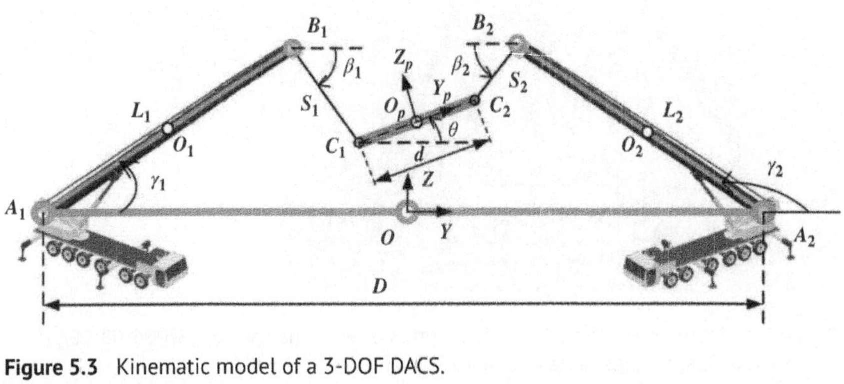

Figure 5.3 Kinematic model of a 3-DOF DACS.

The position vector of joint point A_i in the base frame $\{B\}$ can be expressed as

$$\boldsymbol{r}_{A_1} = \frac{1}{2}[-D \quad 0]^T, \quad \boldsymbol{r}_{A_2} = \frac{1}{2}[D \quad 0]^T \tag{5.1}$$

The position vector of joint point B_i in the base frame $\{B\}$ can be expressed as

$$\boldsymbol{r}_{B_i} = \boldsymbol{r}_{A_i} + L_i \begin{bmatrix} \cos\gamma_i \\ \sin\gamma_i \end{bmatrix}, \quad i = 1, 2 \tag{5.2}$$

Here, L_i and γ_i denote the length and luffing angle of the lifting arm A_iB_i, respectively, while $\boldsymbol{r}_{A_i}$ and $\boldsymbol{r}_{B_i}$ represent the position vectors of the joint points A_i and B_i.

Based on Eqs. (5.1) and (5.2), the position vector of centroid O_i of lifting arm A_iB_i in the base frame $\{B\}$ can be expressed as

$$\boldsymbol{r}_{O_i} = \frac{1}{2}\left(\boldsymbol{r}_{A_i} + \boldsymbol{r}_{B_i}\right), \quad i = 1, 2 \tag{5.3}$$

The position vector of joint point C_i in the moving frame $\{P\}$ can be expressed as

$$\boldsymbol{r}^p_{C_1} = \frac{1}{2}[-d \quad 0]^T, \quad \boldsymbol{r}^p_{C_2} = \frac{1}{2}[d \quad 0]^T \tag{5.4}$$

The position vector of joint point C_i in the base frame $\{B\}$ can be expressed as

$$\boldsymbol{r}_{C_i} = \boldsymbol{r}_{O_p} + \boldsymbol{R}\cdot\boldsymbol{r}^p_{C_i} = \boldsymbol{r}_{B_i} + S_i \begin{bmatrix} \cos\beta_i \\ \sin\beta_i \end{bmatrix}, \quad i = 1, 2 \tag{5.5}$$

where $\boldsymbol{r}_{O_p}$ is the position vector of centroid O_p in the base frame $\{B\}$, and $\boldsymbol{r}_{O_p} = (y \quad z)^T$. S_i and β_i are the length and rotation angle of hoisting rope B_iC_i, respectively. $\boldsymbol{R}$ denotes the rotation matrix from moving frame $\{P\}$ to base frame $\{B\}$, and $\boldsymbol{R} = \begin{bmatrix} \cos\theta & -\sin\theta \\ \sin\theta & \cos\theta \end{bmatrix}$. Here, θ denotes the rotation of $\{P\}$ relative to $\{B\}$, and y and z are the Cartesian coordinates of the origin O_p.

The constraint equation related to the ith hoisting rope can be formulated as

$$\left\|\boldsymbol{r}_{C_i} - \boldsymbol{r}_{B_i}\right\| = S_i, \quad i = 1, 2 \tag{5.6}$$

By combining Eqs. (5.2), (5.4), and (5.5) with Eq. (5.6), the above constraint equation can be expressed as

$$\begin{cases} \left(y - \dfrac{d\cos\theta}{2} + \dfrac{D}{2} - L_1\cos\gamma_1\right)^2 + \left(z - \dfrac{d\sin\theta}{2} - L_1\sin\gamma_1\right)^2 = S_1^2, \\ \left(y + \dfrac{d\cos\theta}{2} - \dfrac{D}{2} - L_2\cos\gamma_2\right)^2 + \left(z + \dfrac{d\sin\theta}{2} - L_2\sin\gamma_2\right)^2 = S_2^2 \end{cases} \tag{5.7}$$

Equation (5.7) can be rewritten as

$$K_{1i}\sin\gamma_i + K_{2i}\cos\gamma_i + K_{3i} = 0,\ i = 1,2 \tag{5.8}$$

Based on Eq. (5.8), the inverse kinematic solution for the DACS can be expressed as

$$\gamma_i = 2\tan^{-1}\frac{-K_{1i} \pm \sqrt{K_{1i}^2 + K_{2i}^2 - K_{3i}^2}}{K_{3i} - K_{2i}},\ i = 1,2 \tag{5.9}$$

where

$$\begin{aligned}
&K_{11} = -2L_1\left(z - \frac{d\sin\theta}{2}\right),\ K_{21} = -2L_1\left(y - \frac{d\cos\theta}{2} + \frac{D}{2}\right)\\
&K_{31} = \left(y - \frac{d\cos\theta}{2} + \frac{D}{2}\right)^2 + \left(z - \frac{d\sin\theta}{2}\right)^2 + L_1^2 - S_1^2\\
&K_{12} = -2L_2\left(z + \frac{d\sin\theta}{2}\right),\ K_{22} = -2L_2\left(y + \frac{d\cos\theta}{2} - \frac{D}{2}\right)\\
&K_{32} = \left(y + \frac{d\cos\theta}{2} - \frac{D}{2}\right)^2 + \left(z + \frac{d\sin\theta}{2}\right)^2 + L_2^2 - S_2^2
\end{aligned} \tag{5.10}$$

The primary goal of the inverse kinematic analysis of the DACS is to determine the luffing angles γ_i $(i = 1, 2)$ given the payload's position and orientation vector $\mathbf{p} = (y\ \ z\ \ \theta)^T$, the length of the lifting arm A_iB_i, and the length of the hoisting rope B_iC_i. A detailed derivation of Eq. (5.1) can be found in [9].

5.2.3 DACS Equilibrium Equation Under Narrowly Random Model

Due to the nonnegative of the luffing angular γ_i $(i = 1, 2)$, Eq. (5.9) can also be written as

$$\tan\frac{\gamma_i}{2} = \frac{\sqrt{K_{1i}^2 + K_{2i}^2 - K_{3i}^2} - K_{1i}}{K_{3i} - K_{2i}},\ i = 1,2 \tag{5.11}$$

so we rewrite Eq. (5.11) in the following form:

$$\boldsymbol{M} = \boldsymbol{T}\boldsymbol{\gamma} \tag{5.12}$$

Here, $\boldsymbol{M}$ and $\boldsymbol{T}$ represent the random vector and random matrix, respectively, while $\boldsymbol{\gamma}$ denotes the luffing angular response vector.

Consequently, the random vector of the DACS $\boldsymbol{M}$ can be expressed as

$$\boldsymbol{M} = \{\boldsymbol{M}_1\ \ \boldsymbol{M}_2\}^T \tag{5.13}$$

where $\boldsymbol{M}_1 = \sqrt{K_{11}^2 + K_{21}^2 - K_{31}^2} - K_{11}$, $\boldsymbol{M}_2 = \sqrt{K_{12}^2 + K_{22}^2 - K_{32}^2} - K_{12}$

Similarly, the random matrix of the DACS $\boldsymbol{T}$ can be expressed as

$$\boldsymbol{T} = \begin{bmatrix} \boldsymbol{T}_1 & 0 \\ 0 & \boldsymbol{T}_2 \end{bmatrix} \tag{5.14}$$

where $\boldsymbol{T}_1 = K_{31} - K_{21}$, $\boldsymbol{T}_2 = K_{32} - K_{22}$.

Similarly, the luffing angular response vector of the DACS $\boldsymbol{\gamma}$ can be expressed as

$$\boldsymbol{\gamma} = \{\gamma_1 \quad \gamma_2\}^T \tag{5.15}$$

where $\gamma_1 = \tan\frac{\gamma_1}{2}$, $\gamma_2 = \tan\frac{\gamma_2}{2}$.

From Eq. (5.12), the luffing angular response equilibrium equation of the DACS under the random model can thus be reformulated as

$$\boldsymbol{M}_i\,(\boldsymbol{K}_i\,(\boldsymbol{X})) = \boldsymbol{T}_i\,(\boldsymbol{K}_i\,(\boldsymbol{X}))\;\gamma_i,\; i = 1, 2 \tag{5.16}$$

Here, $\boldsymbol{M}_i\,(\boldsymbol{K}_i\,(\boldsymbol{X}))$ and $\boldsymbol{T}_i\,(\boldsymbol{K}_i\,(\boldsymbol{X}))$ denote the composite function vector and composite function matrix of the ith crane with respect to the random vector $\boldsymbol{X}$, respectively. Accordingly, the following expressions can be derived:

$$\begin{cases} \boldsymbol{M}_i = \boldsymbol{M}_i\,(\boldsymbol{K}_i\,(\boldsymbol{X})) \\ \boldsymbol{T}_i = \boldsymbol{T}_i\,(\boldsymbol{K}_i\,(\boldsymbol{X})) \end{cases},\; i = 1, 2 \tag{5.17}$$

Based on Eq. (5.9), $\boldsymbol{K}_i\,(\boldsymbol{X})$ stands for the membership function vector of random vector $\boldsymbol{X}$, which can be expressed as

$$\boldsymbol{K}_i\,(\boldsymbol{X}) = \{K_{1i}\,(\boldsymbol{X}), K_{2i}\,(\boldsymbol{X}), K_{3i}\,(\boldsymbol{X})\}^T \tag{5.18}$$

5.2.4 MHRM for Luffing Angular Response Field of the DACS with Narrow Uncertainty

In this subsection, according to the work of Xia et al. [35] and Gao et al. [34], a modified hybrid random method (MHRM) is proposed to handle the prediction of luffing angular response field of the DACS with narrowly random parameters, in which the perturbation-based random composite function method (PRCFM) and the random variable functional moment method (RVFMM) are used to analyze the luffing angular response and its statistical characteristics, respectively.

5.2.4.1 PRCFM for Luffing Angular Response Analysis

Based on the PRCFM and neglecting the higher order terms, the first-order Taylor series expansion of the **composite** function vector $\boldsymbol{M}_i\,(\boldsymbol{K}_i\,(\boldsymbol{X}))$ can be expressed as

$$\boldsymbol{M}_i\left(\boldsymbol{K}_i\left(\boldsymbol{X}\right)\right)=\boldsymbol{M}_i^e+\Delta_1\boldsymbol{M}_i=\boldsymbol{M}_i\left(\boldsymbol{K}_i\left(\boldsymbol{X}^e\right)\right)+\sum_{r=1}^{n}\frac{\partial\boldsymbol{M}_i\left(\boldsymbol{K}_i\left(\boldsymbol{X}\right)\right)}{\partial\boldsymbol{K}_i\left(\boldsymbol{X}\right)}\cdot\frac{\partial\boldsymbol{K}_i\left(\boldsymbol{X}\right)}{\partial x_r}_{|\boldsymbol{X}=\boldsymbol{X}^e}\left(x_r-x_r^e\right),\ i=1,2 \tag{5.19}$$

where $\boldsymbol{X}^e=\{x_1^e,x_2^e,\dots,x_r^e,\dots,x_n^e\}^T$ is the expectation of the random vector $\boldsymbol{X}$; x_r^e is the expectation of x_r.

Similarly, the composite function matrix $\boldsymbol{T}_i\left(\boldsymbol{K}_i\left(\boldsymbol{X}\right)\right)$ can also be expanded at the expectation of the random vector $\boldsymbol{X}$, one obtains

$$\begin{aligned}\boldsymbol{T}_i\left(\boldsymbol{K}_i\left(\boldsymbol{X}\right)\right)&=\boldsymbol{T}_i^e+\Delta_1\boldsymbol{T}_i\\&=\boldsymbol{T}_i\left(\boldsymbol{K}_i\left(\boldsymbol{X}^e\right)\right)+\sum_{r=1}^{n}\frac{\partial\boldsymbol{T}_i\left(\boldsymbol{K}_i\left(\boldsymbol{X}\right)\right)}{\partial\boldsymbol{K}_i\left(\boldsymbol{X}\right)}\cdot\frac{\partial\boldsymbol{K}_i\left(\boldsymbol{X}\right)}{\partial x_r}_{|\boldsymbol{X}=\boldsymbol{X}^e}\left(x_r-x_r^e\right),i=1,2\end{aligned} \tag{5.20}$$

Based on the perturbation theory, by substituting Eqs. (5.19) and (5.20) into Eq. (5.16), one yields

$$\boldsymbol{\gamma}_i=\left(\boldsymbol{T}_i^e+\Delta_1\boldsymbol{T}_i\right)^{-1}\left(\boldsymbol{M}_i^e+\Delta_1\boldsymbol{M}_i\right),\ i=1,2 \tag{5.21}$$

$\left(\boldsymbol{T}_i^e+\Delta_1\boldsymbol{T}_i\right)^{-1}$ can be expanded by using Neumann series expansion

$$\left(\boldsymbol{T}_i^e+\Delta_1\boldsymbol{T}_i\right)^{-1}=\left(\boldsymbol{T}_i^e\right)^{-1}+\sum_{p=1}^{\infty}\left(\boldsymbol{T}_i^e\right)^{-1}\left(-\Delta_1\boldsymbol{T}_i\cdot\left(\boldsymbol{T}_i^e\right)^{-1}\right)^p,\ i=1,2 \tag{5.22}$$

Therefore, the luffing angular response vector $\boldsymbol{\gamma}_i$ can be obtained as

$$\begin{aligned}\boldsymbol{\gamma}_i=&\left(\boldsymbol{T}_i\left(\boldsymbol{K}_i\left(\boldsymbol{X}^e\right)\right)\right)^{-1}\boldsymbol{M}_i\left(\boldsymbol{K}_i\left(\boldsymbol{X}^e\right)\right)+\left(\boldsymbol{T}_i\left(\boldsymbol{K}_i\left(\boldsymbol{X}^e\right)\right)\right)^{-1}\\&\left(\sum_{r=1}^{n}\frac{\partial\boldsymbol{M}_i\left(\boldsymbol{K}_i\left(\boldsymbol{X}\right)\right)}{\partial\boldsymbol{K}_i\left(\boldsymbol{X}\right)}\cdot\frac{\partial\boldsymbol{K}_i\left(\boldsymbol{X}\right)}{\partial x_r}_{|\boldsymbol{X}=\boldsymbol{X}^e}\left(x_r-x_r^e\right)\right)\\&-\left(\boldsymbol{T}_i\left(\boldsymbol{K}_i\left(\boldsymbol{X}^e\right)\right)\right)^{-1}\left(\sum_{r=1}^{n}\frac{\partial\boldsymbol{T}_i\left(\boldsymbol{K}_i\left(\boldsymbol{X}\right)\right)}{\partial\boldsymbol{K}_i\left(\boldsymbol{X}\right)}\cdot\frac{\partial\boldsymbol{K}_i\left(\boldsymbol{X}\right)}{\partial x_r}_{|\boldsymbol{X}=\boldsymbol{X}^e}\left(x_r-x_r^e\right)\right)\\&\left(\boldsymbol{T}_i\left(\boldsymbol{K}_i\left(\boldsymbol{X}^e\right)\right)\right)^{-1}\boldsymbol{M}_i\left(\boldsymbol{K}_i\left(\boldsymbol{X}^e\right)\right),i=1,2\end{aligned} \tag{5.23}$$

5.2.4.2 RVFMM for Luffing Angular Response Field Calculation

In this subsection, the statistical properties of the luffing angular response field are represented by the expectation and variance of the response vector $\boldsymbol{\gamma}_i$. Specifically, the expectation and variance of $\boldsymbol{\gamma}_i$ are expressed as follows:

$$E\left(\boldsymbol{\gamma}_i\right)=\left(\boldsymbol{T}_i\left(\boldsymbol{K}_i\left(\boldsymbol{X}^e\right)\right)\right)^{-1}\boldsymbol{M}_i\left(\boldsymbol{K}_i\left(\boldsymbol{X}^e\right)\right),\ i=1,2 \tag{5.24}$$

$$D(\gamma_i) = \sum_{1}^{r} \left((T_i (K_i (X^e)))^{-1} \left(\left(\frac{\partial M_i (K_i (X))}{\partial K_i (X)} \cdot \frac{\partial K_i (X)}{\partial x_r} \right)_{|X=X^e} - \left(\frac{\partial T_i (K_i (X))}{\partial K_i (X)} \cdot \frac{\partial K_i (X)}{\partial x_r} \right)_{|X=X^e} (T_i (K_i (X^e)))^{-1} M_i (K_i (X^e)) \right) \right)^2 \operatorname{var}(x_r), i = 1, 2 \tag{5.25}$$

where $E(\gamma_i)$ and $D(\gamma_i)$ are the expectation and variance of the luffing angular response vector γ_i, respectively.

5.2.5 Numerical Examples

This subsection presents several examples conducted using MATLAB for the DACS to demonstrate the feasibility and effectiveness of the proposed method. Examples are provided to showcase the applicability of the MHRM in solving the stochastic DACS luffing angular response field problem with narrowly distributed random payload parameters. The Monte Carlo method serves as a benchmark to validate the feasibility and efficiency of the proposed MHRM. The deterministic system parameters for the DACS are provided in Table 5.1.

In the simulation, the centroid of the payload is moved from its initial position and orientation at (0 m, 0 m, 0°) to the terminal coordinate at (−0.72 m, 0.72 m, 30°) along the specified spatial trajectory, as described by

$$\begin{cases} y = -0.5 * \sin\left(\frac{\pi}{2} * t\right), \\ z = 0.5 * \cos\left(\frac{\pi}{2} * t\right), \\ \theta = \pi/6 * t \end{cases} \tag{5.26}$$

Table 5.1 Deterministic parameters of the DACS.

System Parameters	Value
Length of A_1A_2, D (m)	12
Length of payload, d (m)	2.5
Length of lifting arm A_1B_1, L_1 (m)	5
Length of lifting arm A_2B_2, L_2 (m)	5
Length of hoisting rope B_1C_1, S_1 (m)	1
Length of hoisting rope B_2C_2, S_2 (m)	1
Mass of lifting arm A_1B_1, m_1 (kg)	50,000
Mass of lifting arm A_2B_2, m_2 (kg)	50,000
Mass of payload C_1C_2, m_p (kg)	50,000

The spatial trajectory of the payload is a smooth curve shown in Figure 5.4.

To examine the impact of the proposed MHRM outlined in Section 5.2.4 on the stochastic DACS response field problem, the payload parameters (y, z, θ) are treated as narrowly random variables. The characteristics of these random parameters are provided in Table 5.2.

Based on above description, we have

$$\boldsymbol{X} = \{x_1, x_2, x_3\}^T = \{y, z, \theta\}^T \tag{5.27}$$

$$\boldsymbol{X}^e = \{x_1^e, x_2^e, x_3^e\}^T = \{\mu_y, \mu_z, \mu_\theta\}^T \tag{5.28}$$

Therefore, the luffing angular response vector $\boldsymbol{\gamma}_1$ of crane 1 with narrowly random payload parameters can be rewritten as

$$\begin{aligned}\boldsymbol{\gamma}_1 = (\boldsymbol{T}_1^e)^{-1}\boldsymbol{M}_1^e &+ \left[(\boldsymbol{T}_1^e)^{-1}\boldsymbol{M}_{1y} - (\boldsymbol{T}_1^e)^{-2}\boldsymbol{M}_1^e\boldsymbol{T}_{1y}\right]\cdot(y-\mu_y)\\ &+ \left[(\boldsymbol{T}_1^e)^{-1}\boldsymbol{M}_{1z} - (\boldsymbol{T}_1^e)^{-2}\boldsymbol{M}_1^e\boldsymbol{T}_{1z}\right]\cdot(z-\mu_z)\\ &+ \left[(\boldsymbol{T}_1^e)^{-1}\boldsymbol{M}_{1\theta} - (\boldsymbol{T}_1^e)^{-2}\boldsymbol{M}_1^e\boldsymbol{T}_{1\theta}\right]\cdot(\theta-\mu_\theta)\end{aligned} \tag{5.29}$$

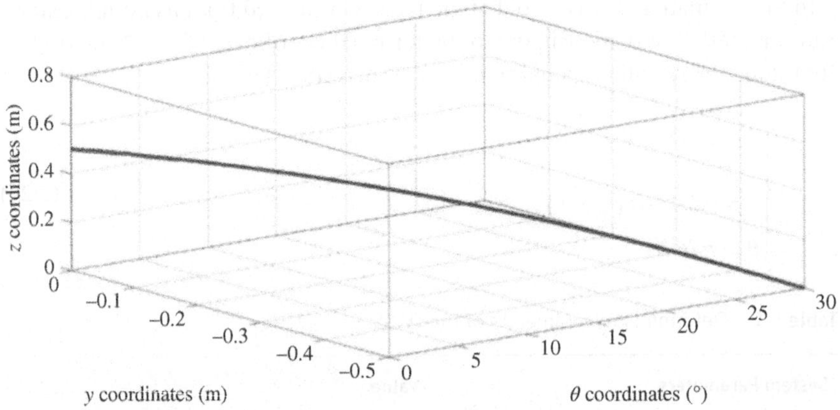

Figure 5.4 Trajectory of the payload.

Table 5.2 Stochastic DACS parameters.

Random Parameters	Value
The expectation of the random variable y, μ_y (m)	0.5
The expectation of the random variable z, μ_z (m)	0.5
The expectation of the random variable θ, μ_θ (rad)	$\pi/6$

Here, let us introduce the coefficients of variation V_y, V_z, and V_θ for random variables y, z, and θ, respectively, and they can be expressed as follows:

$$V_y = \frac{\sigma_y}{\mu_y},\ V_z = \frac{\sigma_z}{\mu_z},\ V_\theta = \frac{\sigma_\theta}{\mu_\theta} \tag{5.30}$$

The expectation and variance of the luffing angular response vector $\boldsymbol{\gamma}_1$ for crane 1, with narrowly distributed random payload parameters, as calculated by the MHRM, can be expressed as follows:

$$E\left(\boldsymbol{\gamma}_1\right) = \left(\boldsymbol{T}_1^e\right)^{-1} \boldsymbol{M}_1^e \tag{5.31}$$

$$\begin{aligned} D\left(\boldsymbol{\gamma}_1\right) = &\left(\left(\boldsymbol{T}_1^e\right)^{-1} \boldsymbol{M}_{1y}\right)^2 \cdot \left(\mu_y \cdot V_y\right)^2 + \left(\left(\boldsymbol{T}_1^e\right)^{-2} \boldsymbol{M}_1^e \boldsymbol{T}_{1y}\right)^2 \cdot \left(\mu_y \cdot V_y\right)^2 \\ &-2\left(\left(\boldsymbol{T}_1^e\right)^{-1} \boldsymbol{M}_{1y}\right)\left(\left(\boldsymbol{T}_1^e\right)^{-2} \boldsymbol{M}_1^e \boldsymbol{T}_{1y}\right) \cdot \left(\mu_y \cdot V_y\right)^2 \\ &+\left(\left(\boldsymbol{T}_1^e\right)^{-1} \boldsymbol{M}_{1z}\right)^2 \cdot \left(\mu_z \cdot V_z\right)^2 + \left(\left(\boldsymbol{T}_1^e\right)^{-2} \boldsymbol{M}_1^e \boldsymbol{T}_{1z}\right)^2 \cdot \left(\mu_z \cdot V_z\right)^2 \\ &-2\left(\left(\boldsymbol{T}_1^e\right)^{-1} \boldsymbol{M}_{1z}\right)\left(\left(\boldsymbol{T}_1^e\right)^{-2} \boldsymbol{M}_1^e \boldsymbol{T}_{1z}\right) \cdot \left(\mu_z \cdot V_z\right)^2 \\ &+\left(\left(\boldsymbol{T}_1^e\right)^{-1} \boldsymbol{M}_{1\theta}\right)^2 \cdot \left(\mu_\theta \cdot V_\theta\right)^2 + \left(\left(\boldsymbol{T}_1^e\right)^{-2} \boldsymbol{M}_1^e \boldsymbol{T}_{1\theta}\right)^2 \cdot \left(\mu_\theta \cdot V_\theta\right)^2 \\ &-2\left(\left(\boldsymbol{T}_1^e\right)^{-1} \boldsymbol{M}_{1\theta}\right)\left(\left(\boldsymbol{T}_1^e\right)^{-2} \boldsymbol{M}_1^e \boldsymbol{T}_{1\theta}\right) \cdot \left(\mu_\theta \cdot V_\theta\right)^2 \end{aligned} \tag{5.32}$$

Similarly, for crane 2, the luffing angular response vector $\boldsymbol{\gamma}_2$, with narrowly distributed random payload parameters, can be expressed as follows:

$$\begin{aligned} \boldsymbol{\gamma}_2 = &\left(\boldsymbol{T}_2^e\right)^{-1} \boldsymbol{M}_2^e + \left[\left(\boldsymbol{T}_2^e\right)^{-1} \boldsymbol{M}_{2y} - \left(\boldsymbol{T}_2^e\right)^{-2} \boldsymbol{M}_2^e \boldsymbol{T}_{2y}\right] \cdot \left(y - \mu_y\right) \\ &+\left[\left(\boldsymbol{T}_2^e\right)^{-1} \boldsymbol{M}_{2z} - \left(\boldsymbol{T}_2^e\right)^{-2} \boldsymbol{M}_2^e \boldsymbol{T}_{2z}\right] \cdot \left(z - \mu_z\right) \\ &+\left[\left(\boldsymbol{T}_2^e\right)^{-1} \boldsymbol{M}_{2\theta} - \left(\boldsymbol{T}_2^e\right)^{-2} \boldsymbol{M}_2^e \boldsymbol{T}_{2\theta}\right] \cdot \left(\theta - \mu_\theta\right) \end{aligned} \tag{5.33}$$

The expectation and variance of the luffing angular response vector $\boldsymbol{\gamma}_2$ for crane 2, with narrowly distributed random payload parameters, as computed by the MHRM, can be expressed as follows:

$$E\left(\boldsymbol{\gamma}_2\right) = \left(\boldsymbol{T}_2^e\right)^{-1} \boldsymbol{M}_2^e \tag{5.34}$$

$$\begin{aligned} D\left(\boldsymbol{\gamma}_2\right) = &\left(\left(\boldsymbol{T}_2^e\right)^{-1} \boldsymbol{M}_{2y}\right)^2 \cdot \left(\mu_y \cdot V_y\right)^2 + \left(\left(\boldsymbol{T}_2^e\right)^{-2} \boldsymbol{M}_2^e \boldsymbol{T}_{2y}\right)^2 \cdot \left(\mu_y \cdot V_y\right)^2 \\ &-2\left(\left(\boldsymbol{T}_2^e\right)^{-1} \boldsymbol{M}_{2y}\right)\left(\left(\boldsymbol{T}_2^e\right)^{-2} \boldsymbol{M}_2^e \boldsymbol{T}_{2y}\right) \cdot \left(\mu_y \cdot V_y\right)^2 \\ &+\left(\left(\boldsymbol{T}_2^e\right)^{-1} \boldsymbol{M}_{2z}\right)^2 \cdot \left(\mu_z \cdot V_z\right)^2 + \left(\left(\boldsymbol{T}_2^e\right)^{-2} \boldsymbol{M}_2^e \boldsymbol{T}_{2z}\right)^2 \cdot \left(\mu_z \cdot V_z\right)^2 \\ &-2\left(\left(\boldsymbol{T}_2^e\right)^{-1} \boldsymbol{M}_{2z}\right)\left(\left(\boldsymbol{T}_2^e\right)^{-2} \boldsymbol{M}_2^e \boldsymbol{T}_{2z}\right) \cdot \left(\mu_z \cdot V_z\right)^2 \\ &+\left(\left(\boldsymbol{T}_2^e\right)^{-1} \boldsymbol{M}_{2\theta}\right)^2 \cdot \left(\mu_\theta \cdot V_\theta\right)^2 + \left(\left(\boldsymbol{T}_2^e\right)^{-2} \boldsymbol{M}_2^e \boldsymbol{T}_{2\theta}\right)^2 \cdot \left(\mu_\theta \cdot V_\theta\right)^2 \\ &-2\left(\left(\boldsymbol{T}_2^e\right)^{-1} \boldsymbol{M}_{2\theta}\right)\left(\left(\boldsymbol{T}_2^e\right)^{-2} \boldsymbol{M}_2^e \boldsymbol{T}_{2\theta}\right) \cdot \left(\mu_\theta \cdot V_\theta\right)^2 \end{aligned} \tag{5.35}$$

We use the deterministic model $V_y = V_z = V_\theta = 0$ as a reference. Simulations for the DACS response field problem, obtained using the MHRM, are conducted with MATLAB R2014a on a 2.5 GHz Intel(R) Core(TM) i7-4710MQ CPU computer.

To explore the impact of random payload parameters (y, z, θ) on the stochastic DACS response field problem, various combinations of random parameters are considered. The computational results obtained using the MHRM are presented in Tables 5.3 and 5.4, respectively.

Table 5.3 Stochastic luffing angular response field for crane 1.

Deterministic and Random Models	$E(\gamma_1)$ (rad)	$D(\gamma_1)$ ($[\mathrm{rad}]^2$)
$V_y = V_z = V_\theta = 0$	0.0759	0
$V_y = 0.1,\ V_z = V_\theta = 0$	0.0759	3.1976E−06
$V_z = 0.1,\ V_y = V_\theta = 0$	0.0759	0
$V_\theta = 0.1,\ V_y = V_z = 0$	0.0759	0
$V_y = 0.2,\ V_z = V_\theta = 0$	0.0759	1.2888E−05
$V_z = 0.2,\ V_y = V_\theta = 0$	0.0759	0
$V_\theta = 0.2,\ V_y = V_z = 0$	0.0759	0
$V_y = V_z = 0.1,\ V_\theta = 0$	0.0759	3.1976E−06
$V_z = V_\theta = 0.1,\ V_y = 0$	0.0759	0
$V_y = V_\theta = 0.1,\ V_z = 0$	0.0759	3.1976E−06
$V_y = V_z = 0.2,\ V_\theta = 0$	0.0759	1.2888E−05
$V_z = V_\theta = 0.2,\ V_y = 0$	0.0759	0
$V_y = V_\theta = 0.2,\ V_z = 0$	0.0759	1.2888E−05
$V_y = V_z = V_\theta = 0.1$	0.0759	3.1976E−06
$V_y = V_z = V_\theta = 0.2$	0.0759	1.2888E−05

Table 5.4 Stochastic luffing angular response field for crane 2.

Deterministic and Random Models	$E(\gamma_2)$ (rad)	$D(\gamma_2)$ ($[\mathrm{rad}]^2$)
$V_y = V_z = V_\theta = 0$	4.4974	0
$V_y = 0.1,\ V_z = V_\theta = 0$	4.4974	1.9929E+02
$V_z = 0.1,\ V_y = V_\theta = 0$	4.4974	0
$V_\theta = 0.1,\ V_y = V_z = 0$	4.4974	0
$V_y = 0.2,\ V_z = V_\theta = 0$	4.4974	7.9961E+02
$V_z = 0.2,\ V_y = V_\theta = 0$	4.4974	0
$V_\theta = 0.2,\ V_y = V_z = 0$	4.4974	0
$V_y = V_z = 0.1,\ V_\theta = 0$	4.4974	1.9929E+02
$V_z = V_\theta = 0.1,\ V_y = 0$	4.4974	0

Deterministic and Random Models	$E(\gamma_2)$ (rad)	$D(\gamma_2)$ $\left([\text{rad}]^2\right)$
$V_y = V_\theta = 0.1,\ V_z = 0$	4.4974	1.9929E+02
$V_y = V_z = 0.2,\ V_\theta = 0$	4.4974	7.9961E+02
$V_z = V_\theta = 0.2,\ V_y = 0$	4.4974	0
$V_y = V_\theta = 0.2,\ V_z = 0$	4.4974	7.9961E+02
$V_y = V_z = V_\theta = 0.1$	4.4974	1.9929E+02
$V_y = V_z = V_\theta = 0.2$	4.4974	7.9961E+02

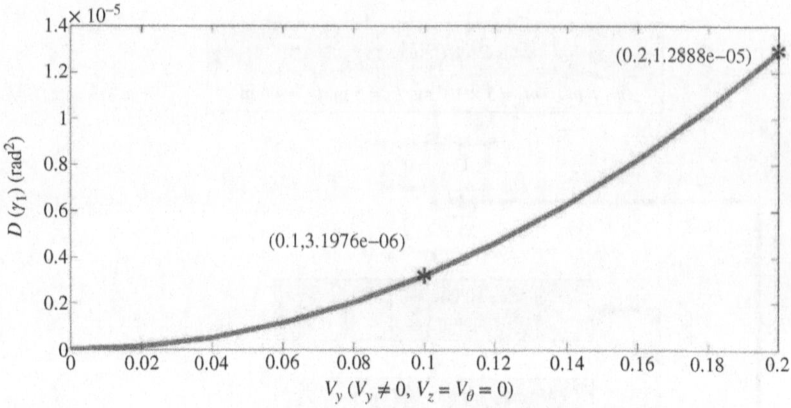

Figure 5.5 The variance of luffing angular response vector of crane 1 calculated by MHRM.

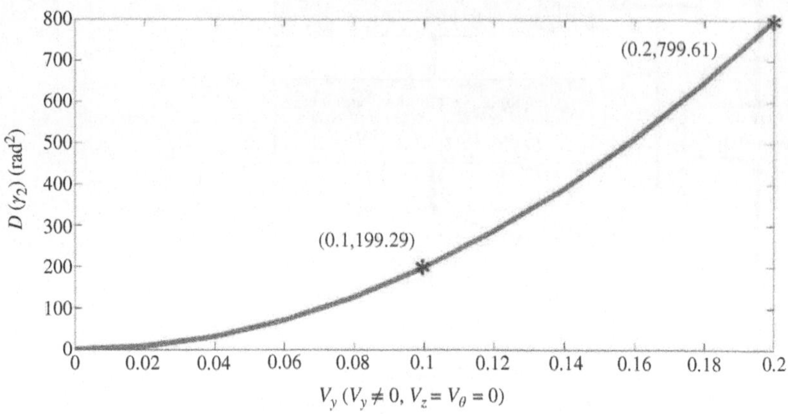

Figure 5.6 The variance of luffing angular response vector of crane 2 calculated by MHRM.

The variances of the luffing angular response field of crane 1 and crane 2 of DACS, as calculated by the MHRM, are illustrated in Figures 5.5 and 5.6.

To further assess the feasibility and effectiveness of the MHRM for the stochastic luffing angular response field of the DACS with narrowly random parameters, we consider a model with three random parameters $(V_y = V_z = V_\theta \neq 0)$ as the research focus, while keeping other system parameters fixed. The HMCM with 10^4 samples is used as a reference solution, and the simulation is carried out in five steps, as shown in the flowchart in Figure 5.7. Consequently, the expectations and variances of the luffing angular response fields for crane 1 and crane 2 of the DACS, calculated by both the MHRM and Monte Carlo simulations (MCM) with the three random parameters, are displayed in Figures 5.8–5.11.

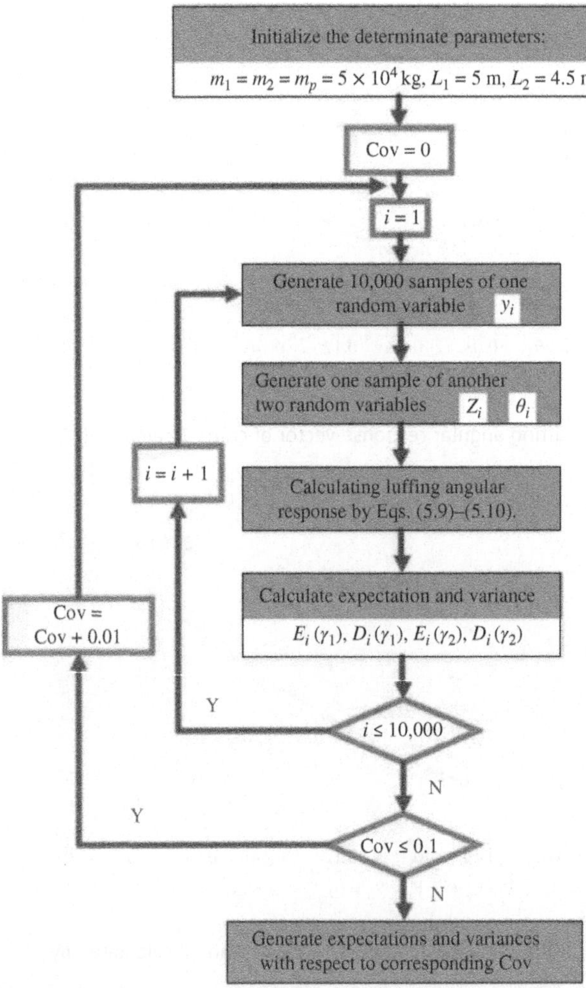

Figure 5.7 Flowchart of the MHRM for the stochastic luffing angular response field analysis.

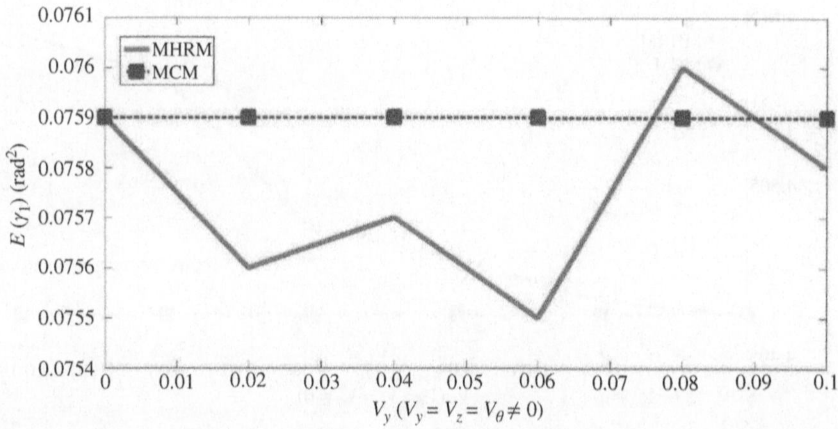

Figure 5.8 The expectation of luffing angular response field of crane 1 calculated by MHRM and MCM.

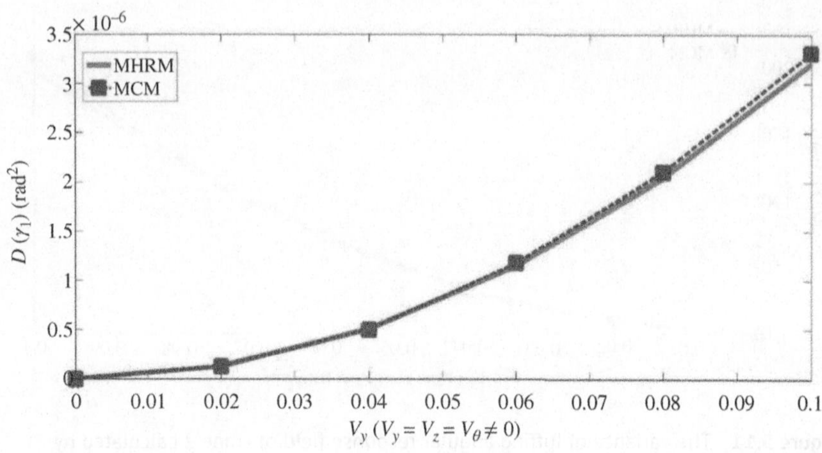

Figure 5.9 The variance of luffing angular response field of crane 1 calculated by MHRM and MCM.

As depicted in Figures 5.8 and 5.10, it is observed that the expectations of the luffing angular response vector for crane 1 and crane 2, calculated using the MHRM, closely match those obtained by the MCM within a small range of V_y. However, as V_y increases, the results from the MHRM show a noticeable deviation from those obtained by the MCM. This indicates that the MHRM is more suitable for predicting the variances of the luffing angular response vector for both crane 1 and crane 2 of the DACS with narrowly random parameters.

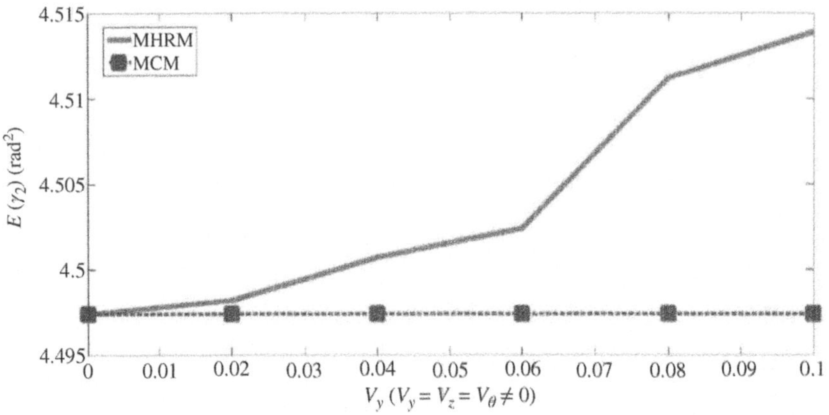

Figure 5.10 The expectation of luffing angular response field of crane 2 calculated by MHRM and MCM.

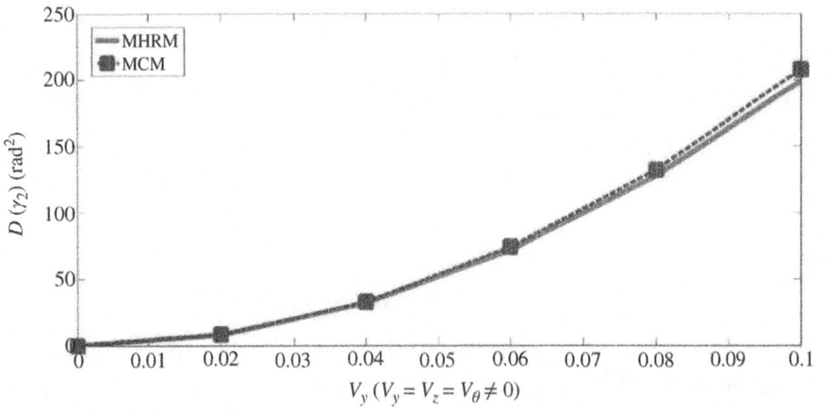

Figure 5.11 The variance of luffing angular response field of crane 2 calculated by MHRM and MCM.

As shown in Figures 5.9 and 5.11, the variances of the luffing angular response vector for crane 1 and crane 2, calculated using the MHRM, closely align with those obtained by the MCM. This suggests that the MHRM can achieve the same level of accuracy as the MCM in predicting the luffing angular response field of the DACS with narrowly random parameters.

5.2.6 Conclusion

This section introduces an MHRM to predict the luffing angular response field of DACS under narrowly distributed random payload parameters. Leveraging inverse kinematic analysis, a deterministic model is first established. Uncertainties in payload parameters (y, z, θ) are then modeled as random variables, and a PRCFM is developed to derive approximate luffing angle expressions using first-order Neumann series expansion. Statistical characteristics of the response fields are efficiently predicted via an RVFMM.

Numerical comparisons with MCM demonstrate MHRM's effectiveness for low-uncertainty scenarios. Sensitivity analysis reveals that payload parameter y exerts the strongest influence on response variability, surpassing z and θ. While the method achieves computational efficiency by neglecting higher-order terms, its accuracy diminishes for large uncertainty levels. Future work should explore extensions to broader parameter uncertainties and diverse DACS configurations. This approach provides a targeted solution for stochastic response prediction in systems with tightly bounded random parameters.

5.3 Kinematic Uncertainty Analysis with Interval Variables

5.3.1 Interval Kinematic Equilibrium Equation

Based on the inverse kinematics in Section 5.2.2, in practical crane engineering, due to uncertain sources such as mechanical errors, unpredictable wind loads, and complex environment, structure parameters with bounded uncertainty are unavoidable. Assume that the interval vector $\boldsymbol{y}^I$ denotes all interval variables in the DCS

$$\boldsymbol{y}^I = \left(\underline{y_r}, \overline{y_r}\right)_n = (y_r^c + \Delta y_r \delta^I)_n, \quad r = 1, 2, \dots, n \tag{5.36}$$

where $\underline{y_r}$ and $\overline{y_r}$ are the lower bound (LB) and upper bound (UB) of y_r^I, respectively; $y_r^c = \left(\underline{y_r} + \overline{y_r}\right)/2$ and $\Delta y_r = \left(\overline{y_r} - \underline{y_r}\right)/2$ are the mean value (MV) and interval radius (IR) of y_r^I, respectively; and δ^I is an unite interval, i.e., $\delta^I = [-1, 1]$. And the interval kinematic equilibrium equation of the DCS can be expressed as

$$\boldsymbol{M}_i\left(\boldsymbol{K}_i\left(\boldsymbol{y}^I\right)\right) = \boldsymbol{T}_i\left(\boldsymbol{K}_i\left(\boldsymbol{y}^I\right)\right)\gamma_i^I\left(\boldsymbol{y}^I\right) \tag{5.37}$$

where $\boldsymbol{M}_i\left(\boldsymbol{K}_i\left(\boldsymbol{y}^I\right)\right)$ and $\boldsymbol{T}_i\left(\boldsymbol{K}_i\left(\boldsymbol{y}^I\right)\right)$ are two interval compound function vectors and $\gamma_i^I\left(\boldsymbol{y}^I\right)$ is the interval LAV.

5.3.2 Hybrid Compound Function/Subinterval Perturbation Method

5.3.2.1 Subinterval Perturbation Method

Based on the SIPM, y_j^I with a large uncertain level is divided into N_j subinterval variables with uniform small uncertain levels

$$y_{k_j}^I = \left[\underline{y_j^I} + \frac{2\left(k_j - 1\right)\Delta y_j}{N_j}, \underline{y_j^I} + \frac{2k_j \Delta y_j}{N_j} \right], \quad j = 1, 2, \dots, n; k_j = 1, 2, \dots, N_j \tag{5.38}$$

where $y_{k_j}^I$ stands for the k_jth subinterval variable; $\underline{y_j^I}$ and Δy_j are the LB and IR of y_j^I, respectively; and N_j is the number of subinterval variables. The total number of all different combinations can be obtained from above subinterval variables,

$$N = N_1 N_2 \cdots N_j \cdots N_n \tag{5.39}$$

where N is the total number of all different combinations.

Based on Eqs. (5.37) and (5.38), the interval kinematic equilibrium equation for every subinterval combination can be rewritten as

$$\begin{gathered} \boldsymbol{M}_i \left(\boldsymbol{K}_i \left(\boldsymbol{y}_{k_1, k_2, \dots, k_j, \dots, k_n}^I \right) \right) = \boldsymbol{T}_i \left(\boldsymbol{K}_i \left(\boldsymbol{y}_{k_1, k_2, \dots, k_j, \dots, k_n}^I \right) \right) \boldsymbol{\gamma}_i^I \left(\boldsymbol{y}_{k_1, k_2, \dots, k_j, \dots, k_n}^I \right), \\ i = 1, 2; \; j = 1, 2, \dots, n; \; k_j = 1, 2, \dots, N_j \end{gathered} \tag{5.40}$$

where $\boldsymbol{y}_{k_1, k_2, \dots, k_j, \dots, k_n}^I$ denotes a subinterval combination vector consisting of the k_1th subinterval of y_1^I and the k_2th subinterval of y_2^I, and up to the k_nth subinterval of y_n^I,

$$\boldsymbol{y}_{k_1, k_2, \dots, k_j, \dots, k_n}^I = \left\{ y_{k_1}^I, y_{k_2}^I, \dots, y_{k_j}^I, \dots, y_{k_n}^I \right\}^T \tag{5.41}$$

For two adjacent subinterval combination vectors $\boldsymbol{y}_{k_1, k_2, \dots, k_j, \dots, k_n}^I$ and $\boldsymbol{y}_{k_1, k_2, \dots, k_{j+1}, \dots, k_n}^I$, based on Eq. (5.40), one has

$$\begin{cases} \boldsymbol{M}_i \left(\boldsymbol{K}_i \left(\boldsymbol{y}_{k_1, k_2, \dots, k_j, \dots, k_n}^I \right) \right) \cap \boldsymbol{M}_i \left(\boldsymbol{K}_i \left(\boldsymbol{y}_{k_1, k_2, \dots, k_j, \dots, k_n}^I \right) \right) \\ \quad = \boldsymbol{M}_i \left(\boldsymbol{K}_i \left(y_{k_1}^I, y_{k_2}^I, \dots, \overline{y_{k_j}^I} = \underline{y_{k_{j+1}}^I}, \dots, y_{k_n}^I \right) \right), \\ \boldsymbol{T}_i \left(\boldsymbol{K}_i \left(\boldsymbol{y}_{k_1, k_2, \dots, k_j, \dots, k_n}^I \right) \right) \cap \boldsymbol{T}_i \left(\boldsymbol{K}_i \left(\boldsymbol{y}_{k_1, k_2, \dots, k_{j+1}, \dots, k_n}^I \right) \right) \\ \quad = \boldsymbol{T}_i \left(\boldsymbol{K}_i \left(y_{k_1}^I, y_{k_2}^I, \dots, \overline{y_{k_j}^I} = \underline{y_{k_{j+1}}^I}, \dots, y_{k_n}^I \right) \right) \end{cases} \tag{5.42}$$

where $\overline{y_{k_j}^I}$ is the UB of the subinterval variable $y_{k_j}^I$ and $\underline{y_{k_{j+1}}^I}$ is the LB of the subinterval variable $y_{k_{j+1}}^I$. Based on the set theory, it can be easily concluded that the intersection of $\boldsymbol{\gamma}_i^I\left(\boldsymbol{K}_i\left(\boldsymbol{y}^I_{k_1,k_2,\ldots,k_j,\ldots,k_n}\right)\right)$ and $\boldsymbol{\gamma}_i^I\left(\boldsymbol{K}_i\left(\boldsymbol{y}^I_{k_1,k_2,\ldots,k_{j+1},\ldots,k_n}\right)\right)$ is a nonempty set,

$$\boldsymbol{\gamma}_i^I\left(\boldsymbol{K}_i\left(\boldsymbol{y}^I_{k_1,k_2,\ldots,k_j,\ldots,k_n}\right)\right) \cap \boldsymbol{\gamma}_i^I\left(\boldsymbol{K}_i\left(\boldsymbol{y}^I_{k_1,k_2,\ldots,k_{j+1},\ldots,k_n}\right)\right) \neq \emptyset \tag{5.43}$$

where $\emptyset$ denotes the empty set. Equation (5.43) shows that the subinterval LAR vector $\boldsymbol{\gamma}_i^I\left(\boldsymbol{K}_i\left(\boldsymbol{y}^I_{k_1,k_2,\ldots,k_j,\ldots,k_n}\right)\right)$ is connected one by one; thus, the $\boldsymbol{\gamma}_i^I\left(\boldsymbol{y}^I\right)$ can be expressed as

$$\begin{aligned}\boldsymbol{\gamma}_i^I\left(\boldsymbol{y}^I\right) &= \bigcup_{\substack{j=1,2,\ldots,n;\\ k_j=1,2,\ldots,N_j}} \boldsymbol{\gamma}_i^I\left(\boldsymbol{K}_i\left(\boldsymbol{y}^I_{k_1,k_2,\ldots,k_j,\ldots,k_n}\right)\right)\\ &= \left[\min_{\substack{j=1,2,\ldots,n;\\ k_j=1,2,\ldots,N_j}} \underline{\boldsymbol{\gamma}}_i\left(\boldsymbol{K}_i\left(\boldsymbol{y}^I_{k_1,k_2,\ldots,k_j,\ldots,k_n}\right)\right),\ \max_{\substack{j=1,2,\ldots,n;\\ k_j=1,2,\ldots,N_j}} \overline{\boldsymbol{\gamma}}_i\left(\boldsymbol{K}_i\left(\boldsymbol{y}^I_{k_1,k_2,\ldots,k_j,\ldots,k_n}\right)\right)\right],\\ & i=1,2\end{aligned} \tag{5.44}$$

5.3.2.2 Hybrid Compound Function/Subinterval Perturbation Method

In this subsection, another interval vector $\boldsymbol{\alpha}^I$ is denoted as all subinterval combination vectors, and the interval vector $\boldsymbol{\alpha}^I$ can be obtained based on Eq. (5.41),

$$\boldsymbol{\alpha}^I = \left\{\boldsymbol{\alpha}_1^I,\ \boldsymbol{\alpha}_2^I,\ \ldots,\ \boldsymbol{\alpha}_s^I,\ \ldots,\ \boldsymbol{\alpha}_N^I\right\}^T,\ s=1,2,\ldots,N \tag{5.45}$$

where $\boldsymbol{\alpha}_s^I$ is the sth subinterval combination vector, and

$$\begin{cases}\boldsymbol{\alpha}_s^I = \left\{\alpha_1^I,\ \alpha_2^I,\ \ldots,\ \alpha_j^I,\ \ldots,\ \alpha_n^I\right\}^T,\ j=1,2,\ldots,n,\\ \boldsymbol{\alpha}_s^I = \boldsymbol{y}^I_{k_1,k_2,\ldots,k_j,\ldots,k_n}\end{cases} \tag{5.46}$$

in which α_j^I is the jth interval variable, and

$$\alpha_j^I = y_{k_j}^I \tag{5.47}$$

Based on Eqs. (5.40) and (5.46), the interval kinematic equilibrium equation of the DCS for the sth subinterval combination can be rewritten as

$$\boldsymbol{M}_i\left(\boldsymbol{K}_i\left(\boldsymbol{\alpha}_s^I\right)\right) = \boldsymbol{T}_i\left(\boldsymbol{K}_i\left(\boldsymbol{\alpha}_s^I\right)\right)\boldsymbol{\gamma}_i^I\left(\boldsymbol{\alpha}_s^I\right),\ i=1,2;\ s=1,2,\ldots,N \tag{5.48}$$

where $\boldsymbol{\gamma}_i^I\left(\boldsymbol{\alpha}_s^I\right)$ is the subinterval LAR vector with respect to the sth subinterval combination vector $\boldsymbol{\alpha}_s^I$. Based on the SIPM, $\boldsymbol{\alpha}_s^I$ is composed of interval variables $\left(\alpha_1^I, \alpha_2^I, \ldots, \alpha_j^I, \ldots, \alpha_n^I\right)$ with small uncertain levels. Expanding interval compound function vectors $\boldsymbol{M}_i\left(\boldsymbol{K}_i\left(\boldsymbol{\alpha}_s^I\right)\right)$ and $\boldsymbol{T}_i\left(\boldsymbol{K}_i\left(\boldsymbol{\alpha}_s^I\right)\right)$ with respect to the MV $\boldsymbol{\alpha}_s^c$ using the first-order Taylor series, respectively, yields

$$\begin{cases} \boldsymbol{M}_i\left(\boldsymbol{K}_i\left(\boldsymbol{\alpha}_s^I\right)\right) = \boldsymbol{M}_i^c + \Delta_1 \boldsymbol{M}_i^I, \\ \boldsymbol{T}_i\left(\boldsymbol{K}_i\left(\boldsymbol{\alpha}_s^I\right)\right) = \boldsymbol{T}_i^c + \Delta_1 \boldsymbol{T}_i^I \end{cases} \tag{5.49}$$

where

$$\begin{cases} \boldsymbol{M}_i^c = \boldsymbol{M}_i\left(\boldsymbol{K}_i\left(\boldsymbol{\alpha}_s^c\right)\right), \\ \boldsymbol{T}_i^c = \boldsymbol{T}_i\left(\boldsymbol{K}_i\left(\boldsymbol{\alpha}_s^c\right)\right) \end{cases} \tag{5.50}$$

$$\begin{cases} \Delta_1 \boldsymbol{M}_i^I = \sum_{j=1}^{n} \dfrac{\partial \boldsymbol{M}_i\left(\boldsymbol{K}_i\left(\boldsymbol{\alpha}_s^c\right)\right)}{\partial \boldsymbol{K}_i\left(\boldsymbol{\alpha}_s^c\right)} \cdot \dfrac{\partial \boldsymbol{K}_i\left(\boldsymbol{\alpha}_s^c\right)}{\partial \alpha_j}\left(\alpha_j^I - \alpha_j^c\right) = \sum_{j=1}^{n} \dfrac{\partial \boldsymbol{M}_i\left(\boldsymbol{K}_i\left(\boldsymbol{\alpha}_s^c\right)\right)}{\partial \boldsymbol{K}_i\left(\boldsymbol{\alpha}_s^c\right)} \cdot \dfrac{\partial \boldsymbol{K}_i\left(\boldsymbol{\alpha}_s^c\right)}{\partial \alpha_j} \Delta\alpha_j \delta_j^I, \\ \Delta_1 \boldsymbol{T}_i^I = \sum_{j=1}^{n} \dfrac{\partial \boldsymbol{T}_i\left(\boldsymbol{K}_i\left(\boldsymbol{\alpha}_s^c\right)\right)}{\partial \boldsymbol{K}_i} \cdot \dfrac{\partial \boldsymbol{K}_i\left(\boldsymbol{\alpha}_s^c\right)}{\partial \alpha_j}\left(\alpha_j^I - \alpha_j^c\right) = \sum_{j=1}^{n} \dfrac{\partial \boldsymbol{T}_i\left(\boldsymbol{K}_i\left(\boldsymbol{\alpha}_s^c\right)\right)}{\partial \boldsymbol{K}_i\left(\boldsymbol{\alpha}_s^c\right)} \cdot \dfrac{\partial \boldsymbol{K}_i\left(\boldsymbol{\alpha}_s^c\right)}{\partial \alpha_j} \Delta\alpha_j \delta_j^I \end{cases} \tag{5.51}$$

in which α_j^c and $\Delta\alpha_j$ are the MV and IR of the α_j^I, respectively; $\widetilde{\boldsymbol{M}}_i^c$ and $\Delta_1 \widetilde{\boldsymbol{M}}_i^I$ are the MV and IR of the $\boldsymbol{M}_i\left(\boldsymbol{K}_i\left(\boldsymbol{\alpha}_s^I\right)\right)$, respectively; and $\widetilde{\boldsymbol{T}}_i^c$ and $\Delta_1 \widetilde{\boldsymbol{T}}_i^I$ are the MV and IR of the $\boldsymbol{T}_i\left(\boldsymbol{K}_i\left(\boldsymbol{\alpha}_s^I\right)\right)$, respectively. Substituting Eq. (5.49) into Eq. (5.48) yields

$$\boldsymbol{\gamma}_i^I\left(\boldsymbol{\alpha}_s^I\right) = \boldsymbol{\gamma}_i^c\left(\boldsymbol{\alpha}_s^I\right) + \Delta_1 \boldsymbol{\gamma}_i^I\left(\boldsymbol{\alpha}_s^I\right) = \left(\widetilde{\boldsymbol{T}}_i^c + \Delta_1 \widetilde{\boldsymbol{T}}_i^I\right)^{-1}\left(\widetilde{\boldsymbol{M}}_i^c + \Delta_1 \widetilde{\boldsymbol{M}}_i^I\right) \tag{5.52}$$

where $\boldsymbol{\gamma}_i^c\left(\boldsymbol{\alpha}_s^I\right)$ and $\Delta_1 \boldsymbol{\gamma}_i^I\left(\boldsymbol{\alpha}_s^I\right)$ are the MV and IR of the $\boldsymbol{\gamma}_i^I\left(\boldsymbol{\alpha}_s^I\right)$, respectively. According to the first-order Neumann series, expanding the $\left(\widetilde{\boldsymbol{T}}_i^c + \Delta_1 \widetilde{\boldsymbol{T}}_i^I\right)^{-1}$ in Eq. (5.52) yields

$$\left(\boldsymbol{T}_i^c + \Delta_1 \boldsymbol{T}_i^I\right)^{-1} = \left(\boldsymbol{T}_i^c\right)^{-1} - \left(\boldsymbol{T}_i^c\right)^{-1} \Delta_1 \boldsymbol{T}_i^I \left(\boldsymbol{T}_i^c\right)^{-1} + \left(\left(\boldsymbol{T}_i^c\right)^{-1} \Delta_1 \boldsymbol{T}_i^I\right)^2 \left(\boldsymbol{T}_i^c\right)^{-1} \tag{5.53}$$

Substituting Eq. (5.53) into Eq. (5.52) yields

$$\begin{aligned} \boldsymbol{\gamma}_i^I\left(\boldsymbol{\alpha}_s^I\right) = \boldsymbol{\gamma}_i^c\left(\boldsymbol{\alpha}_s^I\right) + \Delta_1 \boldsymbol{\gamma}_i^I\left(\boldsymbol{\alpha}_s^I\right) &\approx \left(\boldsymbol{T}_i^c\right)^{-1} \boldsymbol{M}_i^c + \left(\boldsymbol{T}_i^c\right)^{-1} \Delta_1 \boldsymbol{M}_i^I \\ &- \left(\boldsymbol{T}_i^c\right)^{-1} \Delta_1 \boldsymbol{T}_i^I \left(\boldsymbol{T}_i^c\right)^{-1} \boldsymbol{M}_i^c \end{aligned} \tag{5.54}$$

By using interval algorithms, the MV and IR of the $\boldsymbol{\gamma}_i^I\left(\boldsymbol{\alpha}_s^I\right)$ can be expressed as

$$\boldsymbol{\gamma}_i^c\left(\boldsymbol{\alpha}_s^I\right) = \left(\boldsymbol{T}_i^c\right)^{-1} \boldsymbol{M}_i^c \tag{5.55}$$

$$\Delta_1 \boldsymbol{\gamma}_i^I\left(\boldsymbol{\alpha}_s^I\right) = \left(\boldsymbol{T}_i^c\right)^{-1} \Delta_1 \boldsymbol{M}_i^I - \left(\boldsymbol{T}_i^c\right)^{-1} \Delta_1 \boldsymbol{T}_i^I \left(\boldsymbol{T}_i^c\right)^{-1} \boldsymbol{M}_i^c \tag{5.56}$$

respectively. Substituting Eq. (5.51) into Eq. (5.56) yields

$$\begin{aligned}\Delta_1\gamma_i^I(\alpha_s^I) &= (\boldsymbol{T}_i^c)^{-1}\left[\Delta_1\boldsymbol{M}_i^I - \Delta_1\boldsymbol{T}_i^I(\boldsymbol{T}_i^c)^{-1}\boldsymbol{M}_i^c\right]\\ &= (\boldsymbol{T}_i^c)^{-1}\left[\sum_{j=1}^{n}\left(\frac{\partial \boldsymbol{M}_i(\boldsymbol{K}_i(\alpha_s^c))}{\partial \boldsymbol{K}_i(\alpha_s^c)}\cdot\frac{\partial \boldsymbol{K}_i(\alpha_s^c)}{\partial \alpha_j} - \frac{\partial \boldsymbol{T}_i(\boldsymbol{K}_i(\alpha_s^c))}{\partial \boldsymbol{K}_i(\alpha_s^c)}\right.\right.\\ &\quad\left.\left.\cdot\frac{\partial \boldsymbol{K}_i(\alpha_s^c)}{\partial \alpha_j}(\boldsymbol{T}_i^c)^{-1}\boldsymbol{M}_i^c\right)\Delta\alpha_j\delta_j\right]\end{aligned} \tag{5.57}$$

Based on the monotonic technique, the $\Delta_1\gamma_i^I(\alpha_s^I)$ in Eq. (5.57) can be obtained by

$$\begin{aligned}\Delta_1\gamma_i(\alpha_s^I) = \sum_{j=1}^{n}\Bigg|(\boldsymbol{T}_i^c)^{-1}&\left(\frac{\partial \boldsymbol{M}_i(\boldsymbol{K}_i(\alpha_s^c))}{\partial \boldsymbol{K}_i(\alpha_s^c)}\cdot\frac{\partial \boldsymbol{K}_i(\alpha_s^c)}{\partial \alpha_j} - \frac{\partial \boldsymbol{T}_i(\boldsymbol{K}_i(\alpha_s^c))}{\partial \boldsymbol{K}_i(\alpha_s^c)}\right.\\ &\left.\cdot\frac{\partial \boldsymbol{K}_i(\alpha_s^c)}{\partial \alpha_j}(\boldsymbol{T}_i^c)^{-1}\boldsymbol{M}_i^c\right)\Delta\alpha_j\Bigg|\end{aligned} \tag{5.58}$$

Based on Eqs. (5.55) and (5.58), intervals of $\gamma_i^I(\alpha_s^I)$ can be expressed as

$$\underline{\gamma_i^I(\alpha_s^I)} = \gamma_i^c - \Delta_1\gamma_i \tag{5.59}$$

$$\overline{\gamma_i^I(\alpha_s^I)} = \gamma_i^c + \Delta_1\gamma_i \tag{5.60}$$

where $\underline{\gamma_i^I(\alpha_s^I)}$ and $\overline{\gamma_i^I(\alpha_s^I)}$ are the LB and UB of the $\gamma_i^I(\alpha_s^I)$, respectively. Based on Eqs. (5.44), (5.59), and (5.60), intervals of $\gamma_i^I(\alpha^I)$ can be expressed as

$$\underline{\gamma_i^I(\alpha^I)} = \min_{s=1,2,\ldots,N}\underline{\gamma_i^I(\alpha_s^I)} \tag{5.61}$$

$$\overline{\gamma_i^I(\alpha^I)} = \max_{s=1,2,\ldots,N}\overline{\gamma_i^I(\alpha_s^I)} \tag{5.62}$$

where $\underline{\gamma_i^I(\alpha^I)}$ and $\overline{\gamma_i^I(\alpha^I)}$ are the LB and UB of $\gamma_i^I(\alpha^I)$, respectively.

5.3.3 Numerical Examples

In this example, uncertain levels of interval variables are denoted as IR, lengths of A_1A_2 and C_1C_2 are considered as large interval variables, and midpoints of lengths of A_1A_2 and C_1C_2 are 12 and 2.5 m, respectively. The number of subintervals in Eq. (5.38) is $N_1 = N_2 = 2$; thus, the total number of all different combinations in Eq. (5.39) are $N = 4$. Intervals of the $\gamma_i^I(\alpha^I)$ in Eqs. (5.61) and (5.62) with four subinterval combinations by the HCFSPM are calculated in Appendix A of [36].

To investigate the effectiveness of the HCFSPM for large uncertainty problem, a two-interval-variable model (IR = $D_{\mathrm{IR}} = d_{\mathrm{IR}}$), where D_{IR} and d_{IR} are the IR of interval variables D and d, respectively, is constructed. Intervals of $\gamma_i^I\left(\alpha^I\right)$ calculated by the MCM and FCFIPM [9] are used as reference values. From IR = 0.00 to IR = 0.02, intervals of LAVs of cranes 1 and 2 obtained by the MCM, FCFIPM, and HCFSPM are plotted in Figures 5.12 and 5.13, respectively. In these figures, the LB and UB of LAVs yielded by the MCM, FCFIPM, and HCFSPM are named as MCM-LB, MCM-UB, FCFIPM-LB, FCFIPM-UB, HCFSPM-LB, and HCFSPM-UB, respectively.

Figures 5.12 and 5.13 show that intervals of the LAV of cranes 1 and 2 obtained by the HCFSPM are close to those by the MCM, and intervals of the LAVs obtained

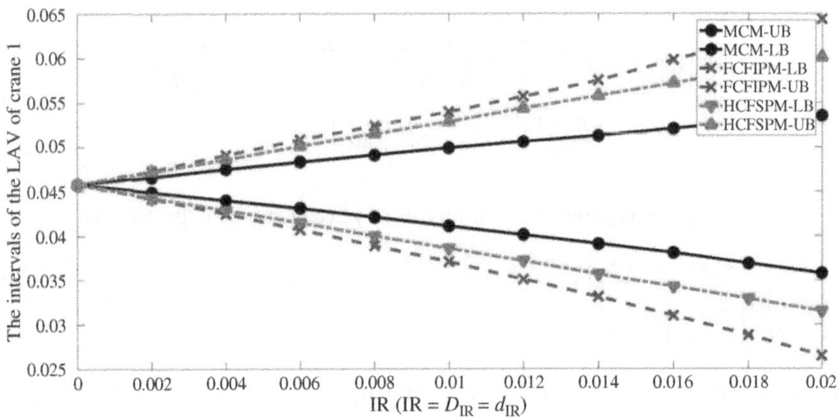

Figure 5.12 Intervals of the LAV of the crane 1.

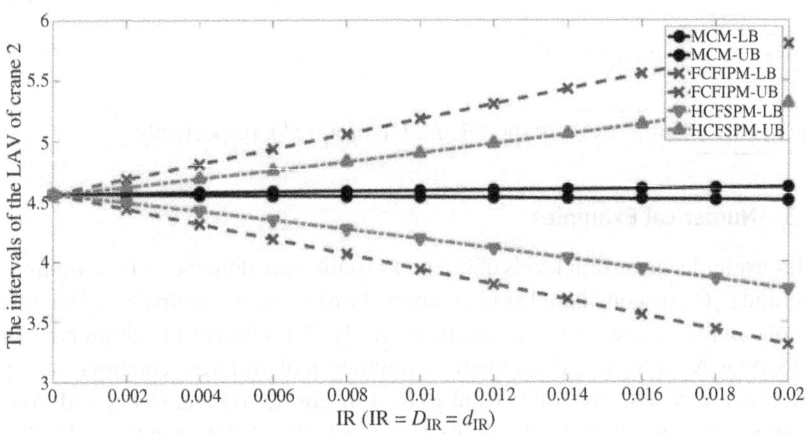

Figure 5.13 Intervals of the LAV of the crane 2.

by the FCFIPM deviate from those by the MCM. However, intervals of the LAVs obtained by both the FCFIPM and HCFSPM significantly differ from those by the MCM with the increase of the IR. The above results show that the accuracy of the HCFSPM is better than that of the FCFIPM when they are applied to conduct the kinematic analysis of the DCS with large bounded uncertainty. It indicates that the results by dividing these large interval variables into several subinterval variables with small uncertainty levels based on the SIPM bring a better UQ than those without using the SIPM.

Relative errors of intervals of LAVs of cranes 1 and crane 2 at IR = 0.00, 0.005, 0.010, 0.015, 0.020% are listed in Tables 5.5 and 5.6, respectively. With the increase of the IR, relative errors of intervals of LAVs between the FCFIPM and MCM, and those between the HCFSPM and MCM all increase, which indicates that the accuracy of the FCFIPM and HCFSPM gradually decrease when the IR increases. Note that the largest relative error of intervals of LAVs between the FCFIPM and MCM, and those between the HCFSPM and MCM are 26.51 and 16.29%, respectively, which shows that the operation of introducing the subinterval theory in the HCFSPM can improve the accuracy greatly, especially when it comes to large uncertain levels. It also indicates that the accuracy of the first-order Taylor and Neumann series in the HCFSPM in Eqs. (5.49), (5.50), (5.51), and (5.53) can be improved greatly by introducing the SIPM. The efficiencies of the FCFIPM and HCFSPM are also compared. From Tables 5.5 and 5.6, calculation times of the HCFSPM and FCFIPM are much shorter than that of the MCM, and the calculation time of the HCFSPM is a little longer than that of the FCFIPM. It is worthy

Table 5.5 Intervals of the LAV of the crane 1.

IR = D_{IR} = d_{IR}	Bounds	MCM (1.6539 s)	FCFIPM (0.0075 s)		HCFSPM (0.0193 s)	
		Value	Value	Error (%)	Value	Error (%)
0.000	LB	0.0458	0.0458	0.00	0.0458	0.00
	UB	0.0458	0.0458	0.00	0.0458	0.00
0.006	LB	0.0431	0.0407	5.57	0.0415	3.71
	UB	0.0483	0.0508	5.18	0.0501	3.73
0.010	LB	0.0411	0.0371	9.73	0.0386	6.08
	UB	0.0499	0.0540	8.22	0.0529	6.01
0.016	LB	0.0381	0.0310	18.64	0.0343	9.97
	UB	0.0521	0.0598	14.78	0.0572	9.79
0.020	LB	0.0358	0.0265	25.98	0.0315	13.65
	UB	0.0535	0.0644	20.37	0.0601	12.34

Table 5.6 Intervals of the LAV of the crane 2.

IR = $D_{IR} = d_{IR}$	Bounds	MCM (1.6539 s)	FCFIPM (0.0075 s)		HCFSPM (0.0193 s)	
		Value	Value	Error (%)	Value	Error (%)
0.000	LB	4.5588	4.5588	0.00	4.5588	0.00
	UB	4.5588	4.5588	0.00	4.5588	0.00
0.005	LB	4.5431	4.1854	7.87	4.3442	4.38
	UB	4.5755	4.9322	7.80	4.7548	3.92
0.010	LB	4.5330	3.9365	13.16	4.1899	7.57
	UB	4.5875	5.1812	12.94	4.9005	6.82
0.016	LB	4.5191	3.5631	21.15	3.9452	13.36
	UB	4.6050	5.5546	20.61	5.1404	11.63
0.020	LB	4.5097	3.3142	26.51	3.7751	16.29
	UB	4.6181	5.8035	25.67	5.3136	15.06

improving the accuracy by taking a little more time in the HCFSPM through the synthesis of the compound function technique and SIPM.

To investigate the number of subintervals on the accuracy of calculating the LAVs with large bounded uncertainty, different numbers of subinterval variables are classified based on Eq. (5.38): (1) one subinterval variable; (2) two subinterval variables; and (3) three subinterval variables. Corresponding subinterval variables of lengths of A_1A_2 and C_1C_2 are shown in Table 5.7. Note that the total number of all different combinations in Eq. (5.39) is $N = 9$.

Intervals of LAVs of cranes 1 and 2 with different numbers of subintervals by the MCM and HCFSPM are plotted in Figures 5.14 and 5.15, respectively, where the HCFSPM1-LB and HCFSPM1-UB denote the LB and UB of LAVs of cranes 1 and 2 calculated by the HCFSPM when $N_1 = N_2 = 1$; the HCFSPM2-LB and HCFSPM2-UB denote the LB and UB of LAVs of cranes 1 and 2 calculated by the HCFSPM when $N_1 = N_2 = 2$; and the HCFSPM3-LB and HCFSPM3-UB denote the LB and UB of LAVs of cranes 1 and 2 calculated by the HCFSPM when $N_1 = N_2 = 3$. Figures 5.14 and 5.15 show that intervals of LAVs yielded by the HCFSPM2 are closer to those yielded by the MCM compared with those by the HCFSPM1, and intervals of LAVs yielded by the HCFSPM3 are closer to those by the MCM compared with those by the HCFSPM3. The above results show that the larger the number of subintervals, the higher the accuracy of the HCFSPM. They also indicate that the operation by dividing these large interval variables into a greater number of subinterval variables can bring more accurate results in the HCFSPM.

Table 5.7 Corresponding subinterval variables of lengths of A_1A_2 and C_1C_2.

One subinterval	Length of A_1A_2, D (mm)	$D^I = D^c\,[1 - \text{IR}, 1 + \text{IR}]$
	Length of C_1C_2, d (mm)	$d^I = d^c\,[1 - \text{IR}, 1 + \text{IR}]$
Two subintervals	Length of A_1A_2, D (mm)	$D_1^I = D^c\,[1 - \text{IR}, 1]$
		$D_2^I = D^c\,[1, 1 + \text{IR}]$
	Length of C_1C_2, d (mm)	$d_1^I = d^c\,[1 - \text{IR}, 1]$
		$d_2^I = d^c\,[1, 1 + \text{IR}]$
Three subintervals	Length of A_1A_2, D (mm)	$D_1^I = D^c\left[1 - \text{IR}, 1 - \frac{1}{3}\text{IR}\right]$
		$D_2^I = D^c\left[1 - \frac{1}{3}\text{IR}, 1 + \frac{1}{3}\text{IR}\right]$
		$D_3^I = D^c\left[1 + \frac{1}{3}\text{IR}, 1 + \text{IR}\right]$
	Length of C_1C_2, d (mm)	$d_1^I = d^c\left[1 - \text{IR}, 1 - \frac{1}{3}\text{IR}\right]$
		$d_2^I = d^c\left[1 - \frac{1}{3}\text{IR}, 1 + \frac{1}{3}\text{IR}\right]$
		$d_3^I = d^c\left[1 + \frac{1}{3}\text{IR}, 1 + \text{IR}\right]$

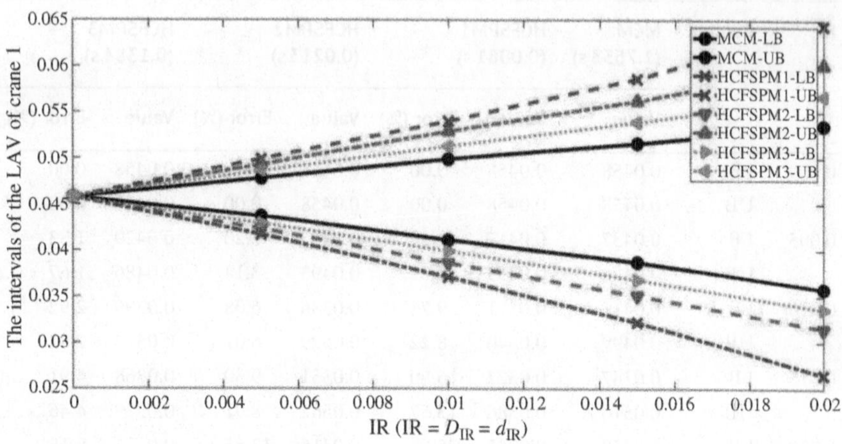

Figure 5.14 Intervals of the LAV of the crane 1 with different numbers of subintervals.

Relative errors of intervals of LAVs of cranes 1 and 2 by the MCM, HCFSPM1, HCFSPM2, and HCFSPM3 at IR = 0.000, 0.005, 0.010, 0.015, 0.020 are listed in Tables 5.8 and 5.9, respectively. Intervals of the LAVs calculated by the MCM are taken as reference solutions. The largest relative errors between intervals of LAVs by the HCFSPM1 and MCM, between those by the HCFSPM2 and MCM, and between those by the HCFSPM3 and MCM are 26.51, 16.29, and 9.46%, respectively. It can be concluded that relative errors between intervals of LAVs

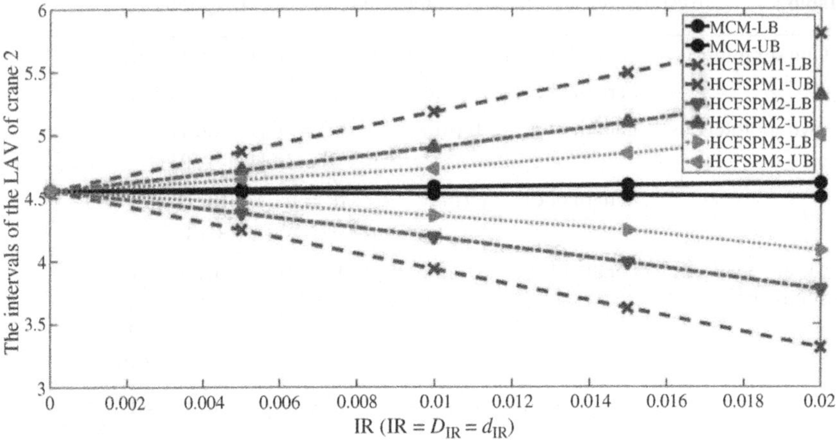

Figure 5.15 Intervals of the LAV of the crane 2 with different numbers of subintervals.

Table 5.8 Intervals of the LAV of the crane 1 with one, two, and three subintervals.

IR	Bounds	MCM (1.7533 s)	HCFSPM1 (0.0081 s)		HCFSPM2 (0.0213 s)		HCFSPM3 (0.1384 s)	
		Value	Value	Error (%)	Value	Error (%)	Value	Error (%)
0.000	LB	0.0458	0.0458	0.00	0.0458	0.00	0.0458	0.00
	UB	0.0458	0.0458	0.00	0.0458	0.00	0.0458	0.00
0.005	LB	0.0437	0.0417	4.58	0.0423	3.20	0.0429	1.83
	UB	0.0478	0.0499	4.39	0.0493	3.14	0.0486	1.67
0.010	LB	0.0411	0.0371	9.73	0.0386	6.08	0.0399	2.92
	UB	0.0499	0.0540	8.22	0.0529	6.01	0.0513	2.81
0.015	LB	0.0387	0.0322	16.80	0.0351	9.30	0.0368	4.91
	UB	0.0516	0.0586	13.57	0.0562	8.91	0.0539	4.46
0.020	LB	0.0358	0.0265	25.98	0.0315	13.65	0.0336	6.15
	UB	0.0535	0.0644	20.37	0.0601	12.34	0.0567	5.98

by the HCFSPM and MCM can be smaller when N becomes larger. It also indicates that solutions calculated by the HCFSPM will converge to referenced solutions calculated by the MCM if the number of the subintervals is infinity. From Tables 5.8 and 5.9, the calculation cost of the HCFSPM increases accordingly with the increase of N. Additional time is implemented to calculate intervals of γ_i^I (α_s^I) from $s = 1$ to $s = N$ in Eqs. (5.59) and (5.60). In other words, the improvement of the accuracy of the HCFSPM based on the SIPM is at the cost of the efficiency.

Table 5.9 Intervals of the LAV of the crane 2 with one, two, and three subintervals.

IR	Bounds	MCM (1.7533 s)	HCFSPM1 (0.0081 s)		HCFSPM2 (0.0213 s)		HCFSPM3 (0.1384 s)	
		Value	Value	Error (%)	Value	Error (%)	Value	Error (%)
0.000	LB	4.5588	4.5588	0.00	4.5588	0.00	4.5588	0.00
	UB	4.5588	4.5588	0.00	4.5588	0.00	4.5588	0.00
0.005	LB	4.5458	4.2478	6.56	4.3813	3.62	4.4641	2.16
	UB	4.5726	4.8699	6.50	4.7216	3.26	4.6481	1.65
0.010	LB	4.5330	3.9365	13.16	4.1899	7.57	4.3623	3.77
	UB	4.5875	5.1812	12.94	4.9005	6.82	4.7331	3.17
0.015	LB	4.5213	3.6254	19.82	3.9869	11.82	4.2449	6.11
	UB	4.6022	5.4924	19.34	5.0990	10.79	4.8517	5.42
0.020	LB	4.5097	3.3142	26.51	3.7751	16.29	4.0832	9.46
	UB	4.6181	5.8035	25.67	5.3136	15.06	4.9950	8.16

5.3.4 Conclusion

The HCFSPM is presented for the kinematic analysis of the DCS with large bounded uncertainty. The SPM is first introduced to transform interval variables with large uncertain levels into several small subinterval variables. The interval kinematic equilibrium equation of the DCS for every subinterval combination is constructed. Interval compound function vectors for every subinterval combination are obtained. Intervals of LAVs of cranes 1 and 2 are obtained based on the interval algorithm and monotonic technique.

Several numerical examples are given to show that: (1) the HCFSPM offers higher accuracy than the FCFIPM and superior efficiency compared to the MCM, especially when dealing with large levels of uncertainty; (2) the accuracy of the HCFSPM increases when the number of subintervals become larger; and (3) the efficiency of the HCFSPM is lower than that of the FCFIPM, but the FCFIPM is limited to small uncertainty problems.

5.4 Kinematic Uncertainty Analysis Based on Evidence Theory

5.4.1 Architecture and Kinematics

In this section, the structure of the CDPR is introduced and the kinematic model is established.

5.4.1.1 Architecture

In this subsection, a CDPR model applied to grasp medicines is established. The 3D model and structural schematic are shown in Figures 5.16 and 5.17, respectively. The designed CDPR is mainly divided into five parts: the fixed base, driving units, pulley blocks, parallel cables, and the end-effector (EE). The EE is connected to driving units through eight parallel cables and the driving units are fixed to the base. These cables connect with the drive units to realize the direction change through the pulley blocks. And pulley blocks adopt fixed type, which can meet the needs of robot operation and simplify the complexity of structure.

The 3D model of the EE is shown in Figure 5.18, including the upper part of the EE, the lower part of the EE, the grip jaw, and springs. The upper and lower parts of the EE can be rotated relatively, and the left and right parts of the grip jaw are fixedly connected with the upper and lower parts of the EE, respectively. The upper and lower parts of the EE keep relatively stationary by giving the spring a tensioning force. The cross-distributed cables can realize the clamping function of the EE, and maintain six degrees of freedom of the CDPR.

5.4.1.2 Kinematics

As shown in Figure 5.17, A_i are attachment points of cables on the base, which are located on center points of trapezoidal distributed pulleys, and B_i are attachment points of cables on the EE, which are symmetrically distributed at the vertex of the EE. A fixed Cartesian frame O–XYZ is attached to the base platform with its

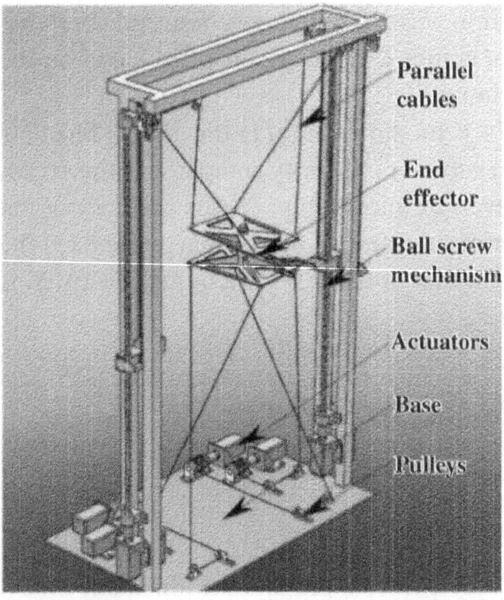

Figure 5.16 The 3D model of the CDPR.

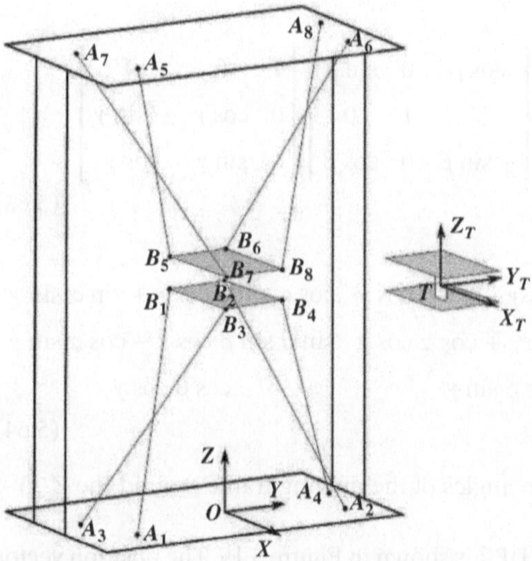

Figure 5.17 Schematic diagram of the CDPR.

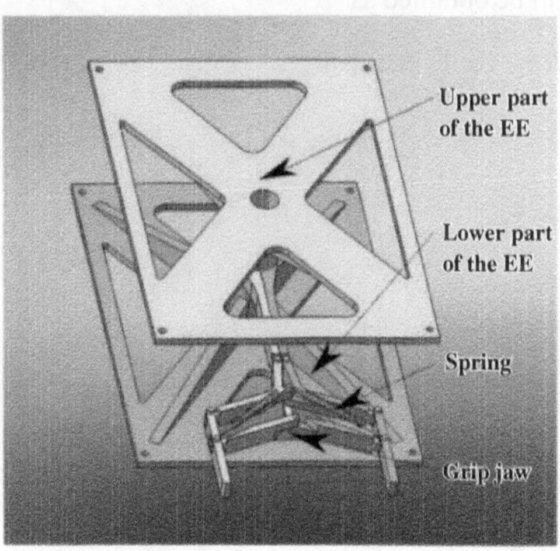

Figure 5.18 The 3D model of the EE.

origin O at the center of the trapezoid $A_1A_2A_3A_4$ and its Y-axis parallel to A_1A_2. A moving Cartesian frame $T\text{–}X_TY_TZ_T$ is defined, the origin T coincides with the center of the EE, and the X-axis is parallel to B_1B_3. Thus, the rotation matrix of the moving frame relative to the fixed frame is expressed as

$$
\begin{aligned}
{}_T^O\boldsymbol{R} &= \boldsymbol{R}_Z(\alpha)\,\boldsymbol{R}_Y(\beta)\,\boldsymbol{R}_X(\gamma) \\
&= \begin{bmatrix} \cos\alpha & -\sin\alpha & 0 \\ \sin\alpha & \cos\alpha & 0 \\ 0 & 0 & 1 \end{bmatrix} \begin{bmatrix} \cos\beta & 0 & \sin\beta \\ 0 & 1 & 0 \\ -\sin\beta & 0 & \cos\beta \end{bmatrix} \begin{bmatrix} 1 & 0 & 0 \\ 0 & \cos\gamma & -\sin\gamma \\ 0 & \sin\gamma & \cos\gamma \end{bmatrix}
\end{aligned} \tag{5.63}
$$

$$
{}_T^O\boldsymbol{R} = \begin{bmatrix} \cos\alpha\cos\beta & \cos\alpha\sin\beta s\gamma - \sin\alpha\cos\gamma & \cos\alpha\sin\beta\cos\gamma + \sin\alpha\sin\gamma \\ \sin\alpha\cos\beta & \sin\alpha\sin\beta s\gamma + \cos\alpha\cos\gamma & \sin\alpha\sin\beta\cos\gamma - \cos\alpha\sin\gamma \\ -\sin\beta & \cos\beta\sin\gamma & \cos\beta\cos\gamma \end{bmatrix} \tag{5.64}
$$

where α, β, and γ are the rotation angles of the moving frame around the Z-, Y-, and X-axes, respectively.

The kinematic diagram of the CDPR is shown in Figure 5.19. The position vector ${}^O\boldsymbol{P}$ is set from O to T, and it can be expressed as ${}^O\boldsymbol{P} = (x\ y\ z)^T$. $\boldsymbol{a}_i$ and $\boldsymbol{b}_i$ denote the position vector of $\boldsymbol{OA}_i$ and $\boldsymbol{TB}_i$, respectively. ${}^O\boldsymbol{b}_i$ denote the position vector of $\boldsymbol{OB}_i$. Thus, the ith cable vector $\boldsymbol{L}_i$ can be obtained as

$$
\boldsymbol{L}_i = \boldsymbol{a}_i - {}^O\boldsymbol{b}_i = \boldsymbol{a}_i - {}^O\boldsymbol{P} - {}_T^O\boldsymbol{R}\boldsymbol{b}_i,\ i = 1,\ 2,\ \dots,\ 8 \tag{5.65}
$$

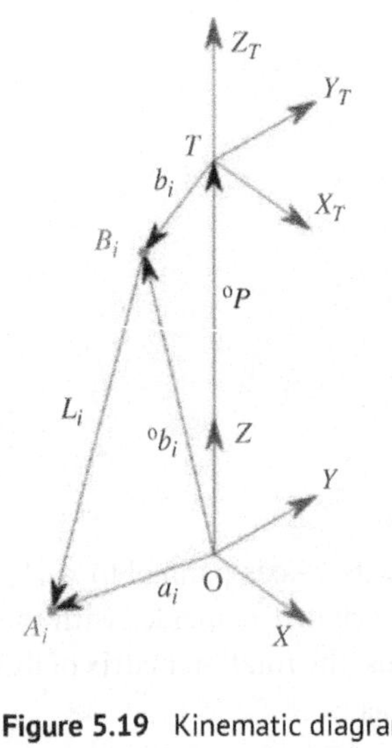

Figure 5.19 Kinematic diagram of the proposed CDPR.

The length of the cable l_i can be calculated by

$$l_i = ||\boldsymbol{L}_i|| = \sqrt{\boldsymbol{L}_i^T \boldsymbol{L}_i} \tag{5.66}$$

Taking the derivative of Eq. (5.65) with respect to time yields

$$\dot{\boldsymbol{L}} = \boldsymbol{J}\dot{\boldsymbol{q}} \tag{5.67}$$

where $\boldsymbol{J}$ is the motion Jacobian matrix of the CDPR; $\dot{\boldsymbol{q}}$ is the motion velocity vector of the EE, and

$$\dot{\boldsymbol{L}} = \begin{pmatrix} \dot{l}_1 & \dot{l}_2 & \cdots & \dot{l}_8 \end{pmatrix}^T \tag{5.68}$$

$$\dot{\boldsymbol{q}} = \begin{pmatrix} \dot{x} & \dot{y} & \dot{z} & \dot{\alpha} & \dot{\beta} & \dot{\gamma} \end{pmatrix}^T \tag{5.69}$$

$$\boldsymbol{J} = -\begin{pmatrix} \frac{\partial \boldsymbol{L}}{\partial x} & \frac{\partial \boldsymbol{L}}{\partial y} & \frac{\partial \boldsymbol{L}}{\partial z} & \frac{\partial \boldsymbol{L}}{\partial \alpha} & \frac{\partial \boldsymbol{L}}{\partial \beta} & \frac{\partial \boldsymbol{L}}{\partial \gamma} \end{pmatrix} \tag{5.70}$$

5.4.2 Error Transfer Model

Based on the kinematic model of the CDPR with considering the pulley radius, this subsection presents the derivative process of error transfer model. The error of the pose of the EE can be expressed as

$$\boldsymbol{de} = \begin{pmatrix} d\boldsymbol{p} \\ d\boldsymbol{\Omega} \end{pmatrix} \tag{5.71}$$

where $d\boldsymbol{p} = \begin{pmatrix} dp_x & dp_y & dp_z \end{pmatrix}^T$ represents the error of the position of the EE; and $d\boldsymbol{\Omega} = \begin{pmatrix} d\alpha & d\beta & d\gamma \end{pmatrix}^T$ represents the error of the posture of the EE.

The unit direction vector along the ith cable is $\boldsymbol{u}_i$, and it can be expressed as

$$\boldsymbol{u}_i = \frac{\boldsymbol{L}_i}{l_i} \tag{5.72}$$

where $\boldsymbol{L}_i$ is the ith cable vector; and l_i is the length of the ith cable vector.

Then we have

$$l_i \boldsymbol{u}_i = \boldsymbol{a}_i - {}^{O}\boldsymbol{P} - {}^{O}_{T}\boldsymbol{R}\boldsymbol{b}_i + \boldsymbol{r}_i \tag{5.73}$$

The differential of Eq. (5.73) can be expressed as

$$dl_i \boldsymbol{u}_i + l_i d\boldsymbol{u}_i = d\boldsymbol{a}_i - d{}^{O}\boldsymbol{P} - d{}^{O}_{T}\boldsymbol{R}\boldsymbol{b}_i - {}^{O}_{T}\boldsymbol{R}d\boldsymbol{b}_i + d\boldsymbol{r}_i \tag{5.74}$$

Left multiplying $\boldsymbol{u}_i^T$ at both sides of Eq. (5.74) deduces

$$\boldsymbol{u}_i^T\, dl_i\boldsymbol{u}_i + \boldsymbol{u}_i^T\, l_i d\boldsymbol{u}_i = \boldsymbol{u}_i^T\, d\boldsymbol{a}_i - \boldsymbol{u}_i^T\, d^O\boldsymbol{P} - \boldsymbol{u}_i^T\, d{}_T^O\boldsymbol{R}\boldsymbol{b}_i - \boldsymbol{u}_i^T\, {}_T^O\boldsymbol{R} d\boldsymbol{b}_i + \boldsymbol{u}_i^T\, d\boldsymbol{r}_i \tag{5.75}$$

where $d\boldsymbol{u}_i$ and $d{}_T^O\boldsymbol{R}$ can be expressed as

$$d\boldsymbol{u}_i = \Delta_{ui}\boldsymbol{u}_i = \begin{pmatrix} 0 & -\partial u_{zi} & \partial u_{yi} \\ \partial u_{zi} & 0 & -\partial u_{xi} \\ -\partial u_{yi} & \partial u_{xi} & 0 \end{pmatrix}\begin{pmatrix} u_{xi} \\ u_{yi} \\ u_{zi} \end{pmatrix} \tag{5.76}$$

$$d{}_T^O\boldsymbol{R} = \Delta_\Omega {}_T^O\boldsymbol{R} = \begin{pmatrix} 0 & -\partial\gamma & \partial\beta \\ \partial\gamma & 0 & -\partial\alpha \\ -\partial\beta & \partial\alpha & 0 \end{pmatrix} {}_T^O\boldsymbol{R} \tag{5.77}$$

where Δ_{ui} and Δ_Ω are different differential factors.

Substituting Eqs. (5.76) and (5.77) into terms $\boldsymbol{u}_i^T\, dl_i\boldsymbol{u}_i$ and $\boldsymbol{u}_i^T\, l_i d\boldsymbol{u}_i$ in Eq. (5.75) respectively yields

$$\boldsymbol{u}_i^T l_i d\boldsymbol{u}_i = l_i \begin{pmatrix} u_{xi} & u_{yi} & u_{zi} \end{pmatrix}\begin{pmatrix} 0 & -\partial u_{zi} & \partial u_{yi} \\ \partial u_{zi} & 0 & -\partial u_{xi} \\ -\partial u_{yi} & \partial u_{xi} & 0 \end{pmatrix}\begin{pmatrix} u_{xi} \\ u_{yi} \\ u_{zi} \end{pmatrix} = 0 \tag{5.78}$$

$$\boldsymbol{u}_i^T d{}_T^O\boldsymbol{R}\boldsymbol{b}_i = \boldsymbol{u}_i^T \Delta_\Omega {}_T^O\boldsymbol{R}\boldsymbol{b}_i = \boldsymbol{u}_i^T d\boldsymbol{\Omega} \times {}_T^O\boldsymbol{R}\boldsymbol{b}_i = \left({}_T^O\boldsymbol{R}\boldsymbol{b}_i \times \boldsymbol{u}_i\right)^T d\boldsymbol{\Omega} \tag{5.79}$$

Substituting Eq. (5.78) into Eq. (5.79) yields

$$\begin{aligned} dl_i &= \boldsymbol{u}_i^T d\boldsymbol{a}_i - \boldsymbol{u}_i^T d\boldsymbol{p} - \left({}_T^O\boldsymbol{R}\boldsymbol{b}_i \times \boldsymbol{u}_i\right)^T d\boldsymbol{\Omega} - \boldsymbol{u}_i^T {}_T^O\boldsymbol{R} d\boldsymbol{b}_i + \boldsymbol{u}_i^T d\boldsymbol{r}_i \\ &= -\begin{pmatrix} \boldsymbol{u}_i^T & \left({}_T^O\boldsymbol{R}\boldsymbol{b}_i \times \boldsymbol{u}_i\right)^T \end{pmatrix}\begin{pmatrix} d\boldsymbol{p} \\ d\boldsymbol{\Omega} \end{pmatrix} + \begin{pmatrix} \boldsymbol{u}_i^T & -\boldsymbol{u}_i^T {}_T^O\boldsymbol{R} & \boldsymbol{u}_i^T \end{pmatrix}\begin{pmatrix} d\boldsymbol{a}_i \\ d\boldsymbol{b}_i \\ d\boldsymbol{r}_i \end{pmatrix} \end{aligned} \tag{5.80}$$

where $dl_i = \boldsymbol{u}_i^T\, dl_i\boldsymbol{u}_i$, and $d\boldsymbol{p} = d^O\boldsymbol{P}$.

Equation (5.80) can be rewritten as

$$d\boldsymbol{e} = -\boldsymbol{J}_u^{-1}\, d\boldsymbol{l} + \boldsymbol{J}_u^{-1}\boldsymbol{J}_d d\boldsymbol{q} \tag{5.81}$$

Replacing the differential symbol d with Δ yields error transfer model of the CDPR:

$$\Delta \boldsymbol{e} = -\boldsymbol{J}_u^{-1}\Delta \boldsymbol{l} + \boldsymbol{J}_u^{-1}\boldsymbol{J}_d\Delta \boldsymbol{q} \tag{5.82}$$

where $\boldsymbol{J}_u^{-1}\boldsymbol{J}_d\Delta \boldsymbol{q}$ refers to errors caused by manufacturing and installation; and $-\boldsymbol{J}_u^{-1}\Delta \boldsymbol{l}$ refers to errors caused by cable lengths.

5.4.3 Uncertainty Analysis Based on Evidence Theory

In order to analyze the movement uncertainty caused by the cable lengths with epistemic uncertainty, the evidence theory-based uncertainty analysis method (ETUAM) based on the evidence theory and error transfer model of the CDPR is proposed.

As mentioned above, cables are not parallel to central planes of fixed pulleys, and with the effects of other factors, uncertain errors of cable lengths are generated. These errors are taken as evidence variables to analyze the kinematic uncertainty of the CDPR. Based on evidence theory and error transfer model, the ETUAM is proposed. The flowchart of the ETUAM is shown in Figure 5.20.

Step 1. Errors of cable lengths $\Delta l_1, \Delta l_2, \Delta l_3, \Delta l_4, \Delta l_5, \Delta l_6, \Delta l_7$, and Δl_8 are taken as evidence variables. And each distribution subinterval of evidence variables is an FD element. Assume that frame elements are $B_i = [b_{il}, b_{ir}]$, $C_i = [c_{il}, c_{ir}]$, $D_i = [d_{il}, d_{ir}]$, $E_i = [e_{il}, e_{ir}]$, $F_i = [f_{il}, f_{ir}]$, $G_i = [g_{il}, g_{ir}]$, $H_i = [h_{il}, h_{ir}]$, and $I_i = [i_{il}, i_{ir}]$, respectively, and the FD is constructed by them.

Step 2. The joint BPA for different evidence sources is established. The joint BPA of a single variable that has different evidence sources is obtained by using Dempster's rule. Then BPAs of all variables are combined to obtain the total joint BPA, which is expressed as $m\,(A)$, and

$$\begin{cases} \boldsymbol{A} = \boldsymbol{B} \cdot \boldsymbol{C} \cdot \boldsymbol{D} \cdot \boldsymbol{E} \cdot \boldsymbol{F} \cdot \boldsymbol{G} \cdot \boldsymbol{H} \cdot \boldsymbol{I}, \\ \sum m\,(\boldsymbol{A}_k) = 1 \end{cases} \tag{5.83}$$

where A is the joint focal element; and k is the number of A.

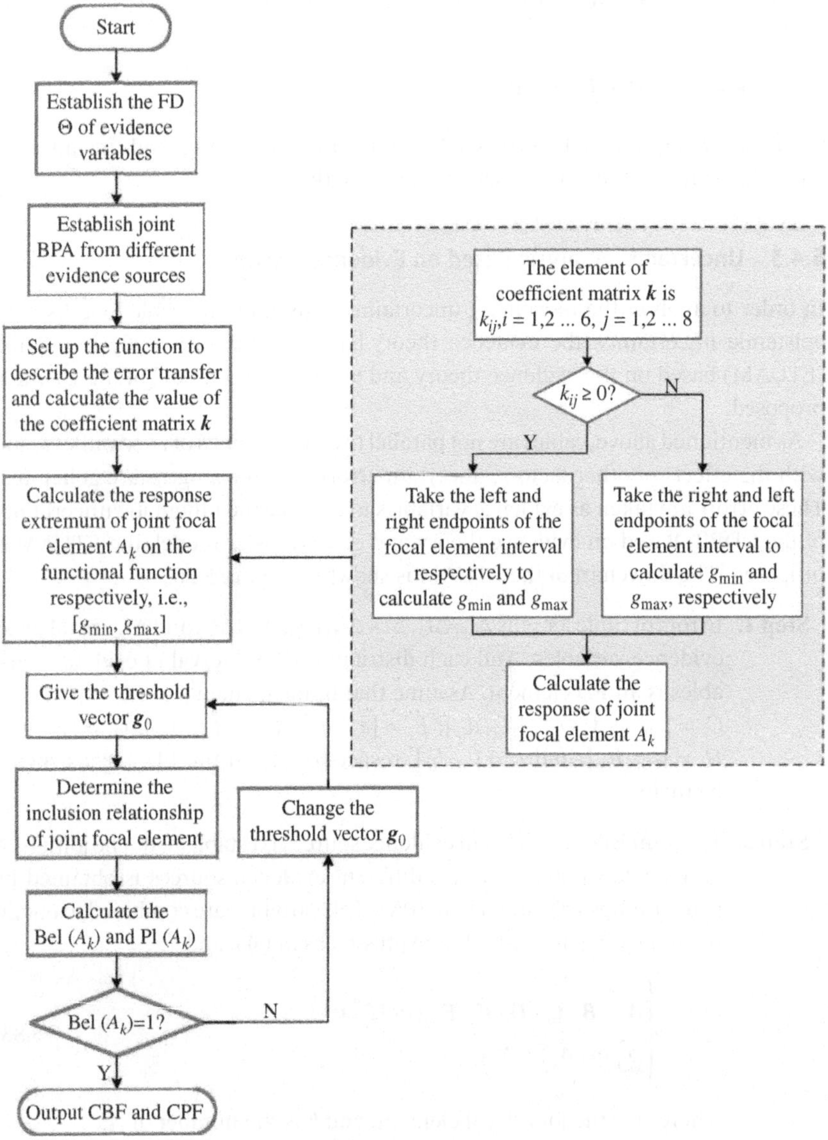

Figure 5.20 Flowchart of the proposed ETUAM.

Step 3. The structural performance function which describes the process of error transmission is established. The structural performance function is obtained by error transfer model, and it can be expressed as

$$g(\Delta \boldsymbol{l}) = \Delta \boldsymbol{e} = \begin{pmatrix} \Delta p_x \\ \Delta p_y \\ \Delta p_z \\ \Delta \alpha \\ \Delta \beta \\ \Delta \gamma \end{pmatrix} = -\boldsymbol{J}_u^{-1} \Delta \boldsymbol{l}$$

$$= -\begin{pmatrix} \boldsymbol{u}_1{}^T & ({}^O_T\boldsymbol{R}\boldsymbol{b}_1 \times \boldsymbol{u}_1)^T \\ \boldsymbol{u}_2{}^T & ({}^O_T\boldsymbol{R}\boldsymbol{b}_2 \times \boldsymbol{u}_2)^T \\ \vdots & \vdots \\ \boldsymbol{u}_8{}^T & ({}^O_T\boldsymbol{R}\boldsymbol{b}_8 \times \boldsymbol{u}_8)^T \end{pmatrix}^{-1} \begin{pmatrix} \Delta l_1 \\ \Delta l_2 \\ \vdots \\ \Delta l_8 \end{pmatrix} \tag{5.84}$$

where $\Delta \boldsymbol{l}$ is the vector composed of evidence variables.

The coefficient matrix $-\boldsymbol{J}_u^{-1}$ can be expressed as

$$\boldsymbol{k} = -\boldsymbol{J}_u^{-1} = -\begin{pmatrix} \boldsymbol{u}_1{}^T & ({}^O_T\boldsymbol{R}\boldsymbol{b}_1 \times \boldsymbol{u}_1)^T \\ \boldsymbol{u}_2{}^T & ({}^O_T\boldsymbol{R}\boldsymbol{b}_2 \times \boldsymbol{u}_2)^T \\ \vdots & \vdots \\ \boldsymbol{u}_8{}^T & ({}^O_T\boldsymbol{R}\boldsymbol{b}_8 \times \boldsymbol{u}_8)^T \end{pmatrix}^{-1} = \begin{pmatrix} k_{11} & k_{12} & \cdots & k_{18} \\ \vdots & \vdots & \ddots & \vdots \\ k_{61} & k_{62} & \cdots & k_{68} \end{pmatrix} \tag{5.85}$$

where k_{ij} is an element of the coefficient matrix.

Step 4. Response extremums of the joint focal element $\boldsymbol{A}_k$ are calculated under the structural performance function.

An element Δp_x of responses is taken as an example to obtain extremums of a joint focal element $\boldsymbol{A}_k$. The calculation function of a response Δp_x can be expressed as

$$\Delta p_x = \boldsymbol{k}_1 \cdot \Delta \boldsymbol{l} \tag{5.86}$$

where $\boldsymbol{k}_1 = (k_{11}, k_{12}, \ldots, k_{18})$ is the first row of the coefficient matrix $\boldsymbol{k}$.

Assuming that elements in $\boldsymbol{k}_1$ satisfy $k_{11} < 0$, $k_{12} \geq 0$, $k_{13} < 0$, $k_{14} \geq 0$, $k_{15} < 0$, $k_{16} \geq 0$, $k_{17} < 0$, and $k_{18} \geq 0$, respectively. According to the values of these elements, the maximum of Δp_x can be obtained when

$$\Delta \boldsymbol{l} = (b_{il}, c_{ir}, d_{il}, e_{ir}, f_{il}, g_{ir}, h_{il}, i_{ir})^T \tag{5.87}$$

where $b_{il}, c_{ir}, d_{il}, e_{ir}, f_{il}, g_{ir}, h_{il}$, and i_{ir} are the bounds of subintervals in BPAs, respectively.

The minimum of Δp_x can be obtained when

$$\Delta \boldsymbol{l} = (b_{ir}, c_{il}, d_{ir}, e_{il}, f_{ir}, g_{il}, h_{ir}, i_{il})^T \tag{5.88}$$

where b_{ir}, c_{il}, d_{ir}, e_{il}, f_{ir}, g_{il}, h_{ir}, and i_{il} are other bounds of subintervals in BPAs, respectively.

The remaining Δp_y, Δp_z, $\Delta\alpha$, $\Delta\beta$, and $\Delta\gamma$ can be calculated successively in the same way. Extremums of the response of the joint focal element $\boldsymbol{A}_k$ are obtained, and

$$[\boldsymbol{g}_{\min}, \boldsymbol{g}_{\max}] = [\min g(\boldsymbol{A}_k), \max g(\boldsymbol{A}_k)] \tag{5.89}$$

Step 5. The threshold vector is given and the attribution of the focal element is judged. The threshold vector can be expressed as

$$\boldsymbol{g}_0 = [p_{x0}, p_{y0}, p_{z0}, \alpha_0, \beta_0, \gamma_0]^T \tag{5.90}$$

where p_{x0}, p_{y0}, p_{z0}, α_0, β_0, and γ_0 are thresholds of the EE moving along the X-, Y-, and Z-axes and rotating around the X-, Y-, and Z-axes, respectively.

Suppose that the event set is $G = \{(\boldsymbol{A}_k) \mid g(\boldsymbol{A}_k) \leq \boldsymbol{g}_0\}$. If a joint focal element $\boldsymbol{A}_k$ satisfies $\boldsymbol{A}_k \subseteq G$, its BPA is counted into $\text{Bel}(G)$, and if it satisfies the $\boldsymbol{A}_k \cap G \neq 0$, its BPA is counted into $\text{Pl}(G)$.

Step 6. Judging whether $\text{Bel}(G)$ reaches 1. When $\text{Bel}(G) = 1$, it means that the threshold vector $\boldsymbol{g}_0$ is large enough, and responses of joint focal elements are within their ranges.

Step 7. When $\text{Bel}(G) < 1$, the threshold vector $\boldsymbol{g}_0$ should be increased and the above steps are repeated until $\text{Bel}(G) = 1$ to get the CBF and CPF graphs.

5.4.4 Simulation and Comparison

In this subsection, to study errors of cable lengths with and without considering pulleys radii, the example of the designed CDPR is simulated. To illustrate the accuracy and efficiency of the ETUAM for calculating the kinematic uncertainty response, the CBF and CPF of the kinematic uncertainty are calculated by the MCM, vertex method, and ETUAM, respectively. The changes of kinematics uncertainty of the CDPR when EE moves are also analyzed.

5.4.4.1 Simulation of Different Modeling Methods

The deterministic structural parameters of the CDPR are listed in Table 5.10.

Three circular moving trajectories of the EE on the X–Y, Y–Z, and X–Z planes are given, respectively. Circular trajectories are shown in Figure 5.21a.

Table 5.10 BPA structures of evidence variables from different sources.

Symbol	Description	Value
a (mm)	The length of A_1A_2	200
b (mm)	The length of A_3A_4	674
c (mm)	The horizontal spacing between A_1A_2 and A_3A_4	836
d (mm)	The vertical spacing between A_1A_2 and A_5A_6	188
e (mm)	The length of B_1B_2	282
f (mm)	The length of B_1B_3	126
h (mm)	The length of B_1B_5	1546
α (rad)	The rotation angle of the moving frame around the *X*-axis	0
β (rad)	The rotation angle of the moving frame around the *Y*-axis	0
γ (rad)	The rotation angle of the moving frame around the *Z*-axis	0
ε (rad)	The angle between the upper and lower parts of the EE	0
r (mm)	The radius of the pulley	12

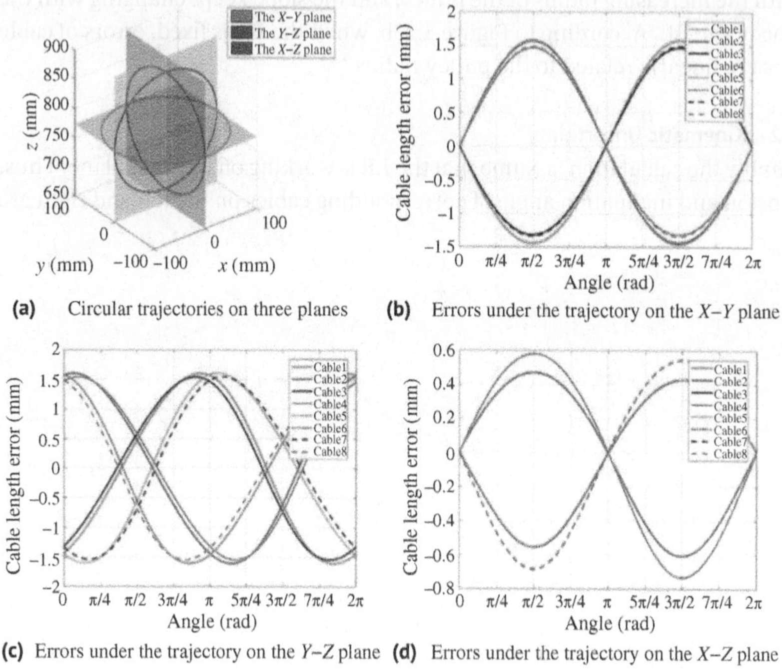

(a) Circular trajectories on three planes **(b)** Errors under the trajectory on the *X*–*Y* plane

(c) Errors under the trajectory on the *Y*–*Z* plane **(d)** Errors under the trajectory on the *X*–*Z* plane

Figure 5.21 Circular trajectories and errors of cable lengths on different planes.

Figure 5.21 shows modeling errors with and without considering pulleys radii when the EE moves on different planes with the pulley radius $r = 12$ mm. As shown in Figure 5.21b,d, when the EE moves on the X–Y, and X–Z plane, symmetry and superposition appear in errors of cable lengths, but not all errors overlap completely. For example, there is a deviation between the errors of cable 1 and cable 3. And the superposition of cable error does not appear in Figure 5.21c. The symmetry and superposition in Figure 5.21b are caused by attachment point A_i being symmetric with the X–Y plane. And the deviations of cables are mainly because of the difference in the position distribution of point A_i. The same is true in Figure 5.21d. When the EE moves on the X–Z plane, the maximum error of the cable length is −0.75 mm. Therefore, the modeling errors have more influence on the motions of the EE on the X–Y plane and Y–Z plane.

To further study the influence of pulleys radii on modeling errors, a spiral moving trajectory of the EE is given in Figure 5.22.

As shown in Figure 5.23a, under the spiral trajectory, overall errors of cable lengths tend to be zero when the EE nears the center of the workspace ($Z = 773$ mm). With the EE moving from bottom to top of the workspace, errors tend to change from positive to negative. And errors of cable lengths are increasing with the increasing radius of the pulley, and the slope keeps changing with the moving of the EE. According to Figure 5.23b, when the EE is fixed, errors of cable lengths are linearly related to the pulley radius.

5.4.4.2 Kinematic Uncertainty

To simplify the calculation, assume that the EE is working on the X–Z plane. Thus, the tension and inclination angle of corresponding cables on the left and right are

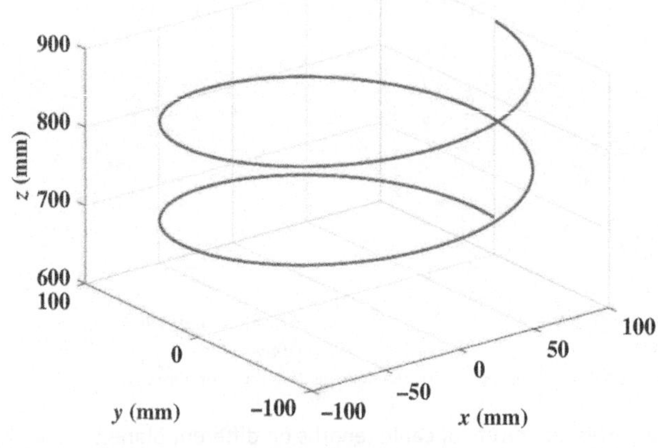

Figure 5.22 The spiral trajectory of the EE.

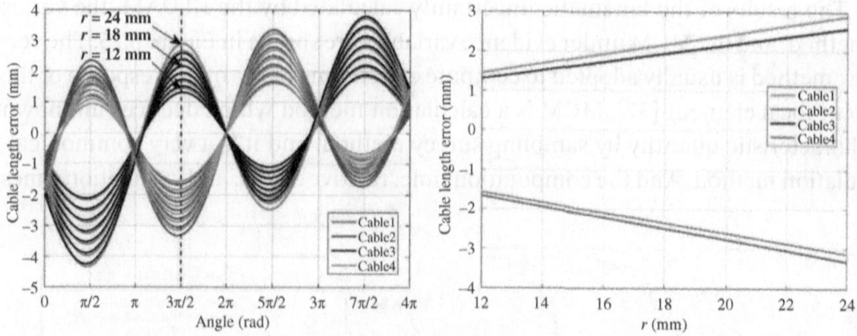

(a) Length errors of four cables

(b) Changes of errors with r when the rotation angle is $3\pi/2$

Figure 5.23 Cable errors under the spiral trajectory.

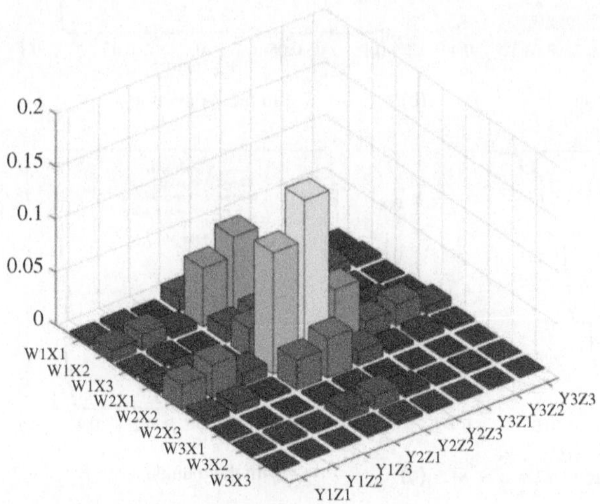

Figure 5.24 The join BPA structures of evidence variables.

the same, and the kinematic uncertainty of the CDPR can be obtained by calculating four cables on one side. When the EE is working on the X–Z plane, the tension and inclination angle of the four cables on the left are different; thus, variables of length errors can be considered that are relatively independent. Assume that the coordinates of the working point of EE on the X–Y plane are (100,0,773).

The joint BPA $m(A)$ is combined by Cartesian product, which is composed of 81 joint elements, and each focal element is a four-dimensional hypercube. Joint BPA structures are shown in Figure 5.24.

The results of the kinematic uncertainty calculated by the ETUAM, the vertex method, and the MCM under evidence variables are shown in Figure 5.25. The vertex method is usually adopted to compute extremums of the of the response of the joint focal element [37]. MCM is a calculation method which deduces unknown characteristic quantity by sampling survey method, and it is a very common calculation method. And the computation time, relative errors, and calculation times

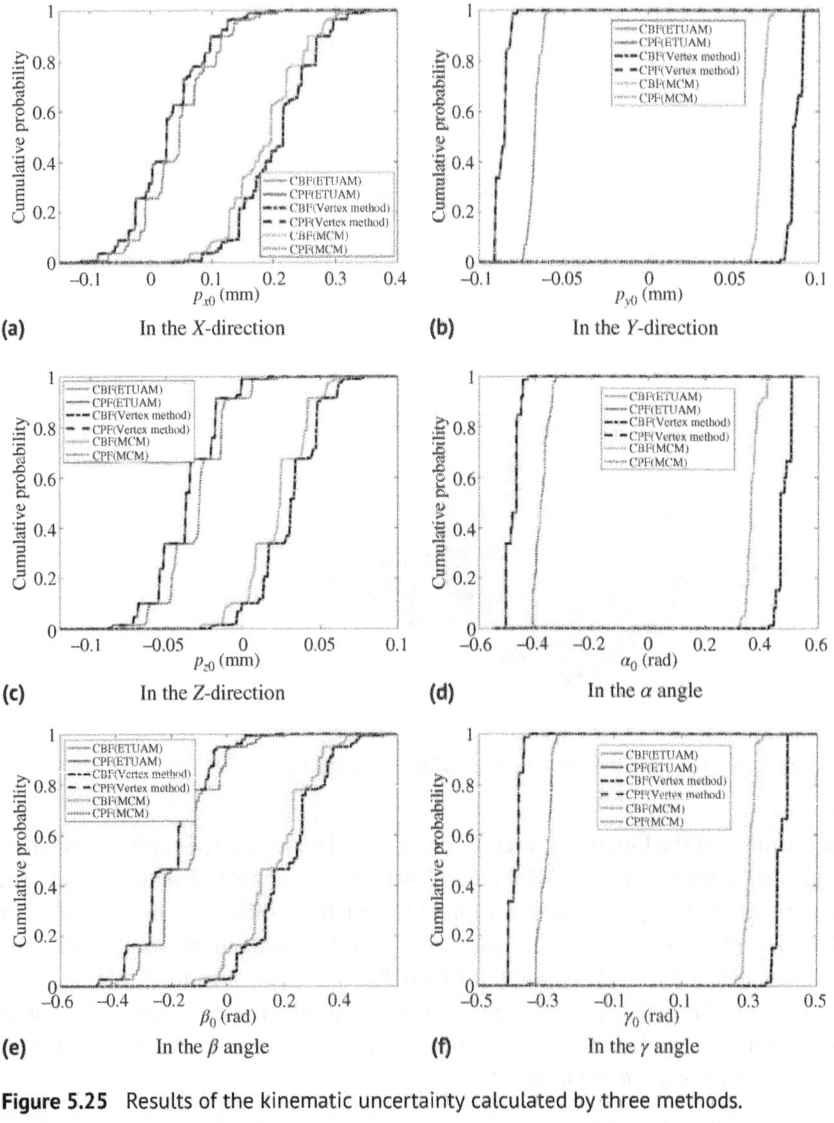

Figure 5.25 Results of the kinematic uncertainty calculated by three methods.

of the kinematic uncertainty by the ETUAM, the vertex method, and the MCM are listed in Tables 5.11 and 5.12.

The CBF and CPF of the kinematic uncertainty calculated by the vertex method, the MCM, and the ETUAM are shown in Figure 5.25. The conventional vertex method can be adopted to provide reference results. The CBF and CPF of the kinematic uncertainty calculated by the vertex method and the ETUAM are completely coincident, which indicates that these two methods all can provide accurate results. However, the CBF and CPF by these two methods significantly differ from those by the MCM. The results indicate that the ETUAM achieves accuracy comparable to that of the vertex method when calculating the CBF and CPF for the kinematic uncertainty, and the ETUAM has much better accuracy than the MCM.

To more clearly illustrate the accuracy and the efficiency of the ETUAM for the kinematic uncertainty calculation of the CDPR with evidence variables, the computational evaluations of the ETUAM, vertex method, and MCM are shown in Tables 5.11 and 5.12. The results listed in the tables show that the ETUAM is more efficient and greatly reduces the computational cost compared with the MCM.

When the EE is at a fixed point, the kinematic uncertainty of the CDPR based on evidence theory has been analyzed. To study the characteristics of the kinematic uncertainty when the EE keeps moving, the simulations are carried out under evidence variables. By changing the X- and Z-coordinates of the EE in the main

Table 5.11 Efficiency and accuracy comparison of different methods.

Parameters	ETUAM	Vertex Method	MCM	
	Computation Time (s)	Computation Time (s)	Computation Time (s)	Error (%)
p_{x0}	<0.1	0.52	7.83	21.64
p_{y0}	<0.1	0.53	7.70	22.61
p_{z0}	<0.1	0.54	7.63	22.90
α_0	<0.1	0.53	7.67	22.95
β_0	<0.1	0.56	7.65	22.50
γ_0	<0.1	0.57	7.58	22.89

Table 5.12 Calculation times of three methods.

	ETUAM	Vertex method	MCM
Calculation times	81	324	50,000

workspace, corresponding changes of thresholds that reflect the changes of uncertainty are obtained when Bel is 1.

As shown in Figures 5.26 and 5.27, when Bel = 1, the thresholds of the response of the kinematic uncertainty vary with the changes of the X- and Z-coordinates of the EE. Changes in kinematic uncertainty can be reflected by changes of thresholds when Bel = 1. As shown in Figure 5.26a,b, thresholds p_{y0} and p_{z0} vary very little with changes of the X-coordinate of the EE. It means uncertainties in the Y- and Z-directions almost do not vary when the EE moves along the X-axis. The threshold p_{x0} varies greatly with changes of the X-coordinate of the EE, and it reaches the maximum when $X = 150$ mm. And the variation of p_{x0} is within 0.45 rad. Uncertainties in α, β, and γ angles vary little when the EE moves along the X-axis, and their variations are within 0.05 rad. As shown in Figure 5.27a,b, kinematic uncertainties in the X-direction, β and γ angles are greatly affected by changes of the

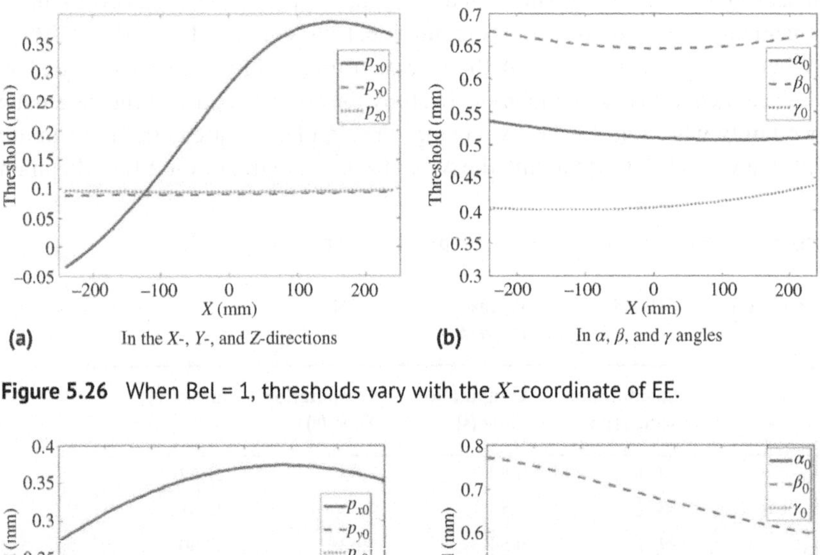

Figure 5.26 When Bel = 1, thresholds vary with the X-coordinate of EE.

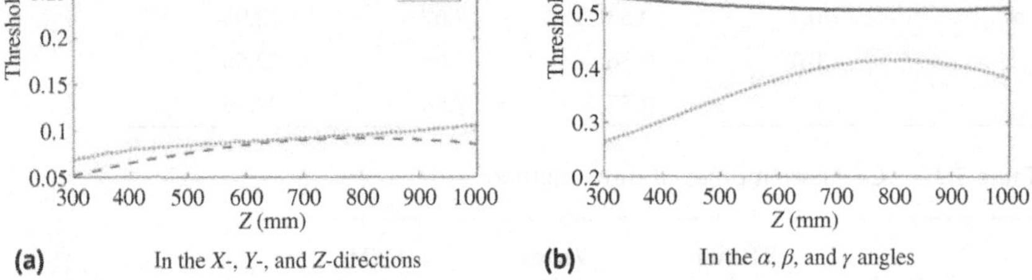

Figure 5.27 When Bel = 1 and Pl = 1, the difference of thresholds varies with the X-coordinate of the EE.

Z-coordinate of the EE. The variation of p_{x0} is within 0.15 mm, and variations of β and γ angles are within 0.2 rad. However, there are few changes in the Y- and Z-directions, and their variations are within 0.05 mm.

Figures 5.28 and 5.29 show the changes of the difference between thresholds when Bel = 1 and Pl = 1 with the changes of the X- and Z-coordinates of the EE. Changes in the degree of epistemic uncertainty of evidential variables can be reflected by the changes of the difference. As shown in Figure 5.28a, the difference of the threshold p_{x0} (in the X-direction) is most affected by changes of the X-coordinate of the EE. When $X = 0$, the difference reaches the maximum. However, with the X-coordinate of the EE increasing, the difference of the threshold in the Y-direction is slightly increased, while in the Z-direction is unchanged. So, the kinematic uncertainty in the X-direction is most affected by the movement of EE along the X-axis. Figure 5.28b shows the changes of difference of angle thresholds with changes of the X-coordinate of the EE. The difference of thresholds in the α

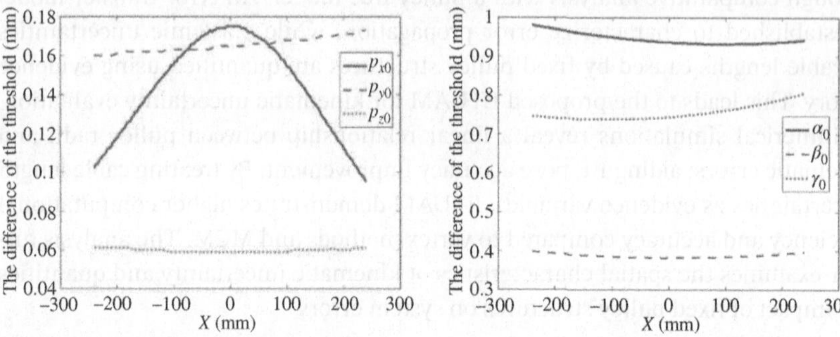

(a) In the X-, Y-, and Z-directions (b) In the α, β, and γ angles

Figure 5.28 When Bel = 1 and Pl = 1, the difference of thresholds varies with the Y-coordinate of the EE.

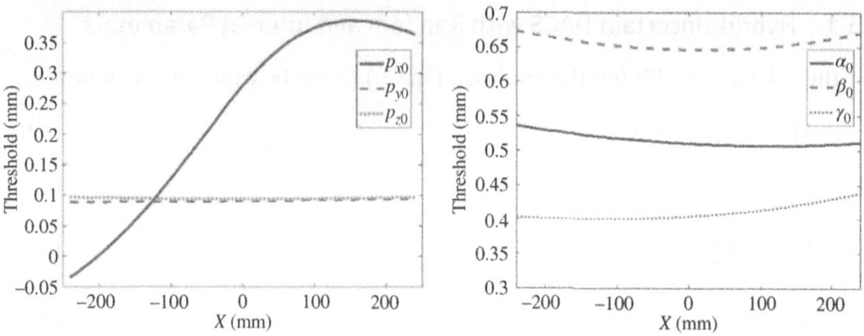

(a) In the X-, Y-, and Z-directions (b) In α, β, and γ angles

Figure 5.29 When Bel = 1 and Pl = 1, the difference of thresholds varies with the Z-coordinate of the EE.

and γ angles varies within 0.1 rad and in the β angle varies slightly. Figure 5.29 shows the changes of difference of thresholds when the EE moves along the Z-axis. The difference of thresholds in the X- and Y-directions and γ angle is greatly affected by changes of the Z-coordinate of the EE, and their variations are within 0.4 mm, 0.8 mm, and 0.3 rad, respectively.

From Figures 5.26–5.29, the kinematic uncertainty of the CDPR based on evidence theory varies with the moving of the EE. And for different responses to the kinematic uncertainty, changes in EE position have different effects on them.

5.4.5 Conclusion

This section presents a CDPR designed for medical capture, featuring eight cross-distributed cables and an EE with grasping functionality enabled by relative rotation between its upper and lower parts. A kinematic model incorporating pulley radii is developed, addressing cable length errors induced by pulley geometry through comparative analysis with a pulley-free model. An error transfer model is established to characterize error propagation, while epistemic uncertainties in cable lengths caused by fixed pulley structures are quantified using evidence theory. This leads to the proposed ETUAM for kinematic uncertainty evaluation.

Numerical simulations reveal a linear relationship between pulley radii and kinematic errors; aiding EE pose accuracy improvement. By treating cable length uncertainties as evidence variables, ETUAM demonstrates higher computational efficiency and accuracy compared to vertex methods and MCM. The analysis further examines the spatial characteristics of kinematic uncertainty and quantifies the impact of fixed pulley structures on system errors.

5.5 Kinematic Uncertainty Analysis with Hybrid Random and Interval Parameters

5.5.1 Hybrid Uncertain DACS with Random and Interval Parameters

For the sake of simplifying the analysis, Eq. (5.11) can be expressed as follows:

$$\boldsymbol{M}_i = \boldsymbol{N}_i \boldsymbol{\gamma}_i,\ i = 1, 2 \tag{5.91}$$

where

$$\boldsymbol{M}_i = \sqrt{{K_{1i}}^2 + {K_{2i}}^2 - {K_{3i}}^2} - K_{1i} \tag{5.92}$$

$$\boldsymbol{N}_i = K_{3i} - K_{2i} \tag{5.93}$$

$$\boldsymbol{\gamma}_i = \tan\frac{\gamma_i}{2} \tag{5.94}$$

where $\boldsymbol{M}_i$ and $\boldsymbol{N}_i$ are two different kinematic vectors; $\boldsymbol{\gamma}_i$ is the response vector.

It should be noted that the kinematic vectors $\boldsymbol{M}_i$ and $\boldsymbol{N}_i$ in Eqs. (5.92)–(5.94) are compound functions of both structural parameters (L_1, L_2, D, d) and payload parameters (y, z, θ). Consequently, the response vector γ_i is also a compound function with respect to these parameters (L_1, L_2, D, d, y, z, θ).

In this section, all random parameters with defined probabilistic distributions form a random vector $\boldsymbol{X}^R$, while all interval parameters with fixed bounds constitute an interval vector $\boldsymbol{Y}^I$.

The property of the random vector $\boldsymbol{X}^R$ can be expressed as

$$\boldsymbol{X}^R = \{x_1, x_2, \dots, x_r, \dots, x_m\}^T \tag{5.95}$$

Thus,

$$\boldsymbol{X}^e = \{x_1^e, x_2^e, \dots, x_r^e, \dots, x_m^e\}^T \tag{5.96}$$

where $\boldsymbol{X}^e$ represents the expectation of the random vector $\boldsymbol{X}^R$; x_r^e denotes the expectation of the random parameter x_r; and m is the total number of random parameters.

The property of the interval vector $\boldsymbol{Y}^I$ can be expressed as

$$\boldsymbol{Y}^I = \{y_1^I, \dots, y_s^I, \dots, y_n^I\}^T = \boldsymbol{Y}^c + \Delta\boldsymbol{Y}\delta^I,\ s = 1, 2, \dots, n \tag{5.97}$$

where

$$\boldsymbol{Y}^c = \{y_1^c, \dots, y_s^c, \dots, y_n^c\}^T \tag{5.98}$$

$$\Delta\boldsymbol{Y} = \{\Delta y_1, \dots, \Delta y_s, \dots, \Delta y_n\}^T \tag{5.99}$$

in which

$$y_s^I = \left[\underline{y_s}, \overline{y_s}\right] = y_s^c + \Delta y_s\delta^I,\ y_s^c = \left(\underline{y_s} + \overline{y_s}\right)/2,\ \Delta y_s = \left(\overline{y_s} - \underline{y_s}\right)/2 \tag{5.100}$$

where $\underline{y_s}$ and $\overline{y_s}$ represent the LB and UB of the interval parameter y_s^I, respectively; y_s^c and Δy_s denote the midpoint (MV) and IR (deviation) of y_s^I, respectively; δ^I indicates a fixed interval, i.e., $\delta^I = [-1, 1]$; n is the total number of the interval parameters; $\boldsymbol{Y}^c$ and $\Delta\boldsymbol{Y}$ represent the midpoint (MV) and IR (deviation) of the interval vector $\boldsymbol{Y}^I$, respectively.

Referring to Eq. (5.91), the hybrid luffing angular equilibrium equation of the DACS, involving the random vector $\boldsymbol{X}^R$ and interval vector $\boldsymbol{Y}^I$, can be expressed as follows:

$$\boldsymbol{M}_i\left(\boldsymbol{K}_i\left(\boldsymbol{X}^R, \boldsymbol{Y}^I\right)\right) = \boldsymbol{N}_i\left(\boldsymbol{K}_i\left(\boldsymbol{X}^R, \boldsymbol{Y}^I\right)\right)\gamma_i\left(\boldsymbol{X}^R, \boldsymbol{Y}^I\right),\ i = 1, 2 \tag{5.101}$$

From Eqs. (5.92) and (5.93), $\boldsymbol{M}_i\left(\boldsymbol{K}_i\left(\boldsymbol{X}^R, \boldsymbol{Y}^I\right)\right)$ and $\boldsymbol{N}_i\left(\boldsymbol{K}_i\left(\boldsymbol{X}^R, \boldsymbol{Y}^I\right)\right)$ can be expressed as

$$\boldsymbol{M}_i\left(\boldsymbol{K}_i\left(\boldsymbol{X}^R, \boldsymbol{Y}^I\right)\right) = \sqrt{K_{1i}\left(\boldsymbol{X}^R, \boldsymbol{Y}^I\right)^2 + K_{2i}\left(\boldsymbol{X}^R, \boldsymbol{Y}^I\right)^2 - K_{3i}\left(\boldsymbol{X}^R, \boldsymbol{Y}^I\right)^2} - K_{1i}\left(\boldsymbol{X}^R, \boldsymbol{Y}^I\right) \tag{5.102}$$

$$\boldsymbol{N}_i\left(\boldsymbol{K}_i\left(\boldsymbol{X}^R, \boldsymbol{Y}^I\right)\right) = K_{3i}\left(\boldsymbol{X}^R, \boldsymbol{Y}^I\right) - K_{2i}\left(\boldsymbol{X}^R, \boldsymbol{Y}^I\right) \tag{5.103}$$

where $\boldsymbol{K}_i\left(\boldsymbol{X}^R, \boldsymbol{Y}^I\right)$ is the hybrid compound subfunction vector.

$\boldsymbol{K}_i\left(\boldsymbol{X}^R, \boldsymbol{Y}^I\right)$ in Eq. (5.101) can be expressed as

$$\boldsymbol{K}_i\left(\boldsymbol{X}^R, \boldsymbol{Y}^I\right) = \left\{K_{1i}\left(\boldsymbol{X}^R, \boldsymbol{Y}^I\right), K_{2i}\left(\boldsymbol{X}^R, \boldsymbol{Y}^I\right), K_{3i}\left(\boldsymbol{X}^R, \boldsymbol{Y}^I\right)\right\}^T \tag{5.104}$$

where $K_{1i}\left(\boldsymbol{X}^R, \boldsymbol{Y}^I\right)$, $K_{2i}\left(\boldsymbol{X}^R, \boldsymbol{Y}^I\right)$, and $K_{3i}\left(\boldsymbol{X}^R, \boldsymbol{Y}^I\right)$ denote the membership expressions with both random vector $\boldsymbol{X}^R$ and interval vector $\boldsymbol{Y}^I$, respectively.

5.5.2 The LAR Analysis of the DACS with Small Uncertainty

To address the hybrid LAR field problem of the DACS with small uncertainties, this section proposes a novel approach called the hybrid perturbation compound function-based moment method, which combines the random interval perturbation compound function-based method with the random interval compound functional moment method. First, the random interval compound functional moment method is extended to LAR analysis. Subsequently, the expectations and variances of the bounds for the hybrid LAR are computed using this method.

5.5.2.1 Random Interval Perturbation Compound Function-based Method

For the hybrid random and interval functional vectors, a random interval perturbation compound function-based method is proposed, building upon the RIPM in [34] and the compound function technique.

For simplicity, all random and interval parameters are assumed to be mutually independent. The Taylor series expansion of $\boldsymbol{f}\left(\boldsymbol{g}_i\left(\boldsymbol{X}^R, \boldsymbol{Y}^I\right)\right)$, retaining terms up to the second order, can be expressed as follows:

$$\begin{aligned}
\boldsymbol{f}\left(\boldsymbol{g}_i\left(\boldsymbol{X}^R, \boldsymbol{Y}^I\right)\right) &= \boldsymbol{f}\left(\boldsymbol{g}_i\left(\boldsymbol{X}^e, \boldsymbol{Y}^c\right)\right) \\
&+ \sum_{i=1}^{N}\sum_{s=1}^{n}\left\{\left.\frac{\partial \boldsymbol{f}\left(\boldsymbol{g}_i\left(\boldsymbol{X}^R, \boldsymbol{Y}^I\right)\right)}{\partial \boldsymbol{g}_i\left(\boldsymbol{X}^R, \boldsymbol{Y}^I\right)}\cdot\frac{\partial \boldsymbol{g}_i\left(\boldsymbol{X}^R, \boldsymbol{Y}^I\right)}{\partial y_s}\right|_{\boldsymbol{X}^e, \boldsymbol{Y}^c}\right\}\Delta y_s + R \\
&+ \sum_{i=1}^{N}\sum_{r=1}^{m}\left(\left.\frac{\partial \boldsymbol{f}\left(\boldsymbol{g}_i\left(\boldsymbol{X}^R, \boldsymbol{Y}^I\right)\right)}{\partial \boldsymbol{g}_i\left(\boldsymbol{X}^R, \boldsymbol{Y}^I\right)}\cdot\frac{\partial \boldsymbol{g}_i\left(\boldsymbol{X}^R, \boldsymbol{Y}^I\right)}{\partial x_r}\right|_{\boldsymbol{X}^e, \boldsymbol{Y}^c}\right. \\
&+ \left.\sum_{s=1}^{n}\left\{\left.\frac{\partial \boldsymbol{f}\left(\boldsymbol{g}_i\left(\boldsymbol{X}^R, \boldsymbol{Y}^I\right)\right)}{\partial \boldsymbol{g}_i\left(\boldsymbol{X}^R, \boldsymbol{Y}^I\right)}\cdot\frac{\partial^2 \boldsymbol{g}_i\left(\boldsymbol{X}^R, \boldsymbol{Y}^I\right)}{\partial x_r \partial y_s}\right|_{\boldsymbol{X}^e, \boldsymbol{Y}^c}\right\}\Delta y_s\right\}\cdot\left(x_r - x_r^e\right)
\end{aligned} \tag{5.105}$$

where R is the remainder term; x_r^e denotes the expectation of the random parameter x_r; Δy_s denotes the IR of the interval parameter y_s.

5.5.2.2 Random Interval Compound Functional Moment Method

Referring to Eq. (5.105), and based on the RIPM in [34] along with the compound function technique, a random interval compound functional moment method is proposed. By neglecting higher-order terms, the expectation $E\left(\boldsymbol{f}\left(\boldsymbol{g}_i\left(\boldsymbol{X}^R,\boldsymbol{Y}^I\right)\right)\right)$ and variance $\mathrm{Var}\left(\boldsymbol{f}\left(\boldsymbol{g}_i\left(\boldsymbol{X}^R,\boldsymbol{Y}^I\right)\right)\right)$ of $\boldsymbol{f}\left(\boldsymbol{g}_i\left(\boldsymbol{X}^R,\boldsymbol{Y}^I\right)\right)$, retaining only the first- and second-order terms, can be expressed as follows:

$$E\left(\boldsymbol{f}\left(\boldsymbol{g}_i\left(\boldsymbol{X}^R,\boldsymbol{Y}^I\right)\right)\right)=\boldsymbol{f}\left(\boldsymbol{g}_i\left(\boldsymbol{X}^e,\boldsymbol{Y}^c\right)\right)+\sum_{i=1}^{N}\sum_{s=1}^{n}\left\{\left.\frac{\partial \boldsymbol{f}\left(\boldsymbol{g}_i\left(\boldsymbol{X}^R,\boldsymbol{Y}^I\right)\right)}{\partial \boldsymbol{g}_i\left(\boldsymbol{X}^R,\boldsymbol{Y}^I\right)}\cdot\frac{\partial \boldsymbol{g}_i\left(\boldsymbol{X}^R,\boldsymbol{Y}^I\right)}{\partial y_s}\right|_{\boldsymbol{X}^e,\boldsymbol{Y}^c}\right\}\Delta y_s$$

$$\begin{aligned}&\mathrm{Var}\left(\boldsymbol{f}\left(\boldsymbol{g}_i\left(\boldsymbol{X}^R,\boldsymbol{Y}^I\right)\right)\right)\\&=\sum_{i=1}^{N}\sum_{r=1}^{m}\left\{\left.\frac{\partial \boldsymbol{f}\left(\boldsymbol{g}_i\left(\boldsymbol{X}^R,\boldsymbol{Y}^I\right)\right)}{\partial \boldsymbol{g}_i\left(\boldsymbol{X}^R,\boldsymbol{Y}^I\right)}\cdot\frac{\partial \boldsymbol{g}_i\left(\boldsymbol{X}^R,\boldsymbol{Y}^I\right)}{\partial x_r}\right|_{\boldsymbol{X}^e,\boldsymbol{Y}^c}\right.\\&\quad\left.+\sum_{s=1}^{n}\left\{\left.\frac{\partial \boldsymbol{f}\left(\boldsymbol{g}_i\left(\boldsymbol{X}^R,\boldsymbol{Y}^I\right)\right)}{\partial \boldsymbol{g}_i\left(\boldsymbol{X}^R,\boldsymbol{Y}^I\right)}\cdot\frac{\partial^2 \boldsymbol{g}_i\left(\boldsymbol{X}^R,\boldsymbol{Y}^I\right)}{\partial x_r\partial y_s}\right|_{\boldsymbol{X}^e,\boldsymbol{Y}^c}\right\}\Delta y_s\right\}\cdot\mathrm{Var}\left(x_r\right)\end{aligned} \tag{5.106}$$

where $\mathrm{Var}\left(x_r\right)$ denotes the variance of the random parameter x_r; m is the total number of random parameters; n is the total number of interval parameters y_s; N is the number of hybrid compound subfunction vector $\boldsymbol{g}_i\left(\boldsymbol{X}^R,\boldsymbol{Y}^I\right)$.

5.5.2.3 Hybrid LAR Analysis of the DACS

The Taylor series expansion of the first hybrid compound function vector $\boldsymbol{M}_i\left(\boldsymbol{K}_i\left(\boldsymbol{X}^R,\boldsymbol{Y}^I\right)\right)$, evaluated at the expectation of the random vector $\boldsymbol{X}^R$ and the midpoint of interval vector $\boldsymbol{Y}^I$, can be expressed by retaining only the first- and second-order terms,

$$\boldsymbol{M}_i\left(\boldsymbol{K}_i\left(\boldsymbol{X}^R,\boldsymbol{Y}^I\right)\right)=\boldsymbol{M}_i^e+\Delta\boldsymbol{M}_i^I \tag{5.107}$$

Similarly, based on Eqs. (5.103)–(5.105), the Taylor series expansion of the second hybrid compound function vector $\boldsymbol{N}_i\left(\boldsymbol{K}_i\left(\boldsymbol{X}^R,\boldsymbol{Y}^I\right)\right)$, evaluated at the expectation of the random vector $\boldsymbol{X}^R$ and the midpoint of interval vector $\boldsymbol{Y}^I$, can be expressed by retaining only the first- and second-order terms,

$$\boldsymbol{N}_i\left(\boldsymbol{K}_i\left(\boldsymbol{X}^R,\boldsymbol{Y}^I\right)\right)=\boldsymbol{N}_i^e+\Delta\boldsymbol{N}_i^I \tag{5.108}$$

Substituting Eqs. (5.107) and (5.108) into Eq. (5.101) yields

$$\gamma_i\left(\boldsymbol{X}^R,\boldsymbol{Y}^I\right)=\left(\boldsymbol{N}_i^e+\Delta\boldsymbol{N}_i^I\right)^{-1}\left(\boldsymbol{M}_i^e+\Delta\boldsymbol{M}_i^I\right),\ i=1,2 \tag{5.109}$$

The hybrid LAR vector $\gamma_i(X^R, Y^I)$ of the DACS can be approximately given as

$$\begin{aligned}
\gamma_i(X^R, Y^I) &= (N_i^e)^{-1} M_i^e + (N_i^e)^{-1} \sum_{t=1i}^{3i} \sum_{s=1}^{n} \left\{ \left. \frac{\partial M_i(K_i(X^R, Y^I))}{\partial K_t(X^R, Y^I)} \cdot \frac{\partial K_t(X^R, Y^I)}{\partial y_s} \right|_{X^R, Y^I = X^e, Y^c} \right\} \Delta y_s^I \\
&+ (N_i^e)^{-1} \sum_{t=1i}^{3i} \sum_{r=1}^{m} \left\{ \left. \frac{\partial M_i(K_i(X^R, Y^I))}{\partial K_t(X^R, Y^I)} \cdot \frac{\partial K_t(X^R, Y^I)}{\partial x_r} \right|_{X^R, Y^I = X^e, Y^c} \right\} \cdot (x_r - x_r^e) \\
&+ (N_i^e)^{-1} \sum_{t=1i}^{3i} \sum_{r=1}^{m} \left\{ \sum_{s=1}^{n} \left\{ \left. \frac{\partial^2 M_i(K_i(X^R, Y^I))}{\partial K_t(X^R, Y^I)} \cdot \frac{\partial K_t(X^R, Y^I)}{\partial x_r \cdot \partial y_s} \right|_{X^R, Y^I = X^e, Y^c} \right\} \cdot \Delta y_s^I \right\} \cdot (x_r - x_r^e) \\
&- (N_i^e)^{-2} M_i^e \sum_{t=2i}^{3i} \sum_{s=1}^{n} \left\{ \left. \frac{\partial N_i(K_i(X^R, Y^I))}{\partial K_t(X^R, Y^I)} \cdot \frac{\partial K_t(X^R, Y^I)}{\partial y_s} \right|_{X^R, Y^I = X^e, Y^c} \right\} \Delta y_s^I \\
&- (N_i^e)^{-2} M_i^e \sum_{t=2i}^{3i} \sum_{r=1}^{m} \left\{ \left. \frac{\partial N_i(K_i(X^R, Y^I))}{\partial K_t(X^R, Y^I)} \cdot \frac{\partial K_t(X^R, Y^I)}{\partial x_r} \right|_{X^R, Y^I = X^e, Y^c} \right\} \cdot (x_r - x_r^e) \\
&- (N_i^e)^{-2} M_i^e \sum_{t=2i}^{3i} \sum_{r=1}^{m} \left\{ \sum_{s=1}^{n} \left\{ \left. \frac{\partial^2 N_i(K_i(X^R, Y^I))}{\partial K_t(X^R, Y^I)} \cdot \frac{\partial K_t(X^R, Y^I)}{\partial x_r \cdot \partial y_s} \right|_{X^R, Y^I = X^e, Y^c} \right\} \cdot \Delta y_s^I \right\} \cdot (x_r - x_r^e)
\end{aligned} \tag{5.110}$$

5.5.3 Hybrid LAR Field Calculation of the DACS

Referring to Eq. (5.110), the hybrid LAR can be rewritten as

$$\gamma_i(X^R, Y^I) = \gamma_i^e + \Delta\gamma_i^I \tag{5.111}$$

$$\boldsymbol{\gamma}_i^e = \left(\boldsymbol{N}_i^e\right)^{-1} \boldsymbol{M}_i^e + \left(\boldsymbol{N}_i^e\right)^{-1} \sum_{t=1i}^{3i} \sum_{r=1}^{m} \left\{ \left. \begin{array}{c} \dfrac{\partial \boldsymbol{M}_i\left(\boldsymbol{K}_i\left(\boldsymbol{X}^R, \boldsymbol{Y}^I\right)\right)}{\partial K_t\left(\boldsymbol{X}^R, \boldsymbol{Y}^I\right)} \\ \cdot \dfrac{\partial K_t\left(\boldsymbol{X}^R, \boldsymbol{Y}^I\right)}{\partial x_r} \end{array} \right|_{\boldsymbol{X}^R, \boldsymbol{Y}^I = \boldsymbol{X}^e, \boldsymbol{Y}^c} \right\} \cdot \left(x_r - x_r^e\right)$$

$$- \left(\boldsymbol{N}_i^e\right)^{-2} \boldsymbol{M}_i^e \sum_{t=2i}^{3i} \sum_{r=1}^{m} \left\{ \left. \begin{array}{c} \dfrac{\partial \boldsymbol{N}_i\left(\boldsymbol{K}_i\left(\boldsymbol{X}^R, \boldsymbol{Y}^I\right)\right)}{\partial K_t\left(\boldsymbol{X}^R, \boldsymbol{Y}^I\right)} \\ \cdot \dfrac{\partial K_t\left(\boldsymbol{X}^R, \boldsymbol{Y}^I\right)}{\partial x_r} \end{array} \right|_{\boldsymbol{X}^R, \boldsymbol{Y}^I = \boldsymbol{X}^e, \boldsymbol{Y}^c} \right\} \cdot \left(x_r - x_r^e\right) \tag{5.112}$$

$$\Delta \boldsymbol{\gamma}_i^I = \left(\boldsymbol{N}_i^e\right)^{-1} \left[\sum_{t=1i}^{3i} \sum_{s=1}^{n} \left\{ \left. \begin{array}{c} \dfrac{\partial \boldsymbol{M}_i\left(\boldsymbol{K}_i\left(\boldsymbol{X}^R, \boldsymbol{Y}^I\right)\right)}{\partial K_t\left(\boldsymbol{X}^R, \boldsymbol{Y}^I\right)} \\ \cdot \dfrac{\partial K_t\left(\boldsymbol{X}^R, \boldsymbol{Y}^I\right)}{\partial y_s} \end{array} \right|_{\boldsymbol{X}^R, \boldsymbol{Y}^I = \boldsymbol{X}^e, \boldsymbol{Y}^c} \right\} \right.$$

$$\left. + \sum_{t=1i}^{3i} \sum_{r=1}^{m} \sum_{s=1}^{n} \left\{ \left. \begin{array}{c} \dfrac{\partial^2 \boldsymbol{M}_i\left(\boldsymbol{K}_i\left(\boldsymbol{X}^R, \boldsymbol{Y}^I\right)\right)}{\partial K_t\left(\boldsymbol{X}^R, \boldsymbol{Y}^I\right)} \\ \cdot \dfrac{\partial K_t\left(\boldsymbol{X}^R, \boldsymbol{Y}^I\right)}{\partial x_r \cdot \partial y_s} \end{array} \right|_{\boldsymbol{X}^R, \boldsymbol{Y}^I = \boldsymbol{X}^e, \boldsymbol{Y}^c} \right\} \cdot \left(x_r - x_r^e\right) \right]$$

$$\cdot \Delta y_s \cdot \delta_s$$

$$- \left(\boldsymbol{N}_i^e\right)^{-2} \boldsymbol{M}_i^e \left[\sum_{t=2i}^{3i} \sum_{s=1}^{n} \left\{ \left. \begin{array}{c} \dfrac{\partial \boldsymbol{N}_i\left(\boldsymbol{K}_i\left(\boldsymbol{X}^R, \boldsymbol{Y}^I\right)\right)}{\partial K_t\left(\boldsymbol{X}^R, \boldsymbol{Y}^I\right)} \\ \cdot \dfrac{\partial K_t\left(\boldsymbol{X}^R, \boldsymbol{Y}^I\right)}{\partial y_s} \end{array} \right|_{\boldsymbol{X}^R, \boldsymbol{Y}^I = \boldsymbol{X}^e, \boldsymbol{Y}^c} \right\} \right. \tag{5.113}$$

$$\left. + \sum_{t=2i}^{3i} \sum_{r=1}^{m} \sum_{s=1}^{n} \left\{ \left. \begin{array}{c} \dfrac{\partial^2 \boldsymbol{N}_i\left(\boldsymbol{K}_i\left(\boldsymbol{X}^R, \boldsymbol{Y}^I\right)\right)}{\partial K_t\left(\boldsymbol{X}^R, \boldsymbol{Y}^I\right)} \\ \cdot \dfrac{\partial K_t\left(\boldsymbol{X}^R, \boldsymbol{Y}^I\right)}{\partial x_r \cdot \partial y_s} \end{array} \right|_{\boldsymbol{X}^R, \boldsymbol{Y}^I = \boldsymbol{X}^e, \boldsymbol{Y}^c} \right\} \cdot \left(x_r - x_r^e\right) \right]$$

$$\cdot \Delta y_s \cdot \delta_s$$

The maximum value of the hybrid LAR vector $\boldsymbol{\gamma}_i\left(\boldsymbol{X}^R, \boldsymbol{Y}^I\right)$ with respect to the interval vector $\boldsymbol{Y}^I$ can be expressed as

$$\max \boldsymbol{\gamma}_i\left(\boldsymbol{X}^R, \boldsymbol{Y}^I\right) = \boldsymbol{\gamma}_i^e + \max_{\delta_s = +1} \Delta \boldsymbol{\gamma}_i^I \tag{5.114}$$

Similarly, the minimum value of the hybrid LAR vector $\boldsymbol{\gamma}_i\left(\mathbf{X}^R, \mathbf{Y}^I\right)$ with respect to the interval vector $\mathbf{Y}^I$ can be expressed as

$$\min \boldsymbol{\gamma}_i\left(\mathbf{X}^R, \mathbf{Y}^I\right) = \boldsymbol{\gamma}_i^e + \min_{\delta_s=-1} \Delta\boldsymbol{\gamma}_i^I \tag{5.115}$$

From Eqs. (5.114) and (5.115), it is evident that the maximum and minimum values of the hybrid LAR vector $\boldsymbol{\gamma}_i\left(\mathbf{X}^R, \mathbf{Y}^I\right)$ are both random functions that depend solely on the random vector $\mathbf{X}^R$.

Referring to Eqs. (5.106) and (5.114), the expectation and variance of $\max \boldsymbol{\gamma}_i\left(\mathbf{X}^R, \mathbf{Y}^I\right)$ can be expressed as

$$\begin{aligned} E\left(\max \boldsymbol{\gamma}_i\left(\mathbf{X}^R, \mathbf{Y}^I\right)\right) &= \left(\mathbf{N}_i^e\right)^{-1} \mathbf{M}_i^e \\ &+ \left| \left(\mathbf{N}_i^e\right)^{-1} \sum_{t=1i}^{3i} \sum_{s=1}^{n} \left\{ \left. \frac{\partial \mathbf{M}_i\left(\mathbf{K}_i\left(\mathbf{X}^R, \mathbf{Y}^I\right)\right)}{\partial K_t\left(\mathbf{X}^R, \mathbf{Y}^I\right)} \cdot \frac{\partial K_t\left(\mathbf{X}^R, \mathbf{Y}^I\right)}{\partial y_s} \right|_{\mathbf{X}^R, \mathbf{Y}^I = \mathbf{X}^e, \mathbf{Y}^c} \right\} \right. \\ &\left. - \left(\mathbf{N}_i^e\right)^{-2} \mathbf{M}_i^e \sum_{t=2i}^{3i} \sum_{s=1}^{n} \left\{ \left. \frac{\partial \mathbf{N}_i\left(\mathbf{K}_i\left(\mathbf{X}^R, \mathbf{Y}^I\right)\right)}{\partial K_t\left(\mathbf{X}^R, \mathbf{Y}^I\right)} \cdot \frac{\partial K_t\left(\mathbf{X}^R, \mathbf{Y}^I\right)}{\partial y_s} \right|_{\mathbf{X}^R, \mathbf{Y}^I = \mathbf{X}^e, \mathbf{Y}^c} \right\} \right| \cdot \Delta y_s \end{aligned} \tag{5.116}$$

$$\begin{aligned} \mathrm{Var}\left(\max \boldsymbol{\gamma}_i\left(\mathbf{X}^R, \mathbf{Y}^I\right)\right) &= \left(\left(\mathbf{N}_i^e\right)^{-1} \sum_{t=1i}^{3i} \sum_{r=1}^{m} \left\{ \left. \frac{\partial \mathbf{M}_i\left(\mathbf{K}_i\left(\mathbf{X}^R, \mathbf{Y}^I\right)\right)}{\partial K_t\left(\mathbf{X}^R, \mathbf{Y}^I\right)} \cdot \frac{\partial K_t\left(\mathbf{X}^R, \mathbf{Y}^I\right)}{\partial x_r} \right|_{\mathbf{X}^R, \mathbf{Y}^I = \mathbf{X}^e, \mathbf{Y}^c} \right\} \right. \\ &\left. - \left(\mathbf{N}_i^e\right)^{-2} \mathbf{M}_i^e \sum_{t=2i}^{3i} \sum_{r=1}^{m} \left\{ \left. \frac{\partial \mathbf{N}_i\left(\mathbf{K}_i\left(\mathbf{X}^R, \mathbf{Y}^I\right)\right)}{\partial K_t\left(\mathbf{X}^R, \mathbf{Y}^I\right)} \cdot \frac{\partial K_t\left(\mathbf{X}^R, \mathbf{Y}^I\right)}{\partial x_r} \right|_{\mathbf{X}^R, \mathbf{Y}^I = \mathbf{X}^e, \mathbf{Y}^c} \right\} \right)^2 \\ &\cdot \mathrm{Var}\left(x_r\right) \\ &+ \left(\left| \left(\mathbf{N}_i^e\right)^{-1} \sum_{t=1i}^{3i} \sum_{r=1}^{m} \sum_{s=1}^{n} \left\{ \left. \frac{\partial^2 \mathbf{M}_i\left(\mathbf{K}_i\left(\mathbf{X}^R, \mathbf{Y}^I\right)\right)}{\partial K_t\left(\mathbf{X}^R, \mathbf{Y}^I\right)} \cdot \frac{\partial K_t\left(\mathbf{X}^R, \mathbf{Y}^I\right)}{\partial x_r \cdot \partial y_s} \right|_{\mathbf{X}^R, \mathbf{Y}^I = \mathbf{X}^e, \mathbf{Y}^c} \right\} \right. \right. \end{aligned}$$

$$\left. - \left(\boldsymbol{N}_i^e\right)^{-2} \boldsymbol{M}_i^e \sum_{t=2i}^{3i} \sum_{r=1}^{m} \sum_{s=1}^{n} \left\{ \begin{array}{l} \dfrac{\partial^2 \boldsymbol{N}_i\left(\boldsymbol{K}_i\left(\boldsymbol{X}^R, \boldsymbol{Y}^I\right)\right)}{\partial K_t\left(\boldsymbol{X}^R, \boldsymbol{Y}^I\right)} \\ \cdot \dfrac{\partial K_t\left(\boldsymbol{X}^R, \boldsymbol{Y}^I\right)}{\partial x_r \cdot \partial y_s} \end{array} \right|_{\boldsymbol{X}^R, \boldsymbol{Y}^I = \boldsymbol{X}^e, \boldsymbol{Y}^c} \right\} \Bigg| \cdot \Delta y_s \Bigg)^2$$

$$\cdot \operatorname{Var}\left(x_r\right)$$

$$+ 2 \cdot \left(\begin{array}{l} \left(\boldsymbol{N}_i^e\right)^{-1} \displaystyle\sum_{t=1i}^{3i} \sum_{r=1}^{m} \left\{ \begin{array}{l} \dfrac{\partial \boldsymbol{M}_i\left(\boldsymbol{K}_i\left(\boldsymbol{X}^R, \boldsymbol{Y}^I\right)\right)}{\partial K_t\left(\boldsymbol{X}^R, \boldsymbol{Y}^I\right)} \\ \cdot \dfrac{\partial K_t\left(\boldsymbol{X}^R, \boldsymbol{Y}^I\right)}{\partial x_r} \end{array} \right|_{\boldsymbol{X}^R, \boldsymbol{Y}^I = \boldsymbol{X}^e, \boldsymbol{Y}^c} \right\} \\ - \left(\boldsymbol{N}_i^e\right)^{-2} \boldsymbol{M}_i^e \displaystyle\sum_{t=2i}^{3i} \sum_{r=1}^{m} \left\{ \begin{array}{l} \dfrac{\partial \boldsymbol{N}_i\left(\boldsymbol{K}_i\left(\boldsymbol{X}^R, \boldsymbol{Y}^I\right)\right)}{\partial K_t\left(\boldsymbol{X}^R, \boldsymbol{Y}^I\right)} \\ \cdot \dfrac{\partial K_t\left(\boldsymbol{X}^R, \boldsymbol{Y}^I\right)}{\partial x_r} \end{array} \right|_{\boldsymbol{X}^R, \boldsymbol{Y}^I = \boldsymbol{X}^e, \boldsymbol{Y}^c} \right\} \end{array} \right) \tag{5.117}$$

$$\cdot \left(\left| \begin{array}{l} \left(\boldsymbol{N}_i^e\right)^{-1} \displaystyle\sum_{t=1i}^{3i} \sum_{r=1}^{m} \sum_{s=1}^{n} \left\{ \begin{array}{l} \dfrac{\partial^2 \boldsymbol{M}_i\left(\boldsymbol{K}_i\left(\boldsymbol{X}^R, \boldsymbol{Y}^I\right)\right)}{\partial K_t\left(\boldsymbol{X}^R, \boldsymbol{Y}^I\right)} \\ \cdot \dfrac{\partial K_t\left(\boldsymbol{X}^R, \boldsymbol{Y}^I\right)}{\partial x_r \cdot \partial y_s} \end{array} \right|_{\boldsymbol{X}^R, \boldsymbol{Y}^I = \boldsymbol{X}^e, \boldsymbol{Y}^c} \right\} \\ - \left(\boldsymbol{N}_i^e\right)^{-2} \boldsymbol{M}_i^e \displaystyle\sum_{t=2i}^{3i} \sum_{r=1}^{m} \sum_{s=1}^{n} \left\{ \begin{array}{l} \dfrac{\partial^2 \boldsymbol{N}_i\left(\boldsymbol{K}_i\left(\boldsymbol{X}^R, \boldsymbol{Y}^I\right)\right)}{\partial K_t\left(\boldsymbol{X}^R, \boldsymbol{Y}^I\right)} \\ \cdot \dfrac{\partial K_t\left(\boldsymbol{X}^R, \boldsymbol{Y}^I\right)}{\partial x_r \cdot \partial y_s} \end{array} \right|_{\boldsymbol{X}^R, \boldsymbol{Y}^I = \boldsymbol{X}^e, \boldsymbol{Y}^c} \right\} \end{array} \right| \cdot \Delta y_s \right)$$

$$\cdot \operatorname{Var}\left(x_r\right)$$

Referring to Eqs. (5.106) and (5.115), the expectation and variance of $\min \gamma_i\left(\boldsymbol{X}^R, \boldsymbol{Y}^I\right)$ can be expressed as

$$E\left(\min \gamma_i\left(\boldsymbol{X}^R, \boldsymbol{Y}^I\right)\right) = \left(\boldsymbol{N}_i^e\right)^{-1} \boldsymbol{M}_i^e$$

$$- \left| \left(\boldsymbol{N}_i^e\right)^{-1} \sum_{t=1i}^{3i} \sum_{s=1}^{n} \left\{ \begin{array}{l} \dfrac{\partial \boldsymbol{M}_i\left(\boldsymbol{K}_i\left(\boldsymbol{X}^R, \boldsymbol{Y}^I\right)\right)}{\partial K_t\left(\boldsymbol{X}^R, \boldsymbol{Y}^I\right)} \\ \cdot \dfrac{\partial K_t\left(\boldsymbol{X}^R, \boldsymbol{Y}^I\right)}{\partial y_s} \end{array} \right|_{\boldsymbol{X}^R, \boldsymbol{Y}^I = \boldsymbol{X}^e, \boldsymbol{Y}^c} \right\}$$

$$
-\left(N_i^e\right)^{-2} M_i^e \sum_{t=2i}^{3i} \sum_{s=1}^{n} \left\{ \left. \frac{\partial N_i\left(K_i\left(X^R, Y^I\right)\right)}{\partial K_t\left(X^R, Y^I\right)} \cdot \frac{\partial K_t\left(X^R, Y^I\right)}{\partial y_s} \right|_{X^R, Y^I = X^e, Y^c} \right\} \Bigg| \cdot \Delta y_s \qquad (5.118)
$$

$$
\begin{aligned}
\mathrm{Var}\left(\min \gamma_i\left(X^R, Y^I\right)\right) = & \left(\left(N_i^e\right)^{-1} \sum_{t=1i}^{3i} \sum_{r=1}^{m} \left\{ \left. \frac{\partial M_i\left(K_i\left(X^R, Y^I\right)\right)}{\partial K_t\left(X^R, Y^I\right)} \cdot \frac{\partial K_t\left(X^R, Y^I\right)}{\partial x_r} \right|_{X^R, Y^I = X^e, Y^c} \right\} \right. \\
& \left. - \left(N_i^e\right)^{-2} M_i^e \sum_{t=2i}^{3i} \sum_{r=1}^{m} \left\{ \left. \frac{\partial N_i\left(K_i\left(X^R, Y^I\right)\right)}{\partial K_t\left(X^R, Y^I\right)} \cdot \frac{\partial K_t\left(X^R, Y^I\right)}{\partial x_r} \right|_{X^R, Y^I = X^e, Y^c} \right\} \right)^2 \\
& \cdot \mathrm{Var}\left(x_r\right) \\
& + \left(\left\| \left(N_i^e\right)^{-1} \sum_{t=1i}^{3i} \sum_{r=1}^{m} \sum_{s=1}^{n} \left\{ \left. \frac{\partial^2 M_i\left(K_i\left(X^R, Y^I\right)\right)}{\partial K_t\left(X^R, Y^I\right)} \cdot \frac{\partial K_t\left(X^R, Y^I\right)}{\partial x_r \cdot \partial y_s} \right|_{X^R, Y^I = X^e, Y^c} \right\} \right. \right. \\
& \left. \left. - \left(N_i^e\right)^{-2} M_i^e \sum_{t=2i}^{3i} \sum_{r=1}^{m} \sum_{s=1}^{n} \left\{ \left. \frac{\partial^2 N_i\left(K_i\left(X^R, Y^I\right)\right)}{\partial K_t\left(X^R, Y^I\right)} \cdot \frac{\partial K_t\left(X^R, Y^I\right)}{\partial x_r \cdot \partial y_s} \right|_{X^R, Y^I = X^e, Y^c} \right\} \right\| \cdot \Delta y_s \right)^2 \\
& \cdot \mathrm{Var}\left(x_r\right) \\
& -2 \cdot \left(\left(N_i^e\right)^{-1} \sum_{t=1i}^{3i} \sum_{r=1}^{m} \left\{ \left. \frac{\partial M_i\left(K_i\left(X^R, Y^I\right)\right)}{\partial K_t\left(X^R, Y^I\right)} \cdot \frac{\partial K_t\left(X^R, Y^I\right)}{\partial x_r} \right|_{X^R, Y^I = X^e, Y^c} \right\} \right. \\
& \left. - \left(N_i^e\right)^{-2} M_i^e \sum_{t=2i}^{3i} \sum_{r=1}^{m} \left\{ \left. \frac{\partial N_i\left(K_i\left(X^R, Y^I\right)\right)}{\partial K_t\left(X^R, Y^I\right)} \cdot \frac{\partial K_t\left(X^R, Y^I\right)}{\partial x_r} \right|_{X^R, Y^I = X^e, Y^c} \right\} \right)
\end{aligned}
$$

$$
\cdot\left(\left|(\boldsymbol{N}_i^e)^{-1}\sum_{t=1i}^{3i}\sum_{r=1}^{m}\sum_{s=1}^{n}\left\{\begin{array}{l}\dfrac{\partial^2\boldsymbol{M}_i(\boldsymbol{K}_i(\boldsymbol{X}^R,\boldsymbol{Y}^I))}{\partial K_t(\boldsymbol{X}^R,\boldsymbol{Y}^I)}\\ \cdot\dfrac{\partial K_t(\boldsymbol{X}^R,\boldsymbol{Y}^I)}{\partial x_r\cdot\partial y_s}\end{array}\right|_{\boldsymbol{X}^R,\boldsymbol{Y}^I=\boldsymbol{X}^e,\boldsymbol{Y}^c}\right\} - (\boldsymbol{N}_i^e)^2\,\boldsymbol{M}_i^e\cdot\sum_{t=2i}^{3i}\sum_{r=1}^{m}\sum_{s=1}^{n}\left\{\begin{array}{l}\dfrac{\partial^2\boldsymbol{N}_i(\boldsymbol{K}_i(\boldsymbol{X}^R,\boldsymbol{Y}^I))}{\partial K_t(\boldsymbol{X}^R,\boldsymbol{Y}^I)}\\ \cdot\dfrac{\partial K_t(\boldsymbol{X}^R,\boldsymbol{Y}^I)}{\partial x_r\cdot\partial y_s}\end{array}\right|_{\boldsymbol{X}^R,\boldsymbol{Y}^I=\boldsymbol{X}^e,\boldsymbol{Y}^c}\right\}\cdot\Delta y_s\right|\right)\cdot\mathrm{Var}(x_r) \tag{5.119}
$$

5.5.4 Numerical Examples

This section presents four numerical examples to showcase the feasibility and effectiveness of the proposed method.

The payload parameters (y, z) are treated as narrowly distributed random variables following a Gaussian distribution, while the structural parameters (D, d) are considered as narrowly bounded interval parameters. All these uncertain parameters are assumed to be mutually independent. The properties of both deterministic and uncertain parameters are summarized in Table 5.13.

Table 5.13 Deterministic and uncertain parameters of the DACS.

System Parameters	Symbol	Description	Constant or Midpoint Values
Deterministic parameters	m_1	Mass of lifting arm A_1B_1	5000 (kg)
	m_2	Mass of lifting arm A_2B_2	5000 (kg)
	m_p	Mass of payload C_1C_2	50,000 (kg)
	S_1	Length of hoisting rope B_1C_1	10 (m)
	S_2	Length of hoisting rope B_2C_2	10 (m)
	L_1	Length of lifting arm A_1B_1	50 (m)
	L_2	Length of lifting arm A_2B_2	45 (m)
	θ	Orientation coordinate of centroid O_p	$\theta = \pi/6\,(°)$
Random parameters	y	Y-axis coordinate of centroid O_p	$\mu_y = 5\,(\mathrm{m})$
	z	Z-axis coordinate of centroid O_p	$\mu_z = 5\,(\mathrm{m})$
Interval parameters	D	Length of A_1A_2	$D^c = 12.75\,(\mathrm{m})$
	d	Length of payload C_1C_2	$d^c = 2.5\,(\mathrm{m})$

To examine the differences in expectation and variance arising from the hybrid random and interval parameters of the DACS, as well as the influence of various uncertain parameters on the LAR field, a HUPM is established,

$$\mathrm{DD1} = V_y = V_z = \Delta D_F = \Delta d_F \tag{5.120}$$

where DD1 denotes the dispersal degree of uncertain variables.

Simulations are performed using MATLAB to compute the LAR field, with DD1 assumed to vary from 0 to 0.01. The LAR field of crane 1, calculated using the proposed algorithm under different HUPMs, is presented in Figures 5.3 and 5.7, respectively.

Figures 5.30–5.34 indicate that the random parameter y influences the variance of the bounds but does not affect their expectation. Figures 5.31 and 5.35 reveal that the random parameter z has a notable effect on the variance of the bounds while leaving the expectation unchanged. Figures 5.32 and 5.36 demonstrate that the interval parameter D significantly impacts the expectation of the bounds but has little effect on their variance. Figures 5.33 and 5.37 show that the interval parameter d slightly affects the expectation of the bounds without influencing the variance.

To demonstrate the effectiveness of the proposed algorithm and to extract the statistical characteristics of the LAR field for crane 1, one random parameter and one interval parameter are treated as uncertain variables, while all other random and interval parameters are fixed. Accordingly, the HUPM is defined as $\mathrm{DD3} = V_y\ (V_z) = \Delta D_F\ (\Delta d_F)$, with the dispersal degree of the uncertain variable DD3 assumed to vary from 0 to 0.01.

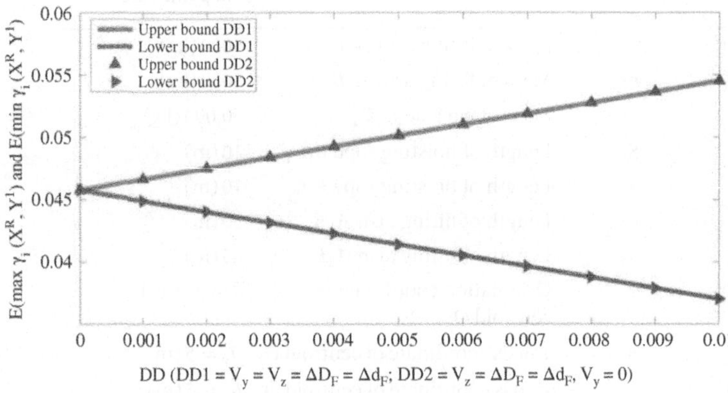

Figure 5.30 Expectation of the bounds for LAR calculated by the proposed algorithm with/without V_y.

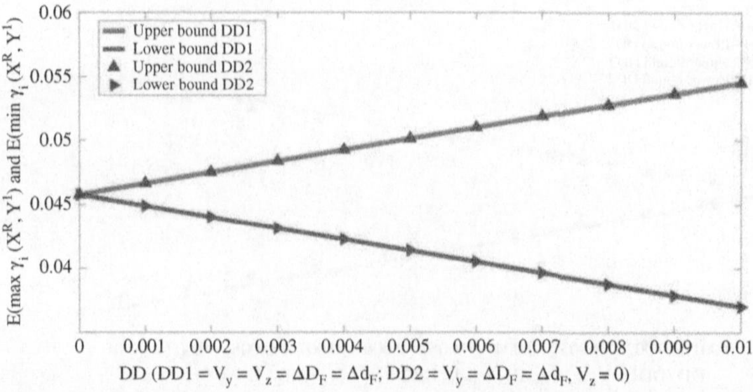

Figure 5.31 Expectation of the bounds for LAR calculated by the proposed algorithm with/without V_z.

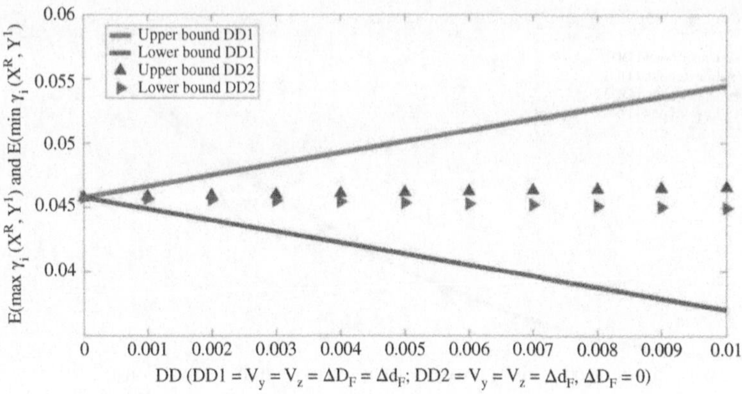

Figure 5.32 Expectation of the bounds for LAR calculated by the proposed algorithm with/without ΔD_F.

The HMCM with 10^8 samples is used as a reference solution, and the simulation is conducted in five steps, following the flowchart depicted in Figure 5.38.

To further assess the accuracy of the proposed algorithm, the relative errors in the expectations and variances of the bounds for the LAR of crane 1, compared with the HMCM results, are presented in Tables 5.14 and 5.15.

As indicated in Tables 5.14 and 5.15, the maximum relative errors remain below 10% when DD3 $\leq$ 0.004, suggesting that the proposed algorithm is suitable for predicting the LAR field of crane 1 under this condition. However, when DD3 increases from 0.004 to 0.01, the LAR field computed by the proposed algorithm deviates significantly from the HMCM results, indicating that even a small

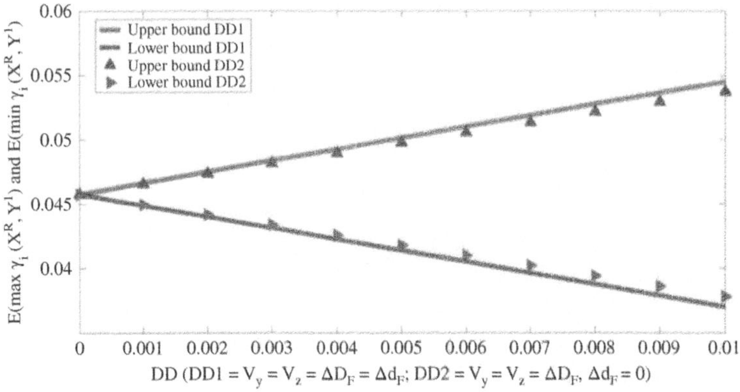

Figure 5.33 Expectation of the bounds for LAR calculated by the proposed algorithm with/without Δd_F.

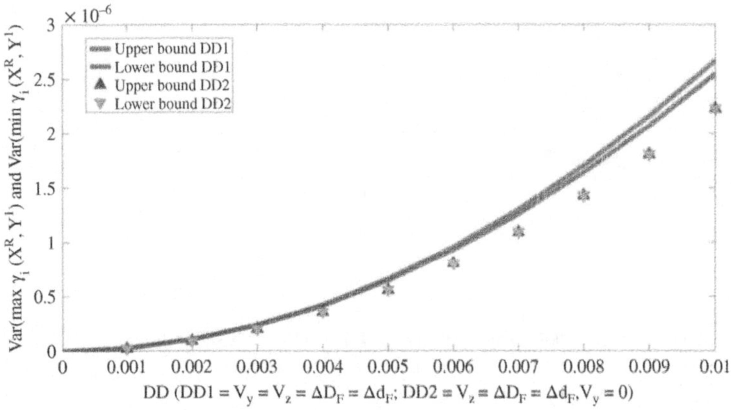

Figure 5.34 Variance of the bounds for LAR calculated by the proposed algorithm with/without V_y.

increase in the dispersal degree of the uncertain parameters can lead to a substantial deviation in the LAR field.

To deeply investigate the impact of stochastic properties in random parameters (y and z) on the LAR field of crane 1, two HUPMs are set as DD4 = V_y, V_z = $\Delta D_F = \Delta d_F = 0.005$, and DD5 = V_z, V_y = $\Delta D_F = \Delta d_F = 0.005$. The variances of the bounds for the LAR field calculated by the proposed algorithm are shown in Figures 5.39–5.41.

From Figure 5.39, it is observed that the randomness of y has a greater effect than that of z when DD ≤ 0.005. Conversely, for DD values exceeding 0.005, the influence of the randomness of z surpasses that of y. Figures 5.40 and 5.41 show

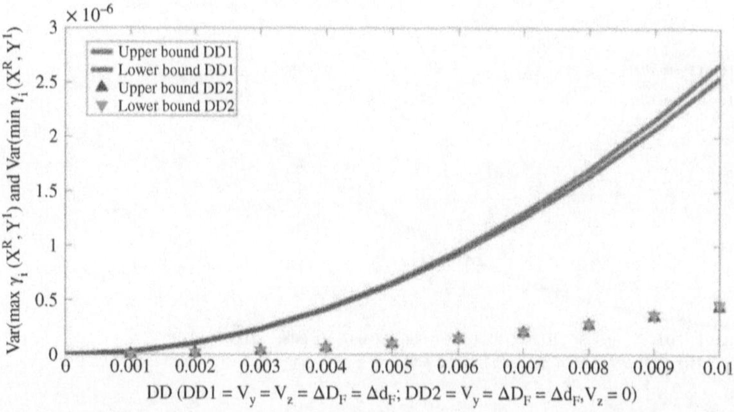

Figure 5.35 Variance of the bounds for LAR calculated by the proposed algorithm with/without V_z.

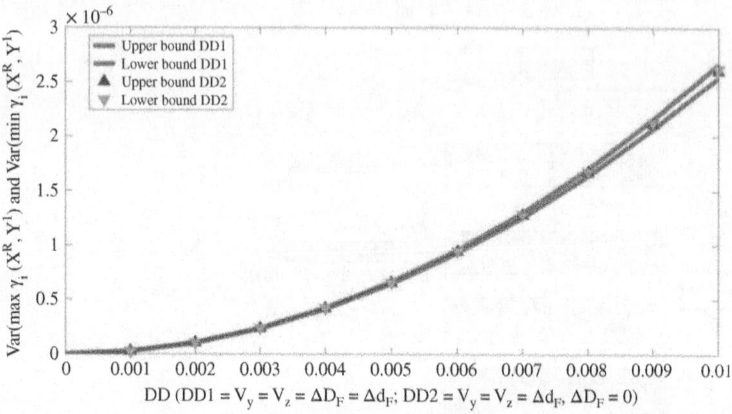

Figure 5.36 Variance of the bounds for LAR calculated by the proposed algorithm with/without ΔD_F.

that the interval width of the variances for the LAR increases with V_y, whereas it remains constant as V_z increases. These results indicate that the random parameter y has a stronger impact than z on the intervals of the variances.

5.5.5 Conclusion

Building on the random interval perturbation compound function-based method and the random interval compound function-based moment method, a hybrid approach—named the hybrid perturbation compound function-based moment

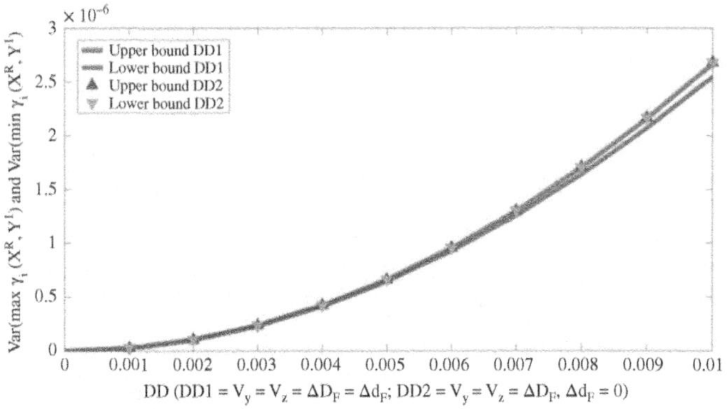

Figure 5.37 Variance of the bounds for LAR calculated by the proposed algorithm with/without Δd_F.

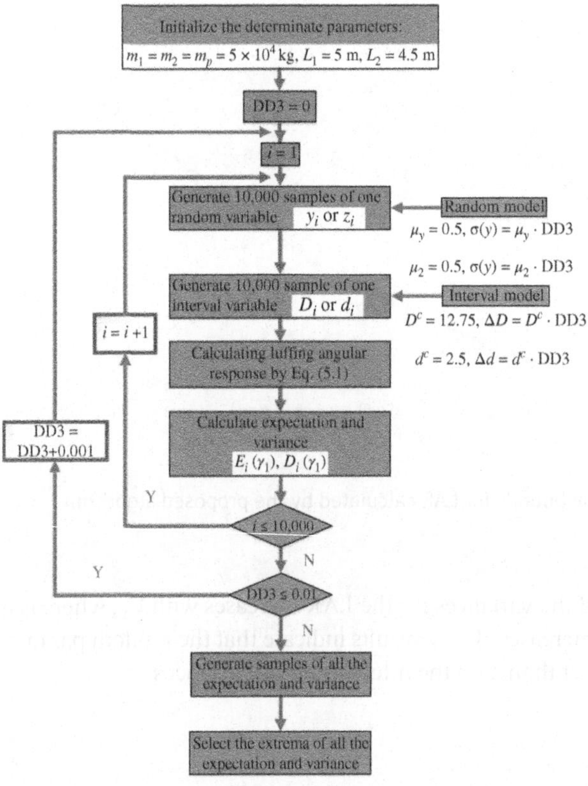

Figure 5.38 Flowchart of the HMCM for the LAR field analysis.

Table 5.14 Expectations of the bounds for the LAR of crane 1.

DD3 = V_y (V_z) = ΔD_F (Δd_F)	Bounds	The Proposed Algorithm		HMCM		Errors (%)
		Time (s)	Value	Time (s)	Value	
0	min $\underline{\gamma}_1$	0.0003	0.0458	1.6798	0.0458	0
	max $\underline{\gamma}_1$		0.0458		0.0458	0
	min $\overline{\gamma}_1$		0.0458		0.0458	0
	max $\overline{\gamma}_1$		0.0458		0.0458	0
0.005	min $\underline{\gamma}_1$	0.0009	0.0410	1.4689	0.0457	10.28
	max $\underline{\gamma}_1$		0.0453		0.0457	0.88
	min $\overline{\gamma}_1$		0.0462		0.0458	0.87
	max $\overline{\gamma}_1$		0.0505		0.0458	10.26
0.01	min $\underline{\gamma}_1$	0.0011	0.0378	1.3089	0.0457	17.29
	max $\underline{\gamma}_1$		0.0450		0.0457	1.53
	min $\overline{\gamma}_1$		0.0466		0.0458	1.75
	max $\overline{\gamma}_1$		0.0537		0.0458	17.25

method—is proposed to predict the LAR field of the DACS with hybrid uncertain parameters. According to the HUPM, two main types of uncertainties are considered, rendering the hybrid uncertain LAR equilibrium equation more representative. In the proposed algorithm, the LAR expression of the DACS is derived using the random interval perturbation compound function-based method combined with a first-order Neumann series expansion. Subsequently, by applying the random interval compound function-based moment method along with monotonic analysis techniques, the probabilistic distribution of the LAR field for the DACS is efficiently obtained.

Compared with the HMCM, the numerical results confirm the feasibility and efficiency of the proposed algorithm in addressing the LAR field problems of the DACS with hybrid uncertainties. By treating random variables as interval variables, the proposed algorithm provides a safer and less conservative prediction of the hybrid LAR response field than the IPM. Regarding the expectation of the bounds for the LAR, the primary influence arises from interval parameters, specifically the lengths of A_1A_2D and C_1C_2d. In contrast, the variance of the bounds is mainly affected by the random parameters, namely the Cartesian coordinates

Table 5.15 Variances of the bounds for the LAR of crane 1.

DD3 = V_y (V_z) = ΔD_F (Δd_F)	Bounds	The Proposed Algorithm		HMCM		Errors (%)
		Time (s)	Value	Time (s)	Value	
0	min $\underline{\gamma_1}$	0.0003	0	1.6798	0	0
	max $\underline{\gamma_1}$		0		0	0
	min $\overline{\gamma_1}$		0		0	0
	max $\overline{\gamma_1}$		0		0	0
0.005	min $\underline{\gamma_1}$	0.0009	1.2711E−7	1.4689	1.405E−7	8.95
	max $\underline{\gamma_1}$		1.3773E−7		1.598E−7	13.83
	min $\overline{\gamma_1}$		1.4220E−7		1.649E−7	13.80
	max $\overline{\gamma_1}$		1.5343E−7		1.834E−7	16.35
0.01	min $\underline{\gamma_1}$	0.0011	3.3022E−7	1.3089	4.241E−7	22.14
	max $\underline{\gamma_1}$		3.7850E−7		4.923E−7	23.13
	min $\overline{\gamma_1}$		3.9919E−7		5.329E−7	25.10
	max $\overline{\gamma_1}$		4.5211E−7		6.165E−7	26.67

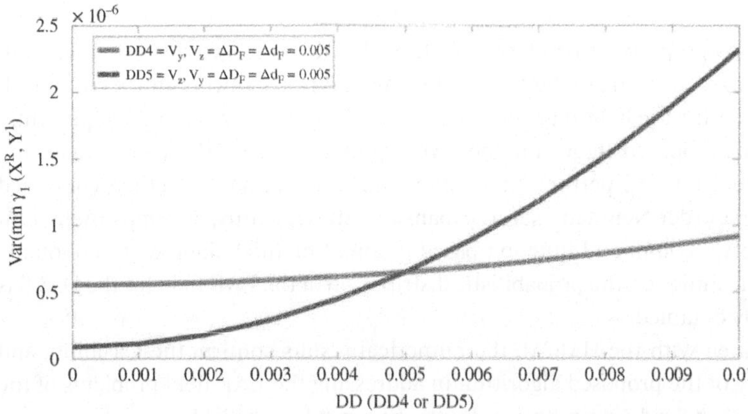

Figure 5.39 Variances of the LB for the LAR field calculated by the proposed algorithm.

y and z of the origin O_p. Additionally, the effect of the randomness of these parameters on the hybrid LAR field of the DACS is examined through numerical simulations, which reveal that the random parameters y and z exert different influences on the hybrid LAR field depending on the values of DD.

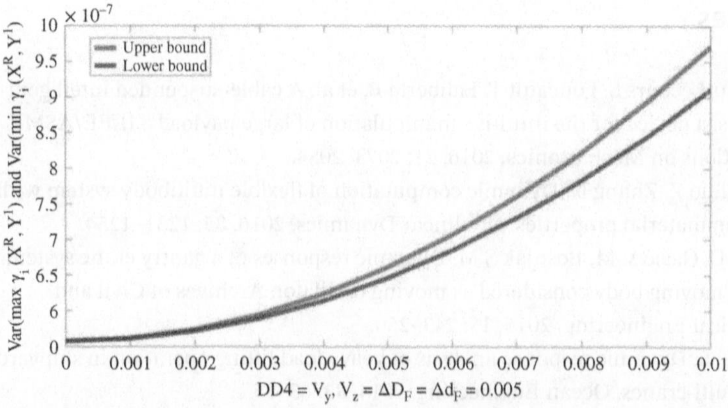

Figure 5.40 Variances of the bounds for the LAR field calculated by the proposed algorithm for DD4.

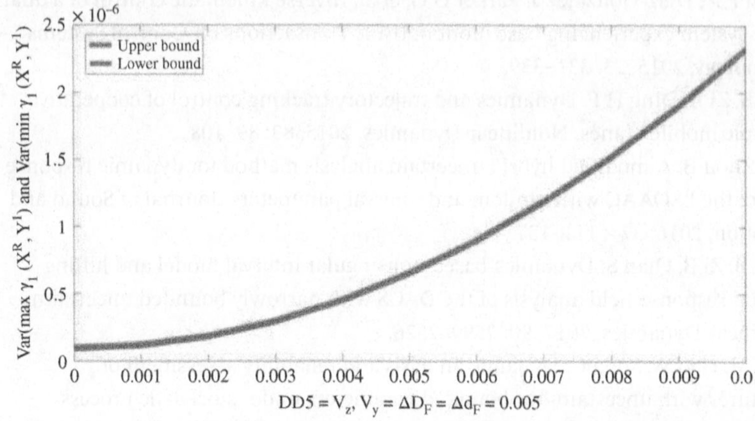

Figure 5.41 Variances of the bounds for the LAR field calculated by the proposed algorithm for DD5.

5.6 Conclusion

This chapter first presents the structural description and kinematic modeling of the dual overhead cranes, followed by the completion of kinematic uncertainty analysis with random parameters, then kinematic uncertainty analysis with interval variables, subsequently kinematic uncertainty analysis based on evidence theory, and finally kinematic uncertainty analysis with hybrid random and interval parameters, providing a reference for future kinematic analysis of rigid–flexible coupling hoisting robots with uncertainty.

References

1 Campeau-Lecours S. Foucault T. Laliberte B, et al. A cable-suspended intelligent crane assist device for the intuitive manipulation of large payloads. IEEE/ASME Transactions on Mechatronics, 2016, 21: 2073–2084.

2 Wu J L, Luo Z, Zhang N. Dynamic computation of flexible multibody system with uncertain material properties. Nonlinear Dynamics, 2016, 85: 1231–1254.

3 Zrnić N D, Gašić V M, Bošnjak S M. Dynamic responses of a gantry crane system due to a moving body considered as moving oscillator. Archives of Civil and Mechanical Engineering, 2015, 15: 243–250.

4 Ku N, Ha S. Dynamic response analysis of heavy load lifting operation in shipyard using multi-cranes. Ocean Engineering, 2014, 83: 63–75.

5 Zi B, Qian S, Ding H, et al. Design and analysis of cooperative cable parallel manipulators for multiple mobile cranes. International Journal of Advanced Robotic Systems, 2012, 9: 1–10.

6 Leban F A, Díaz-Gonzalez J, Parker G G, et al. Inverse kinematic control of a dual crane system experiencing base motion. IEEE Transactions on Control Systems Technology, 2015, 23: 331–339.

7 Qian S, Zi B, Ding H F. Dynamics and trajectory tracking control of cooperative multiple mobile cranes. Nonlinear Dynamics, 2015, 83: 89–108.

8 Zi B. Zhou B. A modified hybrid uncertain analysis method for dynamic response field of the LSOAAC with random and interval parameters. Journal of Sound and Vibration, 2016, 374: 111–137.

9 Zhou B, Zi B, Qian S. Dynamics-based nonsingular interval model and luffing angular response field analysis of the DACS with narrowly bounded uncertainty. Nonlinear Dynamics, 2017, 90: 2599–2626.

10 Do D M, Gao W, Song C. Dynamic analysis and reliability assessment of structures with uncertain-but-bounded parameters under stochastic process excitations. Reliability Engineering and System Safety, 2014, 132: 46–59.

11 Zhu H T, Guo S S. Periodic response of a duffing oscillator under combined harmonic and random excitations. ASME Journal of Vibration and Acoustics, 2015, 137: 041015.1–041015.10.

12 Tabbuso P, Spence S M J, Palizzolo L, et al. An efficient framework for the elasto-plastic reliability assessment of uncertain wind excited systems. Structural Safety, 2016, 58: 69–78.

13 Cui M Y, Liu W, Liu H Z, et al. Extended state observer-based adaptive sliding mode control of differential-driving mobile robot with uncertainties. Nonlinear Dynamics, 2016, 83: 667–683.

14 Sofi A, Romeo E, Barrera O, et al. An interval finite element method for the analysis of structures with spatially varying uncertainties. Advance Engineering Software, 2019, 128: 1–19.

15 Luo Y J, Zhan J J, Xing J, et al. Non-probabilistic uncertainty quantification and response analysis of structures with a bounded field model. Computer Methods in Applied Mechanics and Engineering, 2019, 347: 663–678.

16 Xia B Z, Yu D J. Modified interval and subinterval perturbation methods for the static response analysis of structures with interval parameters. Journal of Structural Engineering, 2014, 140(5): 04013113.

17 Wang C, Qiu Z P, Li Y L. Hybrid uncertainty propagation of coupled structural-acoustic system with large fuzzy and interval parameters. Applied Acoustics, 2016, 102: 62–70.

18 Vieira H L, Fontes J V C, Beck A T, et al. Reliable and failure-free workspaces for motion planning algorithms for parallel manipulators under geometrical uncertainties. Journal of Computational and Nonlinear Dynamics, 2019, 14(2): 021005.

19 Ren B, Lu Z Z, Zhou C C. A novel method for importance measure analysis in the presence of epistemic and aleatory uncertainties. Chinese Journal of Aeronautics, 2014, 27(3): 568–576.

20 Awadallah S K E, Milanovic J V. Quantification of aleatory and epistemic uncertainty in bulk power system reliability evaluation. 2013 IEEE Grenoble PowerTech Conference, 2013: 1–6.

21 Notash L. On the solutions of interval systems for under-constrained and redundant parallel manipulators. Journal of Mechanisms and Robotics, 2017, 9(6): 061009.

22 Yang J, Xu T H, Zhao F. Modified uncertainty measure of rough fuzzy sets from the perspective of fuzzy distance. Mathematical Problems in Engineering, 2018, 2018: 40160905.

23 Zhang Z, Jiang C, Han X, et al. A response surface approach for structural reliability analysis using evidence theory. Advances in Engineering Software, 2014, 69: 37–45.

24 Jiang C, Zhang Z, Han X, et al. A novel evidence-theory-based reliability analysis method for structures with epistemic uncertainty. Computers and Structures, 2013, 129: 1–12.

25 Gao R, Yin S, Xiong F. Response analysis and reliability-based design optimization of structural-acoustic system under evidence theory. Structural and Multidisciplinary Optimization, 2019, 59(3): 959–975.

26 Cao L X, Liu J, Jiang C, et al. Evidence-based structural uncertainty quantification by dimension reduction decomposition and marginal interval analysis. Journal of Mechanical Design, 2020, 142(5): 051701.

27 Yin S W, Yu D J, Yin H, et al. Hybrid evidence theory-based finite element/statistical energy analysis method for mid-frequency analysis of built-up systems with epistemic uncertainties. Mechanical Systems and Signal Processing, 2017, 93: 204–224.

28 Yang D L, Zhang F Y, Miao J J, et al. Dual-rotor misalignment fault quantitative identification based on DBN and improved D-S evidence theory. Mechanics and Industry, 2021, 22: 24.

29 Liang T, Guo H Y, Zhou X Y, et al. Structural damage detection based on cloud model and Dempster-Shafer evidence theory. Journal of Vibroengineering, 2018, 20(2): 909–922.

30 Jian T, Wu Z G, Chao Y. Epistemic uncertainty quantification in flutter analysis using evidence theory. Chinese Journal of Aeronautics, 2015, 28(1): 164–171.

31 Chen L, Diao L, Sang J. Weighted evidence combination rule based on evidence distance and uncertainty measure: an application in fault diagnosis. Mathematical Problems in Engineering, 2018, 2018: 5858272.

32 Liu D. Prediction of network security based on DS evidence theory. ETRI Journal, 2020, 42(5): 802–807.

33 Zhang D, Peng Z, Ning G, et al. Positioning accuracy reliability of industrial robots through probability and evidence theories. Journal of Mechanical Design, 2021, 143(1): 011704.

34 Gao W, Song C M, Tin-Loi F, Probabilistic interval analysis for structures with uncertainty. Structural Safety, 2010, 32: 191–199.

35 Xia B Z, Yu D J. Response probability analysis of random acoustic field based on perturbation stochastic method and change-of-variable technique. ASME Journal of Vibration and Acoustics, 2013, 135: 521–523.

36 Zhou B, Zi B, Li Y, et al. Hybrid compound function/subinterval perturbation method for kinematic analysis of a dual-crane system with large bounded uncertainty, Journal of Computational and Nonlinear Dynamics, 2021, 16(1): 014501.

37 Zhang Z, Jiang C, Wang G G, et al. First and second order approximate reliability analysis methods using evidence theory. Reliability Engineering & System Safety, 2015, 137: 40–49.

6

Dynamic Analysis of Rigid–Flexible Coupling Hoisting Robots with Uncertainty

6.1 Preamble

Over the past few years, the dual crane system (DCS), as a type of cable-driven parallel mechanism, has witnessed rapid development and found extensive applications in transportation, construction, and rescue fields [1–5]. The majority of existing studies have centered on the kinematics, statics, dynamics, and control of the DCS under the assumption of deterministic parameters. Nevertheless, in practical scenarios, uncertain factors such as mechanical errors, joint clearances, and stochastic excitations are inherently present. Without taking these uncertain parameters into account, it becomes challenging to accurately predict the reliable mechanical performances of the DCS [6, 7].

Drawing on the collected statistical data of uncertain parameters, two prevalent categories of uncertainty have been identified: aleatory uncertainty and epistemic uncertainty [8]. Random theory, as a widely used approach, has garnered significant attention for quantifying aleatory uncertainty [9]. However, when it comes to epistemic uncertainty, many random methods struggle to be effectively applied in engineering practices due to the lack of sufficient probabilistic information. In such circumstances, a range of nonprobabilistic methods [10] have been put forward to handle epistemic uncertainty efficiently, including the interval algorithm method [11–14], evidence theory [15–18], and the convex model method [19, 20].

Fuzzy theory represents another nonprobabilistic approach for modeling epistemic uncertainty [21]. In essence, under the level-cut strategy, fuzzy parameters can be regarded as a collection of interval variables accompanied by membership functions. Given this characteristic, modeling fuzzy parameters does not demand strict probabilistic distribution information. As a result, uncertainty methods based on fuzzy theory have been introduced in numerous engineering domains. The fuzzy finite element method, developed by Rao et al. [22, 23], has been employed for uncertainty quantification (UQ) analysis in the heat conduction of mechanical structures, particularly in addressing fuzzy uncertainty in static [24], thermodynamic [25], and natural frequency [26] problems. It is worth noting that fuzzy uncertainty problems can be equivalently converted into interval uncertainty problems. From this perspective, it is feasible to apply various efficient interval analysis methods to fuzzy uncertainty problems, such as classical interval arithmetic [27] and interval and subinterval perturbation methods (IPM) [26, 28]. Generally speaking, most existing fuzzy analysis methods focus on uncertain problems related to structural responses, temperature responses, and the like. However, there has been no research conducted on the static response analysis of the DCS utilizing fuzzy parameters.

In this section, two composite function/fuzzy perturbation methods are put forward to derive the fuzzy static response using fuzzy parameters. In the compound function/fuzzy perturbation method (CFFPM), the fuzzy static equilibrium equation is constructed based on the multiple fuzzy variable (MFV) model. The level-cut technique is utilized to convert the fuzzy static equilibrium equation into a series of interval equations with varying cut levels. To enhance accuracy, in the modified compound function/fuzzy perturbation method (MCFFPM), two innovative approaches are employed: the interval Jacobian matrix and interval virtual work vectors are approximated via the surface generation method (SGM), and the inverse of the interval Jacobian matrix is approximated using the modified Sherman–Morrison–Woodbury (SMW) formula. As a result, these two time-consuming operations are replaced by more efficient oncs.

When analyzing engineering systems or structures, it is advisable to model uncertain parameters with adequate information as random variables, while other uncertain parameters with clear variation ranges can be modeled as interval variables. Therefore, it is of great significance to construct a hybrid uncertainty model that combines the advantages of probabilistic and interval methods. In this hybrid uncertainty model, compared with a single random model, the variation range of responses can be obtained under the condition of considering interval variables. In contrast to a single interval model, the expectation and variance of response bounds can be derived when random variables are taken into account. Gao et al. [29, 30], based on perturbation theory and the random interval moment method, proposed a hybrid probabilistic–interval technique for uncertain bar structures. In hybrid uncertainty analysis, system responses can be obtained through the random IPM, and the expectation and variance of responses can be acquired via the random interval moment method. Based on the hybrid probabilistic–interval model, Xia et al. [31] developed a hybrid perturbation random moment method and a hybrid perturbation inverse mapping method, which are used to efficiently estimate the objective function and component reliability, respectively. Over the past few decades, another hybrid uncertainty model has been put forward to handle certain hybrid uncertainty problems. This model, also known as the interval random model, assumes that uncertain parameters are random variables, while some distribution parameters of these random variables are modeled as interval variables.

From an overall standpoint, research on predicting the dynamic response field of the luffing system of an automobile crane (LSOAAC) using hybrid uncertainty models remains in its initial phase. Firstly, hybrid uncertainty models have not been applied to predict the dynamic response field of the LSOAAC under uncertain information; specifically, no dynamic model incorporating uncertain parameters has been developed for the LSOAAC to date. Secondly, the application of hybrid uncertainty methods to address the uncertain dynamic response problems of the LSOAAC, particularly in predicting its response field, remains unexplored. Moreover, in most fields, hybrid uncertainty methods typically employ the same random vectors and interval vectors on both sides of dynamic equilibrium equations; however, in-depth research on hybrid uncertainty models that involve different random vectors and interval vectors on either side of such equilibrium equations is lacking.

In this section, a modified hybrid uncertainty analysis method (MHUAM) is proposed to predict the dynamic response field of the LSOAAC when both random and interval parameters are present. The process of predicting the dynamic response field is divided into three main steps. The first step involves deriving the dynamic response expression of the LSOAAC under uncertainty by combining the random IPM, first-order Taylor series expansion, and first-order Neumann series. The second step employs monotonic analysis to predict the response field, based on which the extrema of the bounds of the LSOAAC's response are calculated. The third step uses the random interval moment method to obtain the distribution characteristics of the LSOAAC's response field. As a result, the response bounds reflect the interval characteristics of the LSOAAC's response field under uncertainty, while the statistical characteristics of these bounds reflect the random characteristics of the response field. The HMCM and IPM are applied to verify that the MHUAM is more effective and less conservative.

6.2 Static Uncertainty Analysis with Fuzzy Parameters

Static response analysis of a DCS is conducted using fuzzy parameters. The fuzzy static equilibrium equation is established, and two fuzzy perturbation methods, including the CFFPM and MCFFPM, are presented. The CFFPM employs the level-cut technique to transform the fuzzy static equilibrium equation into several interval equations with different cut levels. The interval Jacobian matrix, the first and second interval virtual work vectors, and the inverse of interval Jacobian matrix are approximated by the first-order Taylor series and Neumann series.

The fuzzy static response field for every cut level is obtained by a synthesis of the compound function technique, the IPM, and the fuzzy algorithm. In the MCFFPM, the fuzzy static response field for every cut level is derived based on the surface rail generation method, the modified SMW formula, and the fuzzy theory. Compared with the Monte Carlo method (MCM), numerical examples demonstrate that the MCFFPM has a better accuracy than the CFFPM and both of them bring a higher efficiency than the MCM, especially when it comes to UQ of fuzzy parameters on the static response of the DCS.

6.2.1 Fuzzy Static Equilibrium Equation

As shown in Figure 6.1, the DCS includes the crane 1, the crane 2, and the payload. Based on the principle of virtual work, the static response equation of the DCS without considering masses and elasticity of cables of the cranes can be expressed as [21]

$$\boldsymbol{J}_{\mathrm{DCS}}^{T}\boldsymbol{\tau} + \boldsymbol{W}_{A_iB_i}^{T} + \boldsymbol{W}_{C_1C_2}^{T} = \mathbf{0},\ i = 1, 2 \tag{6.1}$$

where the superscript T denotes transpose of a matrix; $\boldsymbol{\tau}$ is the vector of the driving torque from hydraulic cylinders on lifting arms A_iB_i; $\boldsymbol{J}_{\mathrm{DCS}}$ is the kinematic Jacobian matrix; and $\boldsymbol{W}_{A_iB_i}$ and $\boldsymbol{W}_{C_1C_2}$ are the first virtual work from lifting arms A_iB_i and the second virtual work from the payload C_1C_2, respectively.

The kinematic Jacobian matrix can be expressed as

$$\boldsymbol{J}_{\mathrm{DCS}}^{T} = \boldsymbol{NM} = \begin{bmatrix} N_{11} & N_{21} \\ N_{12} & N_{22} \\ N_{13} & N_{23} \end{bmatrix} \begin{bmatrix} \dfrac{1}{M_{11}} & 0 \\ 0 & \dfrac{1}{M_{22}} \end{bmatrix} \tag{6.2}$$

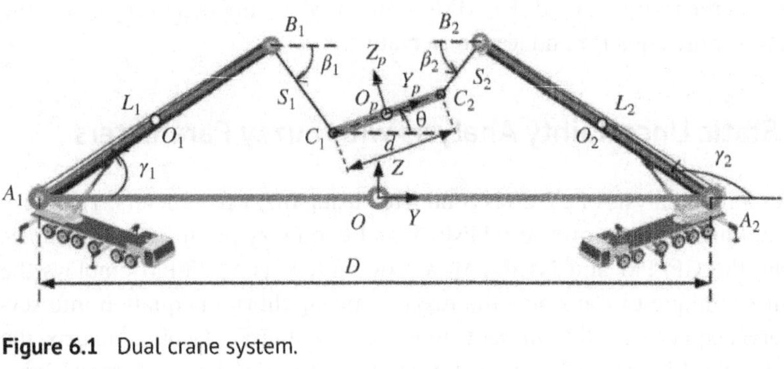

Figure 6.1 Dual crane system.

The first virtual work $\boldsymbol{W}_{A_iB_i}$ can be expressed as

$$\boldsymbol{W}_{A_iB_i} = \sum_{i=1}^{2} \left\{ \Delta_i \begin{bmatrix} \frac{N_{i1}}{M_{ii}} & \frac{N_{i2}}{M_{ii}} & \frac{N_{i3}}{M_{ii}} \end{bmatrix} \right\} \tag{6.3}$$

where $\Delta_i = -\frac{m_i L_i^2}{4}\ddot{\gamma}_i + \frac{m_i g L_i \cos \gamma_i}{2} - \ddot{\gamma}_i \frac{m_i L_i^2}{3}$. The second virtual work $\boldsymbol{W}_{C_1C_2}$ can be expressed as

$$\boldsymbol{W}_{C_1C_2} = \begin{bmatrix} \nabla_1 & \nabla_2 & \nabla_3 \end{bmatrix} \tag{6.4}$$

where

$$\begin{cases} \nabla_1 = -\frac{m_p d^2}{4}\ddot{\theta} \cos 2\theta + \frac{m_p g d \cos \theta}{2} - m_p \ddot{y} \frac{\sin \theta}{2} d - m_p \ddot{z} \frac{\cos \theta}{2} d, \\ \nabla_2 = -m_p \ddot{y} - m_p \frac{d}{2} \ddot{\theta} \sin \theta - m_p \frac{d}{2} \dot{\theta}^2 \cos \theta, \\ \nabla_3 = -m_p \ddot{z} + m_p \frac{d}{2} \ddot{\theta} \cos \theta - m_p \frac{d}{2} \dot{\theta}^2 \sin \theta + m_p g \end{cases} \tag{6.5}$$

Note that a detailed derivation of Eqs. (6.1)–(6.5) can be found in [21].

In practical crane engineering, due to uncertain sources such as mechanical errors, joint clearances, and stochastic excitation, there exists epistemic uncertainty. Assume that the fuzzy vector $\boldsymbol{f}$ represents all uncertain parameters with membership functions based on subjective expert opinions,

$$\boldsymbol{f} = \{f_1, \dots, f_m, \dots, f_n\}^T, m = 1, 2, \dots, n \tag{6.6}$$

where f_m is the mth fuzzy variable and n is the number of fuzzy variables. Based on Eqs. (6.1) and (6.6), the fuzzy static equivalent equation of the DCS can be expressed as

$$\boldsymbol{J}_{\text{DAS}}^T(\boldsymbol{f})\,\boldsymbol{\tau}(\boldsymbol{f}) + \boldsymbol{W}_{A_iB_i}^T(\boldsymbol{f}) + \boldsymbol{W}_{C_1C_2}^T(\boldsymbol{f}) = \boldsymbol{0} \tag{6.7}$$

where $\boldsymbol{J}_{\text{DCS}}(\boldsymbol{f})$ is the fuzzy kinematic Jacobian matrix; $\boldsymbol{W}_{A_iB_i}(\boldsymbol{f})$ and $\boldsymbol{W}_{C_1C_2}(\boldsymbol{f})$ are the first and second fuzzy virtual works, respectively; and $\boldsymbol{\tau}(\boldsymbol{f})$ is the fuzzy static response.

6.2.2 CFFPM

Based on the level-cut technique [32], every fuzzy variable with a defined membership function is composed of interval parameters with corresponding α-cut levels. The property of the fuzzy variable f_m in Eq. (6.6) can be expressed as

$$f_m = f_{m,\alpha_m}^I = \left[\underline{f_{m,\alpha_m}}, \overline{f_{m,\alpha_m}}\right] = f_{m,\alpha_m}^c + \Delta f_{m,\alpha_m} \delta^I \tag{6.8}$$

where $\alpha_m \in [0, 1]$ is a cut level corresponding to f_m; f^I_{m,α_m} is an interval parameter with respect to α_m; $\underline{f}_{m,\alpha_m}$ and $\overline{f}_{m,\alpha_m}$ are the lower bound (LB) and upper bound (UB) of f^I_{m,α_m}, respectively; $f^c_{m,\alpha_m} = (\underline{f}_{m,\alpha_m} + \overline{f}_{m,\alpha_m})/2$ and $\Delta f_{m,\alpha_m} = (\overline{f}_{m,\alpha_m} - \underline{f}_{m,\alpha_m})/2$ are the mean value (MV) and interval radius (IR) of f^I_{m,α_m}, respectively; and $\delta^I = [-1, 1]$ is a unit interval. Based on Eq. (6.8), an MFV model that consists of n fuzzy variables with corresponding cut levels can be expressed as

$$\begin{cases} \boldsymbol{F} = \left\{\left(\boldsymbol{F}^I_1, \boldsymbol{\alpha}_1\right), \dots, \left(\boldsymbol{F}^I_u, \boldsymbol{\alpha}_u\right), \dots, \left(\boldsymbol{F}^I_v, \boldsymbol{\alpha}_v\right)\right\}, \\ \boldsymbol{F}^I_u = \left\{F^I_{u,1}, \dots, F^I_{u,m}, \dots, F^I_{u,n}\right\} \in f_1 \times \dots \times f_m \times \dots \times f_n, \\ m = 1, 2, \dots, n;\ u = 1, 2, \dots, v \end{cases} \tag{6.9}$$

where $F^I_{u,1}, \dots, F^I_{u,m}, \dots, F^I_{u,n}$ are the uth interval parameters from $f_1, \dots, f_m, \dots, f_n$, respectively; $\boldsymbol{F}^I_u$ is the uth interval vector; $\boldsymbol{\alpha}_u$ is the uth cut level vector; and v is the number of cut level vectors. Note that $F^I_{u,m}$ in Eq. (6.9) are assumed to be independent interval variables with small uncertain levels ($\tau \leq 10^{-3}$). The property of $\boldsymbol{\alpha}_u$ can be expressed as

$$\boldsymbol{\alpha}_u = \left\{\alpha_{u,1}, \dots, \alpha_{u,m}, \dots, \alpha_{u,n}\right\} \tag{6.10}$$

Based on Eqs. (6.7), (6.9), and (6.10), the interval static equivalent equation of the DCS with the MFV model with respect to the corresponding cut level vector $\boldsymbol{\alpha}_u$ can be expressed as

$$\boldsymbol{J}^T_{\mathrm{DCS}}\left(\boldsymbol{F}^I_u\right)\boldsymbol{\tau}^I\left(\boldsymbol{F}^I_u\right) + \boldsymbol{W}^T_{A_iB_i}\left(\boldsymbol{F}^I_u\right) + \boldsymbol{W}^T_{C_1C_2}\left(\boldsymbol{F}^I_u\right) = \boldsymbol{0} \tag{6.11}$$

where $\boldsymbol{J}_{\mathrm{DCS}}\left(\boldsymbol{F}^I_u\right)$ is the interval kinematic Jacobian matrix; $\boldsymbol{W}_{A_iB_i}\left(\boldsymbol{F}^I_u\right)$ and $\boldsymbol{W}_{C_1C_2}\left(\boldsymbol{F}^I_u\right)$ are the first and second interval virtual works, respectively; and $\boldsymbol{\tau}\left(\boldsymbol{F}^I_u\right)$ is the interval static response.

From Eqs. (6.2) and (6.11), by using the perturbation theory and compound function technique, the first-order Taylor series expansion of the interval kinematic Jacobian matrix $\boldsymbol{J}^T_{\mathrm{DCS}}\left(\boldsymbol{F}^I_u\right)$ at the MV of $\boldsymbol{F}^I_u$ can be expressed as

$$\boldsymbol{J}^T_{\mathrm{DCS}}\left(\boldsymbol{F}^I_u\right) = \boldsymbol{J}^c + \Delta\boldsymbol{J} \tag{6.12}$$

From Eqs. (6.4) and (6.11), by using the first-order Taylor series to expand the first interval virtual work, $\boldsymbol{W}^T_{A_iB_i}\left(\boldsymbol{F}^I_u\right)$ at the MV of $\boldsymbol{F}^I_u$ can be expressed as

$$\boldsymbol{W}^T_{A_iB_i}\left(\boldsymbol{F}^I_u\right) = \left(\boldsymbol{W}^T_{A_iB_i}\right)^c + \left(\Delta\boldsymbol{W}^T_{A_iB_i}\right)^c \tag{6.13}$$

where $\left(\boldsymbol{W}^T_{A_iB_i}\right)^c$ and $\left(\Delta\boldsymbol{W}^T_{A_iB_i}\right)^c$ are the MV and IR of the first interval virtual work $\boldsymbol{W}^T_{A_iB_i}\left(\boldsymbol{F}^I_u\right)$, respectively, and

$$\left(\boldsymbol{W}^T_{A_iB_i}\right)^c = \boldsymbol{W}^T_{A_iB_i}\left(\boldsymbol{F}^I_u\right)\Big|_{\boldsymbol{F}^I_u = \boldsymbol{F}^c_u} \tag{6.14}$$

$$\left(\Delta W^T_{A_iB_i}\right)^c = \sum_{i=1}^{2}\sum_{m=1}^{n}\left\{\frac{\partial(\Delta_i)}{\partial F_{u,m}}\begin{bmatrix}\frac{N_{i1}}{M_{ii}}\\ \frac{N_{i2}}{M_{ii}}\\ \frac{N_{i3}}{M_{ii}}\end{bmatrix}\Bigg|_{F^I_u=F^c_u} + \Delta_i\begin{bmatrix}\frac{\partial\left(\frac{N_{i1}}{M_{ii}}\right)}{\partial F_{u,m}}\\ \frac{\partial\left(\frac{N_{i2}}{M_{ii}}\right)}{\partial F_{u,m}}\\ \frac{\partial\left(\frac{N_{i3}}{M_{ii}}\right)}{\partial F_{u,m}}\end{bmatrix}\Bigg|_{F^I_u=F^c_u}\right\}\Delta F_{u,m}\delta^I_m \tag{6.15}$$

From Eqs. (6.5) and (6.11), by using the first-order Taylor series to expand the second interval virtual work, $W^T_{C_1C_2}(F^I_u)$ at the MV of F^I_u can be expressed as

$$W^T_{C_1C_2}(F^I_u) = \left(W^T_{C_1C_2}\right)^c + \left(\Delta W^T_{C_1C_2}\right)^c \tag{6.16}$$

where $\left(W^T_{C_1C_2}\right)^c$ and $\left(\Delta W^T_{C_1C_2}\right)^c$ are the MV and IR of the second interval virtual work $W^T_{C_1C_2}(F^I_u)$, respectively, and

$$\left(W^T_{C_1C_2}\right)^c = W^T_{C_1C_2}\left(F^I_u\right)\Big|_{F^I_u=F^c_u} \tag{6.17}$$

$$\left(\Delta W^T_{C_1C_2}\right)^c = \sum_{m=1}^{n}\left\{\begin{bmatrix}\frac{\partial(\nabla_1)}{\partial F_{u,m}}\\ \frac{\partial(\nabla_2)}{\partial F_{u,m}}\\ \frac{\partial(\nabla_3)}{\partial F_{u,m}}\end{bmatrix}\Bigg|_{F^I_u=F^c_u}\right\}\Delta F_{u,m}\delta^I_m \tag{6.18}$$

Substituting Eqs. (6.12), (6.13), and (6.16) into Eq. (6.11) yields

$$\tau^I\left(F^I_u\right) = -\left(J^c+\Delta J\right)^{-1}\left(\left(W^T_{A_iB_i}\right)^c + \left(\Delta W^T_{A_iB_i}\right)^c + W^c_{C_1C_2} + \Delta W_{C_1C_2}\right),\ i=1,2 \tag{6.19}$$

Using the first-order Neumann series to expand $\left(J^c+\Delta J\right)^{-1}$ yields

$$\left(J^c+\Delta J\right)^{-1} = \left(J^c\right)^{-1} - \left(J^c\right)^{-1}\Delta J\left(J^c\right)^{-1} \tag{6.20}$$

Substituting Eq. (6.20) into Eq. (6.21) and using interval algorithms yield

$$\tau^c = -\left(J^c\right)^{-1}\left(\left(W^T_{A_iB_i}\right)^c + \left(W^T_{C_1C_2}\right)^c\right) \tag{6.21}$$

$$\Delta\boldsymbol{\tau}^I = -(\boldsymbol{J}^c)^{-1}\left((\Delta\boldsymbol{W}^T_{A_iB_i})^c + (\Delta\boldsymbol{W}^T_{C_1C_2})^c\right) + (\boldsymbol{J}^c)^{-1}\Delta\boldsymbol{J}(\boldsymbol{J}^c)^{-1}\left((\boldsymbol{W}^T_{A_iB_i})^c + (\boldsymbol{W}^T_{C_1C_2})^c\right) \tag{6.22}$$

where $\boldsymbol{\tau}^c$ and $\Delta\boldsymbol{\tau}^I$ are the MV and IR of the interval static response $\boldsymbol{\tau}^I(\boldsymbol{F}^I_u)$, respectively. Equation (6.22) can be rewritten as

$$\Delta\boldsymbol{\tau}^I = -(\boldsymbol{J}^c)^{-1}\boldsymbol{F}_1\delta^I_m + (\boldsymbol{J}^c)^{-2}\left[(\boldsymbol{W}^T_{A_iB_i})^c + (\boldsymbol{W}^T_{C_1C_2})^c\right]\boldsymbol{F}_2\delta^I_m \tag{6.23}$$

based on Eq. (6.23) and using the monotonic technique, one has

$$\Delta\boldsymbol{\tau} = \left|(\boldsymbol{J}^c)^{-2}\left[(\boldsymbol{W}^T_{A_iB_i})^c + (\boldsymbol{W}^T_{C_1C_2})^c\right]\boldsymbol{F}_2 - (\boldsymbol{J}^c)^{-1}\boldsymbol{F}_1\right| \tag{6.24}$$

where $\Delta\boldsymbol{\tau}$ is the IR of the interval static response $\boldsymbol{\tau}^I(\boldsymbol{F}^I_u)$. From Eqs. (6.21) and (6.24), based on the interval union operation, the LB and UB of the interval static response $\boldsymbol{\tau}(\boldsymbol{F}^I_u)$ can be expressed as

$$\boldsymbol{\tau}^I(\boldsymbol{F}^I_u) = [\underline{\boldsymbol{\tau}}, \overline{\boldsymbol{\tau}}] = [\boldsymbol{\tau}^c - \Delta\boldsymbol{\tau}, \boldsymbol{\tau}^c + \Delta\boldsymbol{\tau}] \tag{6.25}$$

where $\underline{\boldsymbol{\tau}}$ and $\overline{\boldsymbol{\tau}}$ are the LB and UB of the interval static response $\boldsymbol{\tau}^I(\boldsymbol{F}^I_u)$, respectively. From Eqs. (6.10) and (6.25), based on the fuzzy algorithm, the LB and UB of the fuzzy static response $\boldsymbol{\tau}(\boldsymbol{f})$ for every cut level vector $\boldsymbol{\alpha}_u$ can be expressed as

$$\boldsymbol{\tau}(\boldsymbol{f}) = \left[\underline{\boldsymbol{\tau}(\boldsymbol{f})}, \overline{\boldsymbol{\tau}(\boldsymbol{f})}\right] = \left[\bigcup_{\boldsymbol{\alpha}_u\in[0,1]}\boldsymbol{\alpha}_u(\boldsymbol{\tau}^c - \Delta\boldsymbol{\tau}), \bigcup_{\boldsymbol{\alpha}_u\in[0,1]}\boldsymbol{\alpha}_u(\boldsymbol{\tau}^c + \Delta\boldsymbol{\tau})\right] \tag{6.26}$$

where $\underline{\boldsymbol{\tau}(\boldsymbol{f})}$ and $\overline{\boldsymbol{\tau}(\boldsymbol{f})}$ are the LB and UB of the fuzzy static response $\boldsymbol{\tau}(\boldsymbol{f})$, respectively.

6.2.3 MCFFPM

In this section, use of the SGM and modified SMW formula in the MCFFPM further improves accuracy of solving the fuzzy static response, especially when it comes to a higher uncertain level. From Eqs. (6.2) and (6.11), by using the SGM and compound function technique, the expansion of the interval kinematic Jacobian matrix $\boldsymbol{J}^T_{\mathrm{DCS}}(\boldsymbol{F}^I_u)$ at the MV of $\boldsymbol{F}^I_u$ can be expressed as

$$\begin{aligned}\boldsymbol{J}^T_{\mathrm{DACS}}(\boldsymbol{F}^I_u) &= \tilde{\boldsymbol{J}}^c + \Delta\tilde{\boldsymbol{J}} \\ &= \sum_{m=1}^{n}\tilde{\boldsymbol{J}}(\boldsymbol{F}^I_u) - (n-1)\cdot\boldsymbol{J}^T_{\mathrm{DACS}}(\boldsymbol{F}^I_u)\big|_{\boldsymbol{F}^I_u=\boldsymbol{F}^c_u} \\ &= \sum_{m=1}^{n}\tilde{\boldsymbol{J}}(\boldsymbol{F}^c_u) - (n-1)\cdot\boldsymbol{J}^T_{\mathrm{DACS}}(\boldsymbol{F}^I_u)\big|_{\boldsymbol{F}^I_u=\boldsymbol{F}^c_u} \\ &\quad + \sum_{m=1}^{n}\left.\frac{\partial\tilde{\boldsymbol{J}}(\boldsymbol{F}^I_u)}{\partial F_{u,m}}\right|_{\boldsymbol{F}^I_u=\boldsymbol{F}^c_u}\Delta F_{u,m}\delta^I_m\end{aligned} \tag{6.27}$$

From Eqs. (6.4) and (6.11), by using the SGM and compound function technique, the expansion of the first interval virtual work $\boldsymbol{W}_{A_iB_i}^T(\boldsymbol{F}_u^I)$ at the MV of $\boldsymbol{F}_u^I$ can be expressed as

$$
\begin{aligned}
\boldsymbol{W}_{A_iB_i}^T(\boldsymbol{F}_u^I) &= \widetilde{\boldsymbol{W}_{A_iB_i}^T}^c + \Delta\widetilde{\boldsymbol{W}_{A_iB_i}^T} \\
&= \sum_{m=1}^{n} \widetilde{\boldsymbol{W}_{A_iB_i}^T(\boldsymbol{F}_u^I)} - (n-1)\cdot\left(\boldsymbol{W}_{A_iB_i}^T\right)^c \\
&\sum_{m=1}^{n} \widetilde{\boldsymbol{W}_{A_iB_i}^T(\boldsymbol{F}_u^c)} - (n-1)\cdot \boldsymbol{W}_{A_iB_i}^T(\boldsymbol{F}_u^I)\Big|_{\boldsymbol{F}_u^I=\boldsymbol{F}_u^c} \\
&+ \sum_{m=1}^{n} \left.\frac{\partial \widetilde{\boldsymbol{W}_{A_iB_i}^T(\boldsymbol{F}_u^I)}}{\partial F_{u,m}}\right|_{\boldsymbol{F}_u^I=\boldsymbol{F}_u^c} \Delta F_{u,m}\delta_m^I
\end{aligned}
\tag{6.28}
$$

From Eqs. (6.5) and (6.11), by using the SGM and compound function technique, the expansion of the second interval virtual work $\boldsymbol{W}_{C_1C_2}^T(\boldsymbol{F}_u^I)$ at the MV of $\boldsymbol{F}_u^I$ can be expressed as

$$
\begin{aligned}
\boldsymbol{W}_{C_1C_2}^T(\boldsymbol{F}_u^I) &= \widetilde{\boldsymbol{W}_{C_1C_2}^T}^c + \Delta\widetilde{\boldsymbol{W}_{C_1C_2}^T} \\
&= \sum_{m=1}^{n} \widetilde{\boldsymbol{W}_{C_1C_2}^T(\boldsymbol{F}_u^I)} - (n-1)\cdot\left(\boldsymbol{W}_{C_1C_2}^T\right)^c \\
&= \sum_{m=1}^{n} \widetilde{\boldsymbol{W}_{C_1C_2}^T(\boldsymbol{F}_u^c)} - (n-1)\cdot \boldsymbol{W}_{C_1C_2}^T(\boldsymbol{F}_u^I)\Big|_{\boldsymbol{F}_u^I=\boldsymbol{F}_u^c} \\
&+ \sum_{m=1}^{n} \left.\frac{\partial \widetilde{\boldsymbol{W}_{C_1C_2}^T(\boldsymbol{F}_u^I)}}{\partial F_{u,m}}\right|_{\boldsymbol{F}_u^I=\boldsymbol{F}_u^c} \Delta F_{u,m}\delta_m^I
\end{aligned}
\tag{6.29}
$$

Substituting Eqs. (6.27)–(6.29) into Eq. (6.24) yields

$$
\tau^I(\boldsymbol{F}_u^I) = -\left(\widetilde{\boldsymbol{J}}^c + \Delta\widetilde{\boldsymbol{J}}\right)^{-1}\left(\widetilde{\boldsymbol{W}_{A_iB_i}^T}^c + \Delta\widetilde{\boldsymbol{W}_{A_iB_i}^T} + \widetilde{\boldsymbol{W}_{C_1C_2}^T}^c + \Delta\widetilde{\boldsymbol{W}_{C_1C_2}^T}\right), i = 1, 2
\tag{6.30}
$$

Using the first-order Neumann series to expand $\left(\widetilde{\boldsymbol{J}}^c + \Delta\widetilde{\boldsymbol{J}}\right)^{-1}$ yields

$$
\left(\widetilde{\boldsymbol{J}}^c + \Delta\widetilde{\boldsymbol{J}}\right)^{-1} = \left(\widetilde{\boldsymbol{J}}^c\right)^{-1} + \left(\widetilde{\boldsymbol{J}}^c\right)^{-1}\sum_{r=1}^{\infty}\left(-\Delta\widetilde{\boldsymbol{J}}\left(\widetilde{\boldsymbol{J}}^c\right)^{-1}\right)^r
\tag{6.31}
$$

Based on Eq. (6.27), the term $\sum_{r=1}^{\infty}\left(-\Delta\widetilde{\boldsymbol{J}}\left(\widetilde{\boldsymbol{J}}^c\right)^{-1}\right)^r$ can be rewritten as

$$\sum_{r=1}^{\infty}\left(-\Delta\tilde{\boldsymbol{J}}\left(\tilde{\boldsymbol{J}}^{c}\right)^{-1}\right)^{r}=\sum_{r=1}^{\infty}\left(-\sum_{m=1}^{n}\delta_{m}^{I}\tilde{\boldsymbol{J}}_{m}^{c}\right)^{r}$$
$$=\sum_{r=1}^{\infty}\left(-\sum_{m=1}^{n}\left\{\frac{\partial\tilde{\boldsymbol{J}}\left(\boldsymbol{F}_{u}^{I}\right)}{\partial\tilde{\boldsymbol{N}}\left(\boldsymbol{F}_{u}^{I}\right)}\frac{\partial\tilde{\boldsymbol{N}}\left(\boldsymbol{F}_{u}^{I}\right)}{\partial F_{u,m}}\tilde{\boldsymbol{M}}\left(\boldsymbol{F}_{u}^{I}\right)+\tilde{\boldsymbol{N}}\left(\boldsymbol{F}_{u}^{I}\right)\right.\right.$$
$$\left.\left.\frac{\partial\tilde{\boldsymbol{J}}\left(\boldsymbol{F}_{u}^{I}\right)}{\partial\tilde{\boldsymbol{M}}\left(\boldsymbol{F}_{u}^{I}\right)}\frac{\partial\tilde{\boldsymbol{M}}\left(\boldsymbol{F}_{u}^{I}\right)}{\partial F_{u,m}}\tilde{\boldsymbol{M}}\left(\boldsymbol{F}_{u}^{I}\right)\right\}\right|_{\boldsymbol{F}_{u}^{I}=\boldsymbol{F}_{u}^{c}}\left.\Delta F_{u,m}\delta_{m}^{I}\left(\tilde{\boldsymbol{J}}^{c}\right)^{-1}\right)^{r} \tag{6.32}$$

Equation (6.32) can be expressed in the explicit form

$$\begin{aligned}\sum_{r=1}^{\infty}\left(-\Delta\tilde{\boldsymbol{J}}\left(\tilde{\boldsymbol{J}}^{c}\right)^{-1}\right)^{r}&=-\sum_{m=1}^{n}\delta_{m}^{I}\tilde{\boldsymbol{J}}_{m}^{c}+\sum_{m=1}^{n}\left(\delta_{m}^{I}\tilde{\boldsymbol{J}}_{m}^{c}\right)^{2}+\sum_{\substack{m,p=1\\ m\neq p}}^{n}\delta_{m}^{I}\delta_{p}^{I}\tilde{\boldsymbol{J}}_{m}^{c}\tilde{\boldsymbol{J}}_{p}^{c}\\&\quad-\sum_{m=1}^{n}\left(\delta_{m}^{I}\tilde{\boldsymbol{J}}_{m}^{c}\right)^{3}-\sum_{\substack{m,p,q=1\\ m\neq p\neq q}}^{n}\delta_{m}^{I}\delta_{p}^{I}\delta_{q}^{I}\tilde{\boldsymbol{J}}_{m}^{c}\tilde{\boldsymbol{J}}_{p}^{c}\tilde{\boldsymbol{J}}_{q}^{c}\cdots\\&=\sum_{m=1}^{n}\sum_{r=1}^{\infty}\left(-\delta_{m}^{I}\tilde{\boldsymbol{J}}_{m}^{c}\right)^{r}+\sum_{\substack{m,p=1\\ m\neq p}}^{n}\delta_{m}^{I}\delta_{p}^{I}\tilde{\boldsymbol{J}}_{m}^{c}\tilde{\boldsymbol{J}}_{p}^{c}\\&\quad-\sum_{\substack{m,p,q=1\\ m\neq p\neq q}}^{n}\delta_{m}^{I}\delta_{p}^{I}\delta_{q}^{I}\tilde{\boldsymbol{J}}_{m}^{c}\tilde{\boldsymbol{J}}_{p}^{c}\tilde{\boldsymbol{J}}_{q}^{c}\cdots\end{aligned} \tag{6.33}$$

If the norm $\left\|\tilde{\boldsymbol{J}}_{m}^{c}\right\|<1$ is satisfied, one has $\lim_{r\to\infty}\left(-\delta_{m}^{I}\tilde{\boldsymbol{J}}_{m}^{c}\right)^{r}=\boldsymbol{0}$. By using the modified SWM formula, $\left(\tilde{\boldsymbol{J}}^{c}+\Delta\tilde{\boldsymbol{J}}\right)^{-1}$ approximately converges to

$$\begin{aligned}\left(\tilde{\boldsymbol{J}}^{c}+\Delta\tilde{\boldsymbol{J}}\right)^{-1}&=\left(\tilde{\boldsymbol{J}}^{c}\right)^{-1}+\left(\tilde{\boldsymbol{J}}^{c}\right)^{-1}\sum_{m=1}^{n}\sum_{r=1}^{\infty}\left(-\delta_{m}^{I}\tilde{\boldsymbol{J}}_{m}^{c}\right)^{r}\\&=\left(\tilde{\boldsymbol{J}}^{c}\right)^{-1}+\left(\tilde{\boldsymbol{J}}^{c}\right)^{-1}\sum_{m=1}^{n}\frac{-\delta_{m}^{I}\tilde{\boldsymbol{J}}_{m}^{c}\left[I-\left(-\delta_{m}^{I}\tilde{\boldsymbol{J}}_{m}^{c}\right)^{r\to\infty}\right]}{I+\delta_{m}^{I}\tilde{\boldsymbol{J}}_{m}^{c}}\\&\approx\left(\tilde{\boldsymbol{J}}^{c}\right)^{-1}+\left(\tilde{\boldsymbol{J}}^{c}\right)^{-1}\sum_{m=1}^{n}\boldsymbol{K}_{m}^{I}\end{aligned} \tag{6.34}$$

where $\boldsymbol{I}$ is an identity matrix and $\boldsymbol{K}_{m}^{I}=\dfrac{\boldsymbol{I}}{\boldsymbol{I}+\delta_{m}^{I}\tilde{\boldsymbol{J}}_{m}^{c}}$, which can be expressed as

$$\boldsymbol{K}_{m}^{I}=\boldsymbol{K}_{m}^{c}+\Delta\boldsymbol{K}_{m}^{I}=\boldsymbol{K}_{m}^{c}+\Delta\boldsymbol{K}_{m}\delta_{Km}^{I} \tag{6.35}$$

Substituting Eqs. (6.34) and (6.35) into Eq. (6.30) yields

$$\begin{aligned}
\boldsymbol{\tau}^I(\boldsymbol{F}_u^I) &= \boldsymbol{\tau}^c + \Delta\boldsymbol{\tau}^I \\
&= -\left\{\left(\widetilde{\boldsymbol{J}}^c\right)^{-1} + \left(\widetilde{\boldsymbol{J}}^c\right)^{-1}\sum_{m=1}^{n}\left[\boldsymbol{K}_m^c + \Delta\boldsymbol{K}_m^I\right]\right\} \\
&\quad \left(\widetilde{\boldsymbol{W}_{A_iB_i}^T}^c + \Delta\widetilde{\boldsymbol{W}_{A_iB_i}^T} + \widetilde{\boldsymbol{W}_{C_1C_2}^T}^c + \Delta\widetilde{\boldsymbol{W}_{C_1C_2}^T}\right) \\
&= -\left\{\left\{\left(\widetilde{\boldsymbol{J}}^c\right)^{-1} + \left(\widetilde{\boldsymbol{J}}^c\right)^{-1}\sum_{m=1}^{n}\boldsymbol{K}_m^c\right\}\left(\widetilde{\boldsymbol{W}_{A_iB_i}^T}^c + \widetilde{\boldsymbol{W}_{C_1C_2}^T}^c\right)\right\} \\
&\quad -\left\{\left(\widetilde{\boldsymbol{J}}^c\right)^{-1}\sum_{m=1}^{n}\Delta\boldsymbol{K}_m^I\left(\widetilde{\boldsymbol{W}_{A_iB_i}^T}^c + \widetilde{\boldsymbol{W}_{C_1C_2}^T}^c\right)\right\} \\
&\quad -\left\{\left\{\left(\widetilde{\boldsymbol{J}}^c\right)^{-1} + \left(\widetilde{\boldsymbol{J}}^c\right)^{-1}\sum_{m=1}^{n}\boldsymbol{K}_m^c\right\}\left(\Delta\widetilde{\boldsymbol{W}_{A_iB_i}^T} + \Delta\widetilde{\boldsymbol{W}_{C_1C_2}^T}\right)\right\} \\
&\quad -\left\{\left(\widetilde{\boldsymbol{J}}^c\right)^{-1}\sum_{m=1}^{n}\Delta\boldsymbol{K}_m^I\left(\Delta\widetilde{\boldsymbol{W}_{A_iB_i}^T} + \Delta\widetilde{\boldsymbol{W}_{C_1C_2}^T}\right)\right\}
\end{aligned} \tag{6.36}$$

where

$$\boldsymbol{\tau}^c = -\left\{\left(\widetilde{\boldsymbol{J}}^c\right)^{-1} + \left(\widetilde{\boldsymbol{J}}^c\right)^{-1}\sum_{m=1}^{n}\boldsymbol{K}_m^c\right\}\left(\widetilde{\boldsymbol{W}_{A_iB_i}^T}^c + \widetilde{\boldsymbol{W}_{C_1C_2}^T}^c\right) \tag{6.37}$$

$$\begin{aligned}
\Delta\boldsymbol{\tau}^I = &\left\{\left(\widetilde{\boldsymbol{J}}^c\right)^{-1}\sum_{m=1}^{n}\Delta\boldsymbol{K}_m^I\left(\widetilde{\boldsymbol{W}_{A_iB_i}^T}^c + \widetilde{\boldsymbol{W}_{C_1C_2}^T}^c\right)\right\} \\
&\left\{\left\{\left(\widetilde{\boldsymbol{J}}^c\right)^{-1} + \left(\widetilde{\boldsymbol{J}}^c\right)^{-1}\sum_{m=1}^{n}\boldsymbol{K}_m^c\right\}\left(\Delta\widetilde{\boldsymbol{W}_{A_iB_i}^T} + \Delta\widetilde{\boldsymbol{W}_{C_1C_2}^T}\right)\right\}
\end{aligned} \tag{6.38}$$

Substituting Eqs. (6.31), (6.32), and (6.35) into Eq. (6.38) yields

$$\begin{aligned}
\Delta\boldsymbol{\tau}^I = &-\left\{\left(\widetilde{\boldsymbol{J}}^c\right)^{-1}\sum_{m=1}^{n}\Delta\boldsymbol{K}_m\delta_{Km}^I\left(\widetilde{\boldsymbol{W}_{A_iB_i}^T}^c + \widetilde{\boldsymbol{W}_{C_1C_2}^T}^c\right)\right\} \\
&-\left\{\left\{\left(\widetilde{\boldsymbol{J}}^c\right)^{-1} + \left(\widetilde{\boldsymbol{J}}^c\right)^{-1}\sum_{m=1}^{n}\boldsymbol{K}_m^c\right\}\left(\sum_{m=1}^{n}\left\{\frac{\partial\widetilde{\boldsymbol{W}_{A_iB_i}^T}(\boldsymbol{F}_u^I)}{\partial F_{u,m}}\right.\right.\right. \\
&\left.\left.\left.+\frac{\partial\widetilde{\boldsymbol{W}_{C_1C_2}^T}(\boldsymbol{F}_u^I)}{\partial F_{u,m}}\right\}\right|_{\boldsymbol{F}_u^I=\boldsymbol{F}_u^c}\Delta F_{u,m}\delta_m^I\right)\right\}
\end{aligned} \tag{6.39}$$

Based on Eq. (6.39) and by using the monotonic technique, one has

$$\begin{aligned}\Delta\tau = &\left|(\boldsymbol{J}^c)^{-1}\sum_{K_m=1}^{n}\Delta\boldsymbol{K}_m\left(\widetilde{\boldsymbol{W}_{A_iB_i}^T}^c+\widetilde{\boldsymbol{W}_{C_1C_2}^T}^c\right)\right| \\ &+\left|\left\{(\boldsymbol{J}^c)^{-1}+(\boldsymbol{J}^c)^{-1}\sum_{m=1}^{n}\boldsymbol{K}_m^c\right\}\right. \\ &\left.\left(\sum_{m=1}^{n}\left\{\frac{\partial\widetilde{\boldsymbol{W}_{A_iB_i}^T}(\boldsymbol{F}_u^I)}{\partial F_{u,m}}+\frac{\partial\widetilde{\boldsymbol{W}_{C_1C_2}^T}(\boldsymbol{F}_u^I)}{\partial F_{u,m}}\right\}\bigg|_{\boldsymbol{F}_u^I=\boldsymbol{F}_u^c}\Delta F_{u,m}\right)\right|\end{aligned} \tag{6.40}$$

From Eqs. (6.37) and (6.40) and based on the interval union operation, the LB and UB of the interval static response $\boldsymbol{\tau}\left(\boldsymbol{F}_u^I\right)$ can be expressed as

$$\tau^I\left(\boldsymbol{F}_u^I\right)=\left[\underline{\tau},\bar{\tau}\right]=\left[\tau^c-\Delta\tau,\tau^c+\Delta\tau\right] \tag{6.41}$$

From Eqs. (6.11) and (6.41) and based on the fuzzy algorithm, the LB and UB of the fuzzy static response $\boldsymbol{\tau}\left(\boldsymbol{f}\right)$ for every cut level vector $\boldsymbol{\alpha_u}$ can be expressed as

$$\tau\left(\boldsymbol{f}\right)=\left[\underline{\tau\left(\boldsymbol{f}\right)},\overline{\tau\left(\boldsymbol{f}\right)}\right]=\left[\bigcup_{\boldsymbol{\alpha_u}\in[0,1]}\boldsymbol{\alpha_u}\left(\tau^c-\Delta\tau\right),\bigcup_{\boldsymbol{\alpha_u}\in[0,1]}\boldsymbol{\alpha_u}\left(\tau^c+\Delta\tau\right)\right] \tag{6.42}$$

6.2.4 Numerical Examples

To verify the effectiveness of the CFFPM and MCFFPM for obtaining the fuzzy static response of the DCS, some examples are simulated.

All examples use uncertain structural parameters (D, d, L_1, L_2) as narrowly fuzzy variables, which can be expressed as triangular fuzzy numbers: $D = (12.75\,(1-t), 12.75, 12.75\,(1+t))$ m, $d = (2.5\,(1-t), 2.5, 2.5\,(1+t))$ m, $L_1 = (5\,(1-t), 5, 5\,(1+t))$ m, and $L_2 = (4.5\,(1-t), 4.5, 4.5\,(1+t))$ m. Uncertain levels of fuzzy variables are denoted as $t(0 \leq t \leq 1)$. Cut levels of fuzzy variables are denoted as $\alpha(0\leq \alpha_u \leq 1)$.

6.2.4.1 Effects of Fuzzy Parameters on the Fuzzy Static Response

To investigate effects of different fuzzy variables on the fuzzy static response by the CFFPM, an MFV model $t_0(t_0 = t_D = t_d = t_{L_1} = t_{L_2})$, where t_D, t_d, t_{L_1}, and t_{L_2} denote uncertain levels of fuzzy variables D, d, L_1, and L_2, respectively, is constructed. Four other MFV models are also constructed as follows: (1) t_1 $(t_1 = t_d = t_{L_1} = t_{L_2}, t_D = 0)$; (2) $t_2(t_2 = t_D = t_{L_1} = t_{L_2}, t_d = 0)$; (3) $t_3(t_3 = t_D = t_d = t_{L_2}, t_{L_1} = 0)$; and (4) $t_4(t_4 = t_D = t_d = t_{L_1}, t_{L_2} = 0)$. To clearly compare different degrees of influence on the fuzzy static response from fuzzy variables, the LB and UB of the fuzzy static response of the crane 1 with different MFV models obtained by the CFFPM are shown in Figures 6.2–6.5.

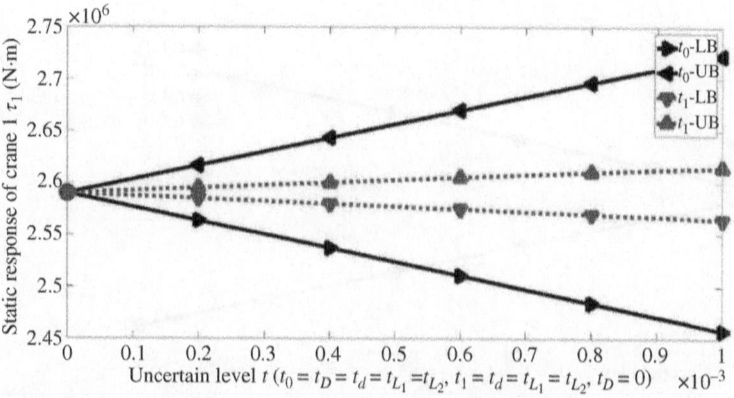

Figure 6.2 LB and UB of the fuzzy static response of the crane 1 with t_0 and t_1.

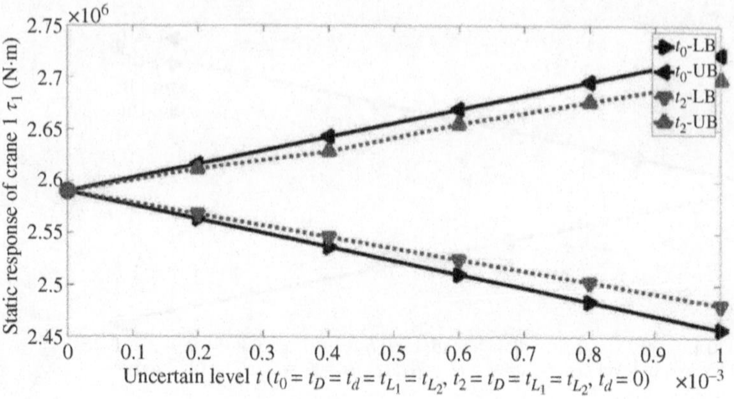

Figure 6.3 LB and UB of the fuzzy static response of the crane 1 with t_0 and t_2.

Figure 6.2 shows the LB and UB of the fuzzy static response of the crane 1 with two MFV models t_0 and t_1. One finds that the two dashed curves and two solid curves have different increasing trends and their intervals obviously increase, which indicate that the fuzzy parameter D has a significant impact on the fuzzy static response. The above results also show that the fuzzy parameter D is a key factor to evaluate the fuzzy static response. Figure 6.3 shows the LB and UB of the fuzzy static response of the crane 1 with two MFV models t_0 and t_2. One finds that the two dashed curves and two solid curves have different increasing trends and their intervals increase a little, which indicate that the fuzzy parameter d has a slight impact on the fuzzy static response. The above results also show that the fuzzy parameter d is also a key factor to evaluate the

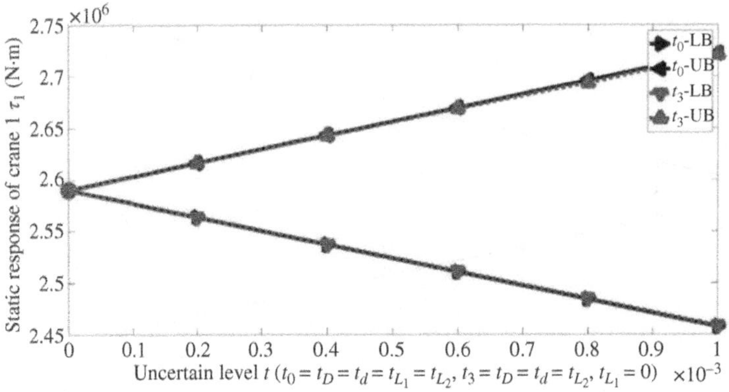

Figure 6.4 LB and UB of the fuzzy static response of the crane 1 with t_0 and t_3.

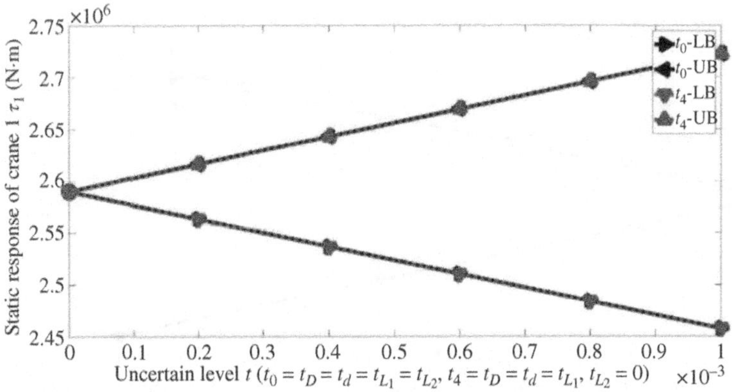

Figure 6.5 LB and UB of the fuzzy static response of the crane 1 with t_0 and t_4.

fuzzy static response. Figure 6.4 shows the LB and UB of the fuzzy static response of the crane 1 with two MFV models t_0 and t_3. One finds that the two dashed curves and two solid curves have the same trends and their intervals are zero, which indicate the fuzzy parameter L_1 has no impact on the fuzzy static response. Figure 6.5 shows the LB and UB of the fuzzy static response of the crane 1 with two MFV models t_0 and t_4. One finds that the two dashed curves and two solid curves have the same trends and their intervals are zero, which indicate that the fuzzy parameter L_2 has no impact on the fuzzy static response.

6.2.4.2 Accuracy and Efficiency of the CFFPM and MCFFPM

To illustrate the accuracy and efficiency of the CFFPM and MCFFPM for calculating the fuzzy static response, the LB and UB of the fuzzy static response with the MFV model $t_0(t_0 = t_D = t_d = t_{L_1} = t_{L_2})$ are calculated by the MCM, CFFPM, and MCFFPM, respectively. The LB and UB of the fuzzy static response of the crane 1 with the MFV model t_0 under cut levels $\alpha = 0.4$ and $\alpha = 0.8$ are shown in Figures 6.6 and 6.7, respectively.

Figures 6.6 and 6.7 show the LB and UB of the fuzzy static response of the crane 1 with the MFV model t_0 under cut levels $\alpha = 0.4$ and $\alpha = 0.8$ calculated by the MCM, CFFPM, and MCFFPM, respectively.

The relative errors between the LB and UB of the fuzzy static response with the MFV model t_0 under $\alpha = 0.4$ and $\alpha = 0.8$ by the CFFPM and MCM, and between those by the MCFFPM and MCM are listed in Tables 6.1 and 6.2, respectively. As shown in Table 6.1, the maximum relative errors are no more than 10.20%, which

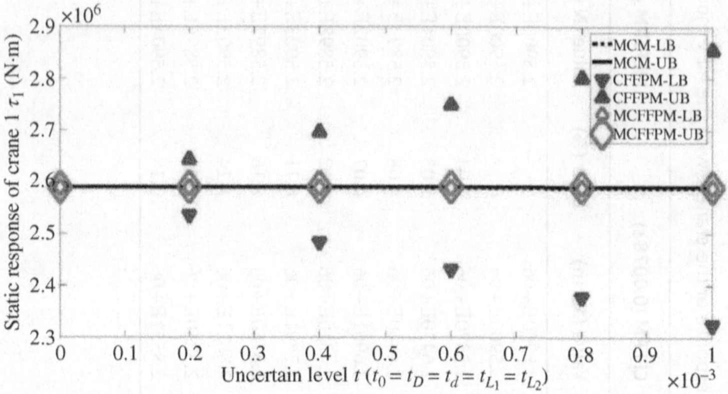

Figure 6.6 LB and UB of the fuzzy static response of the crane 1 with the MFV model t_0 under $\alpha = 0.4$.

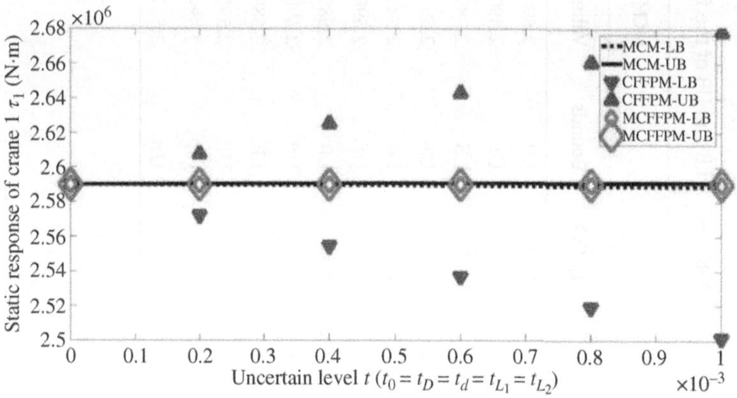

Figure 6.7 LB and UB of the fuzzy static response of the crane 1 with the MFV model t_0 under $\alpha = 0.8$.

Table 6.1 LB and UB of the fuzzy static response of the crane 1 with the MFV model t_0 under $\alpha = 0.4$.

		MCM (515.7 s)	CFFPM (0.0078 s)		MCFFPM (0.0143 s)	
t_0 (%)	Bounds	Value (N · m)	Value (N · m)	Error (%)	Value (N · m)	Error (%)
0.00	LB	2.5900E+06	2.5900E+06	0	2.5900E+06	0
	UB	2.5900E+06	2.5900E+06	0	2.5900E+06	0
0.02	LB	2.5899E+06	2.5370E+06	2.04	2.5900E+06	0
	UB	2.5904E+06	2.6430E+06	2.03	2.5901E+06	0.01
0.04	LB	2.5897E+06	2.4840E+06	4.08	2.5899E+06	0
	UB	2.5906E+06	2.6961E+06	4.07	2.5902E+06	0.02
0.06	LB	2.5894E+06	2.4310E+06	6.12	2.5898E+06	0.02
	UB	2.5907E+06	2.7491E+06	6.11	2.5903E+06	0.02
0.08	LB	2.5893E+06	2.3780E+06	8.16	2.5897E+06	0.02
	UB	2.5911E+06	2.8021E+06	8.14	2.5903E+06	0.03
0.10	LB	2.5890E+06	2.3250E+06	10.20	2.5896E+06	0.02
	UB	2.5913E+06	2.8551E+06	10.18	2.5904E+06	0.03

Table 6.2 LB and UB of the fuzzy static response of the crane 1 with the MFV model t_0 under $\alpha = 0.8$.

		MCM (515.7 s)	CFFPM (0.0017 s)		MCFFPM (0.0065 s)	
t_0 (%)	Bounds	Value (N · m)	Value (N · m)	Error (%)	Value (N · m)	Error (%)
0.00	LB	2.5900E+06	2.5900E+06	0	2.5900E+06	0
	UB	2.5900E+06	2.5900E+06	0	2.5900E+06	0
0.02	LB	2.5899E+06	2.5724E+06	0.68	2.5900E+06	0
	UB	2.5904E+06	2.6077E+06	0.68	2.5901E+06	0.01
0.04	LB	2.5897E+06	2.5547E+06	1.37	2.5900E+06	0.01
	UB	2.5906E+06	2.6254E+06	1.36	2.5901E+06	0.02
0.06	LB	2.5894E+06	2.5370E+06	2.05	2.5900E+06	0.02
	UB	2.5907E+06	2.6430E+06	2.03	2.5901E+06	0.02
0.08	LB	2.5893E+06	2.5194E+06	2.72	2.5899E+06	0.02
	UB	2.5911E+06	2.6607E+06	2.71	2.5901E+06	0.04
0.10	LB	2.5890E+06	2.5017E+06	3.40	2.5899E+06	0.03
	UB	2.5913E+06	2.6784E+06	3.39	2.5902E+06	0.04

indicates that the CFFPM and MCFFPM are appropriate for calculation of the fuzzy static response with small uncertain levels. The maximum relative errors between the LB and UB of the fuzzy static response with the MFV model t_0 under $\alpha = 0.4$ by the CFFPM and MCM, and between those by the MCFFPM and MCM are 10.20 and 0.03%, respectively. It indicates that the accuracy of the MCFFPM for calculating the fuzzy static response is much larger than that of the CFFPM. As shown in Table 6.2, the maximum relative errors between the LB and UB of the fuzzy static response with the MFV model t_0 under $\alpha = 0.8$ by the CFFPM and MCM are 3.39%. It indicates that the accuracy of the CFFPM can be improved by choosing a larger cut level for fuzzy parameters, which leads to smaller relative interval parameters.

The efficiencies of the CFFPM and MCFFPM are also compared in Tables 6.1 and 6.2. Calculation times of the MCM, CFFPM, and MCFFPM are 515.7, 0.0078, and 0.0143 sec, respectively. The CFFPM is 66,114 times faster than the MCM and the MCFFPM is 36,061 times faster than the MCM. Note that while there is a little loss of efficiency, the accuracy of the MCFFPM is greatly improved.

Membership functions of the fuzzy static response of the crane 1 with different uncertain levels are calculated by the MCM, CFFPM, and MCFFPM and shown in Figures 6.8–6.10, respectively. Three uncertain levels are considered: (1) $t_0 = 0.0002$; (2) $t_0 = 0.0006$; and (3) $t_0 = 0.0010$. Figures 6.6–6.8 show that all membership functions of the fuzzy static response of the crane 1 yielded by the MCM, CFFPM, and MCFFPM almost satisfy the triangular distribution. Figures 6.11 and 6.12 show membership functions of the fuzzy static response of the crane 1 with different uncertain levels by the MCM and CFFPM, and the MCM and MCFFPM, respectively. From Figures 6.11 and 6.12, one finds that three dashed curves of membership functions of the fuzzy static response of the crane

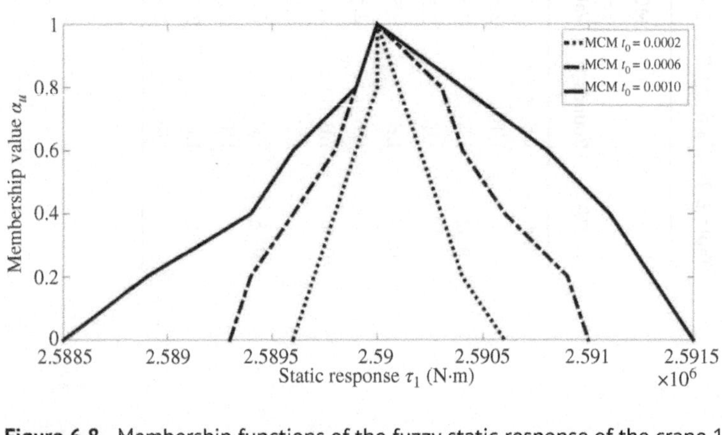

Figure 6.8 Membership functions of the fuzzy static response of the crane 1 yielded by the MCM.

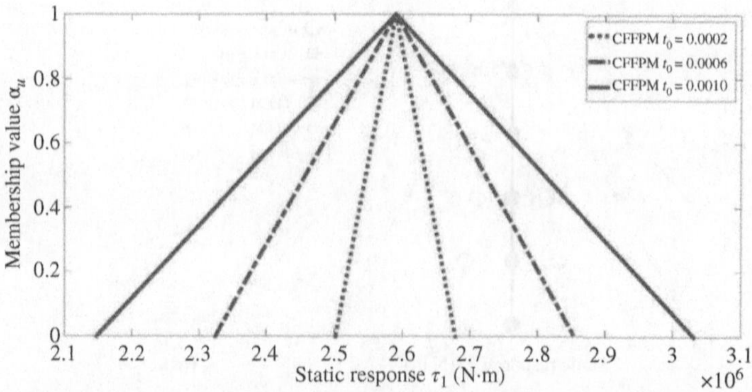

Figure 6.9 Membership functions of the fuzzy static response of the crane 1 yielded by the CFFPM.

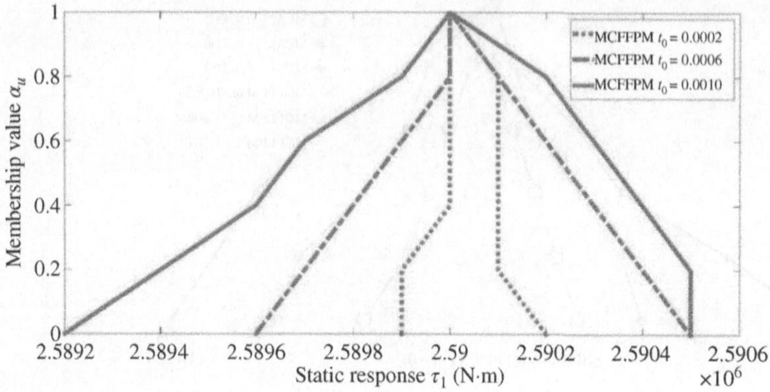

Figure 6.10 Membership functions of the fuzzy static response of the crane 1 yielded by the MCFFPM.

1 obtained by the MCFFPM are much closer to three solid curves obtained by the MCM when they are compared with three dashed curves obtained by the CFFPM. The results show that the accuracy of the MCFFPM is higher than that of the CFFPM for calculating the fuzzy static response, and consideration of higher-order terms in the MCFFPM can greatly improve the accuracy. Note that intervals between MCFFPM dashed curves and MCM solid curves increase from $t_0 = 0.0002$ to $t_0 = 0.0010$, and intervals between CFFPM dashed curves and MCM solid curves almost have no change from $t_0 = 0.0002$ to $t_0 = 0.0010$. The results show that the CFFPM is suitable for calculating the fuzzy static response with narrowly fuzzy parameters, and the MCFFPM is suitable for calculating the fuzzy static response with fuzzy parameters with different uncertain levels.

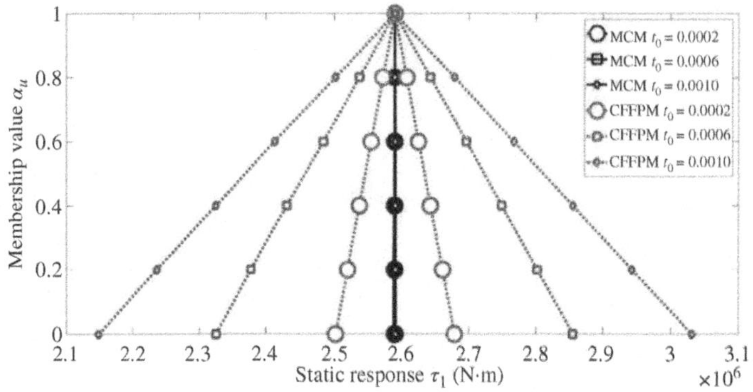

Figure 6.11 Membership functions of the fuzzy static response of the crane 1 yielded by the MCM and CFFPM.

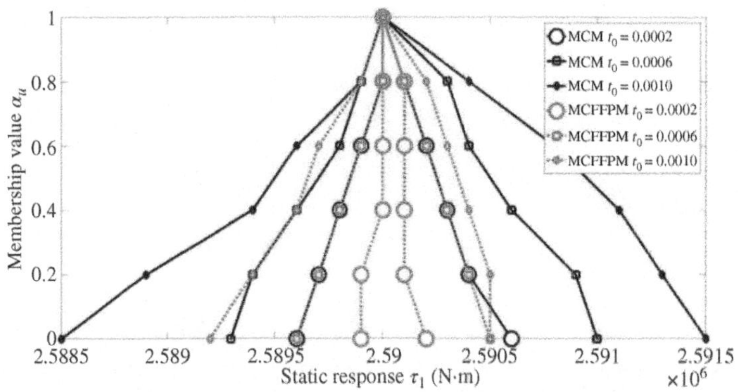

Figure 6.12 Membership functions of the fuzzy static response of the crane 1 yielded by the MCM and MCFFPM.

6.2.5 Conclusion

Two fuzzy perturbation methods, the CFFPM and MCFFPM, are introduced to obtain the fuzzy static response using fuzzy parameters. The fuzzy static equilibrium equation with the MFV model is constructed. The level-cut technique is applied to transform the fuzzy static equilibrium equation into several interval equations with different cut levels. The accuracy of the CFFPM based on the first-order Taylor series and Neumann series is low. To improve the accuracy, in the MCFFPM, the interval Jacobian matrix and interval virtual work vectors are approximated by the SGM, and the inverse of the interval Jacobian matrix is approximated by the modified SMW formula. Thus, two time-consuming operations are replaced by more efficient approaches. Finally, the fuzzy static response for every cut level can be calculated by a combination of the compound function technique, the IPM, and the fuzzy algorithm.

Several numerical examples are given to show that: (1) the fuzzy parameter D achieves a larger impact on the fuzzy static response of the crane 1 than the fuzzy parameter d, and fuzzy parameters L_1 and L_2 have no impact on the fuzzy static response of the crane 1. (2) Both the CFFPM and MCFFPM are appropriate for obtaining the fuzzy static response of the crane 1 with narrowly fuzzy uncertainty. Note that the MCFFPM yields a better accuracy and a lower efficiency than the CFFPM to obtain the fuzzy static response, and both the CFFPM and MCFFPM are more efficient than the MCM.

6.3 Dynamic Uncertainty Analysis with Hybrid Random and Interval Parameters

To predict the dynamic response field of the LSOAAC with random and interval parameters, a hybrid uncertain model is introduced. In this model, parameters with known probability distributions are treated as random variables, while those with specified LB and UB are represented as interval variables rather than precise values. Based on this hybrid uncertain model, the hybrid uncertain dynamic response equilibrium equation is formulated, incorporating both random and interval parameters simultaneously in the input and output terms. A modified hybrid uncertain analysis method (MHUAM) is then proposed. Compared with the hybrid MCM (HMCM) and the IPM, numerical results demonstrate the feasibility and efficiency of the MHUAM in solving hybrid LSOAAC problems.

6.3.1 LSOAAC Equilibrium Equation Under the Hybrid Uncertain Model

Consider the three-dimensional model of the LSOAAC illustrated in Figure 6.13, which consists of five components: the lifting arm, luffing cylinder, rotating table, hoisting rope, and suspended weight. A simplified schematic diagram of the LSOAAC is presented in Figure 6.14. The lower end of the lifting arm is connected to the rotating table via hinge point A, while its upper end is linked to the hoisting rope through hinge point B. For the luffing cylinder, its lower end is hinged to the rotating table at point C, and its upper end is connected to the lifting arm at hinge point D. Additionally, OO_1 serves as the slewing axis of the rotating table.

To simplify the analysis, the following assumptions are adopted [33]:

1. The interplay between the hoisting rope and the suspended weight is not to be disregarded.
2. The mass of the hoisting rope is considered negligible in comparison to the tension it is subjected to.
3. Given the slow luffing speed, the inertia force generated by the mobile crane's body is disregarded.

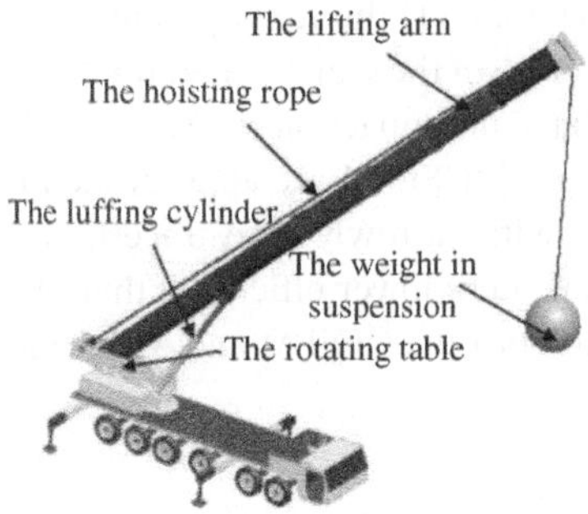

Figure 6.13 Three-dimensional model of the LSOAAC.

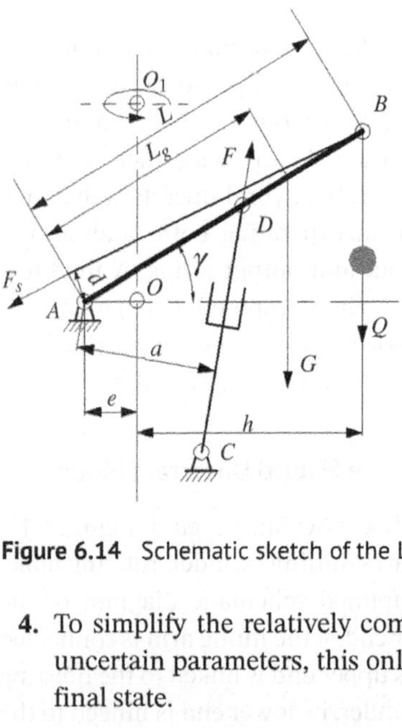

Figure 6.14 Schematic sketch of the LSOAAC.

4. To simplify the relatively complex process of analyzing the LSOAAC with uncertain parameters, this only examines the response of the LSOAAC at its final state.

Based on the moment equilibrium equation at hinge joint A, the dynamic response equation of the LSOAAC during the process of hoisting the suspended weight can be expressed as follows:

$$Fa = Q\,(L\cos\gamma) + G\,(L_g\cos\gamma) - F_s d \tag{6.43}$$

Here, γ denotes the angle of response between the lifting arm and the horizontal plane; F and F_s represent the thrust of the luffing cylinder and the tension of the hoisting rope, respectively; Q and G stand for the gravity of the suspended weight and the gravity of the lifting arm, respectively; L and L_g refer to the length of the lifting arm and the distance from the lower end of the lifting arm to its center of gravity, respectively; a and d indicate the distance from the lower end of the lifting arm to the luffing cylinder and the distance from the lower end of the lifting arm to the hoisting rope, respectively.

Equation (6.43) can also be written as

$$\cos\gamma = \frac{Fa + F_s d}{QL + GL_g} \tag{6.44}$$

To streamline the process of analyzing the dynamic response equation of the LSOAAC involving both random and interval parameters, a hybrid uncertainty model is established. Equation (6.43) is thus rephrased into the following form:

$$\boldsymbol{M} = \boldsymbol{T}\varepsilon \tag{6.45}$$

Here, $\boldsymbol{M}$ and $\boldsymbol{T}$ represent the input moment and output moment of the LSOAAC, respectively; ε denotes the cosine of the response angle γ. These variables can be expressed as follows:

$$\boldsymbol{M} = Fa + F_s d,\ \boldsymbol{T} = QL + GL_g,\ \varepsilon = \cos\gamma \tag{6.46}$$

The dynamic equilibrium equation (Eq. [6.45]) of the LSOAAC, which incorporates the hybrid uncertainty model, can be re-expressed as follows:

$$\boldsymbol{M}\left(\vec{\boldsymbol{X}}_{\boldsymbol{M}}^{R}, \vec{\boldsymbol{Y}}_{\boldsymbol{M}}^{I}\right) = \boldsymbol{T}\left(\vec{\boldsymbol{X}}_{\boldsymbol{T}}^{R}, \vec{\boldsymbol{Y}}_{\boldsymbol{T}}^{I}\right)\varepsilon \tag{6.47}$$

Here, $\boldsymbol{M}\left(\vec{\boldsymbol{X}}_{\boldsymbol{M}}^{R}, \vec{\boldsymbol{Y}}_{\boldsymbol{M}}^{I}\right)$ and $\boldsymbol{T}\left(\vec{\boldsymbol{X}}_{\boldsymbol{T}}^{R}, \vec{\boldsymbol{Y}}_{\boldsymbol{T}}^{I}\right)$ respectively denote the known input moment vector and known output moment vector of the LSOAAC, which contain distinct random parameters and interval parameters. Specifically, the random vector $\vec{\boldsymbol{X}}_{\boldsymbol{M}}^{R}$ stands for all input random parameters, while the random vector $\vec{\boldsymbol{X}}_{\boldsymbol{T}}^{R}$ represents all output random parameters. As for interval vectors, $\vec{\boldsymbol{Y}}_{\boldsymbol{M}}^{I}$ corresponds to all input interval parameters, and $\vec{\boldsymbol{Y}}_{\boldsymbol{T}}^{I}$ refers to all output interval parameters. These vectors can be expressed in the following forms:

$$\begin{aligned} &\vec{\boldsymbol{X}}_{\boldsymbol{M}}^{R} = \left(\vec{X}_{M1}^{R}, \vec{X}_{M2}^{R}, \ldots, \vec{X}_{Mk}^{R}, \ldots, \vec{X}_{Ms}^{R}\right), \vec{\boldsymbol{X}}_{\boldsymbol{T}}^{R} = \left(\vec{X}_{T1}^{R}, \vec{X}_{T2}^{R}, \ldots, \vec{X}_{Ti}^{R}, \ldots, \vec{X}_{Tn}^{R}\right) \\ &\vec{\boldsymbol{Y}}_{\boldsymbol{M}}^{I} = \left(\vec{Y}_{M1}^{I}, \vec{Y}_{M2}^{I}, \ldots, \vec{Y}_{Ml}^{I}, \ldots, \vec{Y}_{Mw}^{I}\right), \vec{\boldsymbol{Y}}_{\boldsymbol{T}}^{I} = \left(\vec{Y}_{T1}^{I}, \vec{Y}_{T2}^{I}, \ldots, \vec{Y}_{Tj}^{I}, \ldots, \vec{Y}_{Tm}^{I}\right) \end{aligned} \tag{6.48}$$

Clearly, the cosine of the response angle ε is also an unknown hybrid variable that simultaneously involves distinct random parameters and interval parameters. Here, we define the dynamic response field as the quantitative influence of uncertain parameters on the LSOAAC's response, which can be expressed through the mathematical properties of the extrema of the dynamic response bounds.

6.3.2 MHUAM for the Dynamic Response Analysis of LSOAAC

In this section, building upon the research by Gao et al. [29, 30] and Wang et al. [34, 35], an MHUAM is proposed. This method is designed to address the prediction of the dynamic response field of the LSOAAC when it involves different random parameters and interval parameters at the same time. Specifically, the random IPM is employed to analyze the dynamic response intervals, while the random interval moment method is used to examine the probabilistic distribution characteristics of these intervals.

6.3.2.1 Random IPM and Random Interval Moment Method

In [29, 30, 34–36], the random IPM (RIPM) and the random interval moment method (RIMM) have been widely applied to engineering problems for predicting response fields. A brief introduction to these two methods is provided as follows:

Without loss of generation, let random interval variable $\mathbf{z}^{\mathbf{RI}}$ be a function of multiple random variables and interval variables, which are respectively represented by the random vector $\vec{\boldsymbol{X}}^{R} = (x_1^R, x_2^R, \ldots, x_i^R, \ldots, x_n^R)$ and the interval vector $\vec{\boldsymbol{Y}}^{I} = \left(y_1^I, y_2^I, \ldots, y_j^I, \ldots, y_m^I\right)$, respectively. For the sake of simplicity, it is assumed that all variables are independent of one another. By applying the first-order Taylor series expansion and perturbation theory, the random interval variable $\boldsymbol{z}^{RI}$ can be expressed as

$$
\begin{aligned}
z^{RI} &= f\left(\vec{X}^R, \vec{Y}^I\right) = f\left(E\left(\vec{X}^R\right), \vec{Y}^C\right) + \sum_{j=1}^{m} \left\{ \left. \frac{\partial f\left(\vec{X}^R, \vec{Y}^I\right)}{\partial y_j^I} \right|_{E\left(\vec{X}^R\right), \vec{Y}^C} \right\} \Delta y_j^I \\
&\quad + \sum_{i=1}^{n} \left\{ \left. \frac{\partial f\left(\vec{X}^R, \vec{Y}^I\right)}{\partial x_i^R} \right|_{E\left(\vec{X}^R\right), \vec{Y}^C} + \sum_{j=1}^{m} \left\{ \left. \frac{\partial^2 f\left(\vec{X}^R, \vec{Y}^I\right)}{\partial x_i^R \partial y_j^I} \right|_{E\left(\vec{X}^R\right), \vec{Y}^C} \right\} \Delta y_j^I \right\} \\
&\quad (x_i^R - E\,(x_i^R)) + R
\end{aligned}
\tag{6.49}
$$

Here, R denotes the remainder term. $\vec{\boldsymbol{Y}}^{C} = \left(y_1^c, y_2^c, \ldots, y_j^c, \ldots, y_m^c\right)$ represents the interval midpoint vector, where y_j^c and Δy_j^I respectively stand for the interval midpoint and IR of the interval variable y_j^I.

By employing the random interval moment method and disregarding higher-order terms, the expectation and variance of the random interval variable $\boldsymbol{z}^{RI}$ can be expressed as

$$E\left(z^{RI}\right)=f\left(E\left(\vec{X}^{R}\right),\vec{Y}^{C}\right)+\sum_{j=1}^{m}\left\{\left.\frac{\partial f\left(\vec{X}^{R},\vec{Y}^{I}\right)}{\partial y_j^I}\right|_{E\left(\vec{X}^{R}\right),\vec{Y}^{C}}\right\}\Delta y_j^I$$

$$\mathrm{Var}\left(z^{RI}\right)=\sum_{i=1}^{n}\left\{\left.\frac{\partial f\left(\vec{X}^{R},\vec{Y}^{I}\right)}{\partial x_i^R}\right|_{E\left(\vec{X}^{R}\right),\vec{Y}^{C}}+\sum_{j=1}^{m}\left\{\left.\frac{\partial^2 f\left(\vec{X}^{R},\vec{Y}^{I}\right)}{\partial x_i^R\partial y_j^I}\right|_{E\left(\vec{X}^{R}\right),\vec{Y}^{C}}\right\}\Delta y_j^I\right\}^2\cdot\mathrm{Var}\left(x_i^R\right) \tag{6.50}$$

Here, E and Var represent the expectation operator and variance operator, respectively.

6.3.2.2 Dynamic Response Analysis of the LSOAAC

Neglecting the higher order terms, the first-order Taylor expansions of input moment vector $\boldsymbol{M}\left(\vec{X}_{M}^{R},\vec{Y}_{M}^{I}\right)$ around the expectation of the known input moment vector at $\left(E\left(\vec{X}_{M}^{R}\right),\vec{Y}_{M}^{C}\right)$ can be expressed as

$$\boldsymbol{M}\left(\vec{X}_{M}^{R},\vec{Y}_{M}^{I}\right)=\boldsymbol{M}_{\mathrm{m}}+\Delta_1\boldsymbol{M} \tag{6.51}$$

where

$$\boldsymbol{M}_{\mathrm{m}}=\boldsymbol{M}\left(E\left(\vec{X}_{M}^{R}\right),\vec{Y}_{M}^{C}\right)=\boldsymbol{M}\left(\vec{X}_{M}^{R},\vec{Y}_{M}^{I}\right)\Big|_{\vec{X}_{M}^{R},\vec{Y}_{M}^{I}=E\left(\vec{X}_{M}^{R}\right),\vec{Y}_{M}^{C}} \tag{6.52}$$

$$\begin{aligned}\Delta_1\boldsymbol{M}=&\sum_{l=1}^{w}\left\{\left.\frac{\partial\boldsymbol{M}\left(\vec{X}_{M}^{R},\vec{Y}_{M}^{I}\right)}{\partial y_{Ml}^{I}}\right|_{\vec{X}_{M}^{R},\vec{Y}_{M}^{I}=E\left(\vec{X}_{M}^{R}\right),\vec{Y}_{M}^{C}}\right\}\cdot\Delta y_{Ml}^{I}\\&+\sum_{k=1}^{s}\left\{\left.\frac{\partial\boldsymbol{M}\left(\vec{X}_{M}^{R},\vec{Y}_{M}^{I}\right)}{\partial x_{Mk}^{R}}\right|_{\vec{X}_{M}^{R},\vec{Y}_{M}^{I}=E\left(\vec{X}_{M}^{R}\right),\vec{Y}_{M}^{C}}\right\}\cdot\left(x_{Mk}^{R}-E\left(x_{Mk}^{R}\right)\right)\\&+\sum_{l=1}^{w}\sum_{k=1}^{s}\left\{\left.\frac{\partial^2\boldsymbol{M}\left(\vec{X}_{M}^{R},\vec{Y}_{M}^{I}\right)}{\partial x_{Mk}^{R}\partial y_{Ml}^{I}}\right|_{\vec{X}_{M}^{R},\vec{Y}_{M}^{I}=E\left(\vec{X}_{M}^{R}\right),\vec{Y}_{M}^{C}}\right\}\cdot\left(x_{Mk}^{R}-E\left(x_{Mk}^{R}\right)\right)\cdot\Delta y_{Ml}^{I}\end{aligned} \tag{6.53}$$

where Δy_{Ml}^{I} is the IR of interval variable $\vec{Y}_{Ml}^{I}$.

Likewise, by applying the first-order Taylor series expansion, the output moment vector $\boldsymbol{T}\left(\vec{\boldsymbol{X}}_T^R, \vec{\boldsymbol{Y}}_T^I\right)$ evaluated at the expectation of the known output moment vector, i.e., at $\left(E\left(\vec{\boldsymbol{X}}_T^R\right), \vec{\boldsymbol{Y}}_T^C\right)$, can be expressed as

$$\boldsymbol{T}\left(\vec{\boldsymbol{X}}_T^R, \vec{\boldsymbol{Y}}_T^I\right) = \boldsymbol{T}_{\mathrm{m}} + \Delta_1 \boldsymbol{T} \tag{6.54}$$

where

$$\boldsymbol{T}_{\mathrm{m}} = \boldsymbol{T}\left(E\left(\vec{\boldsymbol{X}}_T^R\right), \vec{\boldsymbol{Y}}_T^C\right) = \boldsymbol{T}\left(\vec{\boldsymbol{X}}_T^R, \vec{\boldsymbol{Y}}_T^I\right)\Bigg|_{\vec{\boldsymbol{X}}_T^R, \vec{\boldsymbol{Y}}_T^I = E\left(\vec{\boldsymbol{X}}_T^R\right), \vec{\boldsymbol{Y}}_T^C} \tag{6.55}$$

$$\begin{aligned} \Delta_1 \boldsymbol{T} = & \sum_{j=1}^{m} \left\{ \frac{\partial \boldsymbol{T}\left(\vec{\boldsymbol{X}}_T^R, \vec{\boldsymbol{Y}}_T^I\right)}{\partial y_{Tj}^I} \Bigg|_{\vec{\boldsymbol{X}}_T^R, \vec{\boldsymbol{Y}}_T^I = E\left(\vec{\boldsymbol{X}}_T^R\right), \vec{\boldsymbol{Y}}_T^C} \right\} \cdot \Delta y_{Tj}^i \\ & + \sum_{i=1}^{n} \left\{ \frac{\partial \boldsymbol{T}\left(\vec{\boldsymbol{X}}_T^R, \vec{\boldsymbol{Y}}_T^I\right)}{\partial x_{Ti}^R} \Bigg|_{\vec{\boldsymbol{X}}_T^R, \vec{\boldsymbol{Y}}_T^I = E\left(\vec{\boldsymbol{X}}_T^R\right), \vec{\boldsymbol{Y}}_T^C} \right\} \cdot \left(x_{Ti}^R - E\left(x_{Ti}^R\right)\right) \\ & + \sum_{j=1}^{m} \sum_{i=1}^{n} \left\{ \frac{\partial^2 \boldsymbol{T}\left(\vec{\boldsymbol{X}}_T^R, \vec{\boldsymbol{Y}}_T^I\right)}{\partial x_{Ti}^R \partial y_{Tj}^I} \Bigg|_{\vec{\boldsymbol{X}}_T^R, \vec{\boldsymbol{Y}}_T^I = E\left(\vec{\boldsymbol{X}}_T^R\right), \vec{\boldsymbol{Y}}_T^C} \right\} \cdot \left(x_{Ti}^R - E\left(x_{Ti}^R\right)\right) \cdot \Delta y_{Tj}^i \end{aligned} \tag{6.56}$$

where Δy_{Tj}^I is the IR of interval variable $\vec{Y}_{Tj}^I$.

Applying perturbation theory and substituting Eqs. (6.51) and (6.54) into Eq. (6.47) results in

$$\varepsilon = \left(\boldsymbol{T}_{\mathrm{m}} + \Delta_1 \boldsymbol{T}\right)^{-1} \left(\boldsymbol{M}_{\mathrm{m}} + \Delta_1 \boldsymbol{M}\right) \tag{6.57}$$

For most engineering problems, the improvement in accuracy achieved by using higher-order Taylor series expansions is relatively minor, while the computational effort increases correspondingly. Therefore, if the spectral radius of $\left(\boldsymbol{T}_{\mathrm{m}}\right)^{-1} \Delta_1 \boldsymbol{T}$ is less than 1, specifically, $\left\|\left(\boldsymbol{T}_{\mathrm{m}}\right)^{-1} \Delta_1 \boldsymbol{T}\right\| < 1$, the inverse $\left(\boldsymbol{T}_{\mathrm{m}} + \Delta_1 \boldsymbol{T}\right)^{-1}$ can be expanded using the Neumann series [14]

$$\left(\boldsymbol{T}_{\mathrm{m}} + \Delta_1 \boldsymbol{T}\right)^{-1} = \left(\boldsymbol{T}_{\mathrm{m}}\right)^{-1} - \left(\boldsymbol{T}_{\mathrm{m}}\right)^{-1} \Delta_1 \boldsymbol{T} \left(\boldsymbol{T}_{\mathrm{m}}\right)^{-1} + \left(\left(\boldsymbol{T}_{\mathrm{m}}\right)^{-1} \Delta_1 \boldsymbol{T}\right)^2 \left(\boldsymbol{T}_{\mathrm{m}}\right)^{-1} \tag{6.58}$$

Substituting Eq. (6.58) into Eq. (6.57) and ignoring higher-order terms, the following result can be obtained:

$$\varepsilon = \left(\boldsymbol{T}_{\mathrm{m}}\right)^{-1} \boldsymbol{M}_{\mathrm{m}} + \left(\boldsymbol{T}_{\mathrm{m}}\right)^{-1} \Delta_1 \boldsymbol{M} - \left(\boldsymbol{T}_{\mathrm{m}}\right)^{-1} \Delta_1 \boldsymbol{T} \left(\boldsymbol{T}_{\mathrm{m}}\right)^{-1} \boldsymbol{M}_{\mathrm{m}} \tag{6.59}$$

Substituting Eqs. (6.52), (6.53), (6.55), and (6.56) into Eq. (6.59) yields

$$
\begin{aligned}
\varepsilon &= T_{\mathrm{m}}^{-1} M_{\mathrm{m}} \\
&+T_{\mathrm{m}}^{-1}\left\{\sum_{l=1}^{w}\left\{\left.\frac{\partial M\left(\vec{X}_M^R, \vec{Y}_M^I\right)}{\partial y_{Ml}^I}\right|_{\vec{X}_M^R,\vec{Y}_M^I=E\left(\vec{X}_M^R\right),\vec{Y}_M^C}\right\}\Delta y_{Ml}^I\right\} \\
&+T_{\mathrm{m}}^{-1}\left\{\sum_{k=1}^{s}\left\{\left.\frac{\partial M\left(\vec{X}_M^R, \vec{Y}_M^I\right)}{\partial x_{Mk}^R}\right|_{\vec{X}_M^R,\vec{Y}_M^I=E\left(\vec{X}_M^R\right),\vec{Y}_M^C}\right\}\left(x_{Mk}^R-E\left(x_{Mk}^R\right)\right)\right\} \\
&+T_{\mathrm{m}}^{-1}\left\{\sum_{l=1}^{w}\sum_{k=1}^{s}\left\{\left.\frac{\partial^2 M\left(\vec{X}_M^R, \vec{Y}_M^I\right)}{\partial x_{Mk}^R\partial y_{Ml}^I}\right|_{\vec{X}_M^R,\vec{Y}_M^I=E\left(\vec{X}_M^R\right),\vec{Y}_M^C}\right\}\left(x_{Mk}^R-E\left(x_{Mk}^R\right)\right)\Delta y_{Ml}^I\right\} \\
&-T_{\mathrm{m}}^{-2}M_{\mathrm{m}}\left\{\sum_{j=1}^{m}\left\{\left.\frac{\partial T\left(\vec{X}_T^R, \vec{Y}_T^I\right)}{\partial y_{Tj}^I}\right|_{\vec{X}_T^R,\vec{Y}_T^I=E\left(\vec{X}_T^R\right),\vec{Y}_T^C}\right\}\Delta y_{Tj}^i\right\} \\
&-T_{\mathrm{m}}^{-2}M_{\mathrm{m}}\left\{\sum_{i=1}^{n}\left\{\left.\frac{\partial T\left(\vec{X}_T^R, \vec{Y}_T^I\right)}{\partial x_{Ti}^R}\right|_{\vec{X}_T^R,\vec{Y}_T^I=E\left(\vec{X}_T^R\right),\vec{Y}_T^C}\right\}\left(x_{Ti}^R-E\left(x_{Ti}^R\right)\right)\right\} \\
&-T_{\mathrm{m}}^{-2}M_{\mathrm{m}}\left\{\sum_{j=1}^{m}\sum_{i=1}^{n}\left\{\left.\frac{\partial^2 T\left(\vec{X}_T^R, \vec{Y}_T^I\right)}{\partial x_{Ti}^R\partial y_{Tj}^I}\right|_{\vec{X}_T^R,\vec{Y}_T^I=E\left(\vec{X}_T^R\right),\vec{Y}_T^C}\right\}\left(x_{Ti}^R-E\left(x_{Ti}^R\right)\right)\Delta y_{Tj}^i\right\}
\end{aligned}
\tag{6.60}
$$

6.3.3 Hybrid LSOAAC Response Field Calculation

The expectation and variance of the hybrid uncertain response field respectively represent the MV and dispersion degree of this field. Consequently, the probabilistic distribution characteristics of the hybrid LSOAAC response field can be characterized by the mathematical properties of the extrema of the bounds of the LSOAAC response.

6.3.3.1 Monotonic Analysis for Hybrid Uncertain Response Field

To determine the interval ranges of the dynamic response field under the hybrid uncertainty model, the following steps are adopted: First, monotonic analysis is employed to derive the intervals of the dynamic response bounds $\overline{\varepsilon}^I$ and $\underline{\varepsilon}^I$ with respect to the interval variables δ_{Ml} and δ_{Tl}, respectively. Subsequently, based on the random interval moment method, the probabilistic distribution characteristics of the hybrid uncertain response field with respect to the input random variable x_{Mk}^{R} and output random variable x_{Ti}^{R} can be readily obtained, respectively.

Based on Eq. (6.60), it can be rewritten into the following alternative form:

$$
\begin{aligned}
\varepsilon = \boldsymbol{S}_0 + \Delta \boldsymbol{S} = (\boldsymbol{T}_\mathrm{m})^{-1} \boldsymbol{M}_\mathrm{m} \\
&+ (\boldsymbol{T}_\mathrm{m})^{-1} \left\{ \sum_{k=1}^{s} \left\{ \left. \frac{\partial \boldsymbol{M}\left(\vec{\boldsymbol{X}}_M^R, \vec{\boldsymbol{Y}}_M^I\right)}{\partial x_{Mk}^R} \right|_{\vec{X}_M^R, \vec{Y}_M^I = E\left(\vec{X}_M^R\right), \vec{Y}_M^C} \right\} (x_{Mk}^R - E(x_{Mk}^R)) \right\} \\
&- (\boldsymbol{T}_\mathrm{m})^{-2} \boldsymbol{M}_\mathrm{m} \left\{ \sum_{i=1}^{n} \left\{ \left. \frac{\partial \boldsymbol{T}\left(\vec{\boldsymbol{X}}_T^R, \vec{\boldsymbol{Y}}_T^I\right)}{\partial x_{Ti}^R} \right|_{\vec{X}_T^R, \vec{Y}_T^I = E\left(\vec{X}_T^R\right), \vec{Y}_T^C} \right\} (x_{Ti}^R - E(x_{Ti}^R)) \right\} \\
&+ (\boldsymbol{T}_\mathrm{m})^{-1} \left[\sum_{l=1}^{w} \left\{ \left. \frac{\partial \boldsymbol{M}\left(\vec{\boldsymbol{X}}_M^R, \vec{\boldsymbol{Y}}_M^I\right)}{\partial y_{Ml}^I} \right|_{\vec{X}_M^R, \vec{Y}_M^I = E\left(\vec{X}_M^R\right), \vec{Y}_M^C} \right\} \right. \\
&\left. + \sum_{l=1}^{w} \sum_{k=1}^{s} \left\{ \left. \frac{\partial^2 \boldsymbol{M}\left(\vec{\boldsymbol{X}}_M^R, \vec{\boldsymbol{Y}}_M^I\right)}{\partial x_{Mk}^R \partial y_{Ml}^I} \right|_{\vec{X}_M^R, \vec{Y}_M^I = E\left(\vec{X}_M^R\right), \vec{Y}_M^C} \right\} (x_{Mk}^R - E(x_{Mk}^R)) \right] \cdot \Delta y_{Ml}^{l} \\
&- (\boldsymbol{T}_\mathrm{m})^{-2} \boldsymbol{M}_\mathrm{m} \left[\sum_{j=1}^{m} \left\{ \left. \frac{\partial \boldsymbol{T}\left(\vec{\boldsymbol{X}}_T^R, \vec{\boldsymbol{Y}}_T^I\right)}{\partial y_{Tj}^I} \right|_{\vec{X}_T^R, \vec{Y}_T^I = E\left(\vec{X}_T^R\right), \vec{Y}_T^C} \right\} \right. \\
&\left. + \sum_{j=1}^{m} \sum_{i=1}^{n} \left\{ \left. \frac{\partial^2 \boldsymbol{T}\left(\vec{\boldsymbol{X}}_T^R, \vec{\boldsymbol{Y}}_T^I\right)}{\partial x_{Ti}^R \partial y_{Tj}^I} \right|_{\vec{X}_T^R, \vec{Y}_T^I = E\left(\vec{X}_T^R\right), \vec{Y}_T^C} \right\} (x_{Ti}^R - E(x_{Ti}^R)) \right] \cdot \Delta y_{Tj}^{i}
\end{aligned}
\tag{6.61}
$$

where

$$
\begin{aligned}
\boldsymbol{S}_0 = (\boldsymbol{T}_\mathrm{m})^{-1} \boldsymbol{M}_\mathrm{m} \\
&+ (\boldsymbol{T}_\mathrm{m})^{-1} \left\{ \sum_{k=1}^{s} \left\{ \left. \frac{\partial \boldsymbol{M}\left(\vec{\boldsymbol{X}}_M^R, \vec{\boldsymbol{Y}}_M^I\right)}{\partial x_{Mk}^R} \right|_{\vec{X}_M^R, \vec{Y}_M^I = E\left(\vec{X}_M^R\right), \vec{Y}_M^C} \right\} (x_{Mk}^R - E(x_{Mk}^R)) \right\} \\
&- (\boldsymbol{T}_\mathrm{m})^{-2} \boldsymbol{M}_\mathrm{m} \left\{ \sum_{i=1}^{n} \left\{ \left. \frac{\partial \boldsymbol{T}\left(\vec{\boldsymbol{X}}_T^R, \vec{\boldsymbol{Y}}_T^I\right)}{\partial x_{Ti}^R} \right|_{\vec{X}_T^R, \vec{Y}_T^I = E\left(\vec{X}_T^R\right), \vec{Y}_T^C} \right\} (x_{Ti}^R - E(x_{Ti}^R)) \right\}
\end{aligned}
\tag{6.62}
$$

$$
\begin{aligned}
\Delta S = &(T_{\mathrm{m}})^{-1}\left[\sum_{l=1}^{w}\left\{\left.\frac{\partial M\left(\vec{X}_M^R,\vec{Y}_M^I\right)}{\partial y_{Ml}^I}\right|_{\vec{X}_M^R,\vec{Y}_M^I=E\left(\vec{X}_M^R\right),\vec{Y}_M^C}\right\}\right.\\
&\left.+\sum_{l=1}^{w}\sum_{k=1}^{s}\left\{\left.\frac{\partial^2 M\left(\vec{X}_M^R,\vec{Y}_M^I\right)}{\partial x_{Mk}^R\partial y_{Ml}^I}\right|_{\vec{X}_M^R,\vec{Y}_M^I=E\left(\vec{X}_M^R\right),\vec{Y}_M^C}\right\}\left(x_{Mk}^R-E\left(x_{Mk}^R\right)\right)\right]\cdot\Delta y_{Ml}^I\\
&-(T_{\mathrm{m}})^{-2}M_{\mathrm{m}}\left[\sum_{j=1}^{m}\left\{\left.\frac{\partial T\left(\vec{X}_T^R,\vec{Y}_T^I\right)}{\partial y_{Tj}^I}\right|_{\vec{X}_T^R,\vec{Y}_T^I=E\left(\vec{X}_T^R\right),\vec{Y}_T^C}\right\}\right.\\
&\left.+\sum_{j=1}^{m}\sum_{i=1}^{n}\left\{\left.\frac{\partial^2 T\left(\vec{X}_T^R,\vec{Y}_T^I\right)}{\partial x_{Ti}^R\partial y_{Tj}^I}\right|_{\vec{X}_T^R,\vec{Y}_T^I=E\left(\vec{X}_T^R\right),\vec{Y}_T^C}\right\}\left(x_{Ti}^R-E\left(x_{Ti}^R\right)\right)\right]\cdot\Delta y_{Tj}^i
\end{aligned}
\tag{6.63}
$$

Based on the interval algorithm, we have

$$
\Delta y_{Ml}^l = \Delta y_{Ml}\cdot\delta_{Ml},\ \Delta y_{Tj}^i = \Delta y_{Tl}\cdot\delta_{Tl}
\tag{6.64}
$$

In this context, since $\delta_{Ml}\in[-1,1]$ and $\delta_{Tj}\in[-1,1]$ represent fixed intervals, Eq. (6.63) may be re-expressed as follows

$$
\begin{aligned}
\Delta S = &(T_{\mathrm{m}})^{-1}\left[\sum_{l=1}^{w}\left\{\left.\frac{\partial M\left(\vec{X}_M^R,\vec{Y}_M^I\right)}{\partial y_{Ml}^I}\right|_{\vec{X}_M^R,\vec{Y}_M^I=E\left(\vec{X}_M^R\right),\vec{Y}_M^C}\right\}\right.\\
&\left.+\sum_{l=1}^{w}\sum_{k=1}^{s}\left\{\left.\frac{\partial^2 M\left(\vec{X}_M^R,\vec{Y}_M^I\right)}{\partial x_{Mk}^R\partial y_{Ml}^I}\right|_{\vec{X}_M^R,\vec{Y}_M^I=E\left(\vec{X}_M^R\right),\vec{Y}_M^C}\right\}\left(x_{Mk}^R-E\left(x_{Mk}^R\right)\right)\right]\\
&\cdot\Delta y_{Ml}\cdot\delta_{Ml}\\
&-(T_{\mathrm{m}})^{-2}M_{\mathrm{m}}\left[\sum_{j=1}^{m}\left\{\left.\frac{\partial T\left(\vec{X}_T^R,\vec{Y}_T^I\right)}{\partial y_{Tj}^I}\right|_{\vec{X}_T^R,\vec{Y}_T^I=E\left(\vec{X}_T^R\right),\vec{Y}_T^C}\right\}\right.\\
&\left.+\sum_{j=1}^{m}\sum_{i=1}^{n}\left\{\left.\frac{\partial^2 T\left(\vec{X}_T^R,\vec{Y}_T^I\right)}{\partial x_{Ti}^R\partial y_{Tj}^I}\right|_{\vec{X}_T^R,\vec{Y}_T^I=E\left(\vec{X}_T^R\right),\vec{Y}_T^C}\right\}\left(x_{Ti}^R-E\left(x_{Ti}^R\right)\right)\right]\\
&\cdot\Delta y_{Tj}\cdot\delta_{Tj}
\end{aligned}
\tag{6.65}
$$

It is evident that both $\partial\Delta S/\partial\Delta y_{Ml}^l$ and $\partial\Delta S/\partial\Delta y_{Tj}^i$ are constant values, and they have no correlation with the interval variables Δy_{Ml}^l $(l = 1, 2, \dots, w)$ and Δy_{Tj}^i $(j = 1, 2, \dots, m)$. Thus, by virtue of the interval union operation and relying on the monotonicity of $\Delta \boldsymbol{S}$ with respect to the standard interval variables δ_{Ml} and δ_{Tj}, the maximum and minimum values of the bounds of $\Delta \boldsymbol{S}$ can be derived by

$$\max_{\delta_{Tj}=+1} / \min_{\delta_{Tj}=-1} \max_{\delta_{Ml}=+1} \Delta \boldsymbol{S} = (\boldsymbol{T}_{\mathrm{m}})^{-1} \left| \sum_{l=1}^{w} \left\{ \left. \frac{\partial \boldsymbol{M}\left(\vec{X}_M^R, \vec{Y}_M^I\right)}{\partial y_{Ml}^I} \right|_{\vec{X}_M^R, \vec{Y}_M^I = E\left(\vec{X}_M^R\right), \vec{Y}_M^C} \right\} + \sum_{l=1}^{w} \sum_{k=1}^{s} \left\{ \left. \frac{\partial^2 \boldsymbol{M}\left(\vec{X}_M^R, \vec{Y}_M^I\right)}{\partial x_{Mk}^R \partial y_{Ml}^I} \right|_{\vec{X}_M^R, \vec{Y}_M^I = E\left(\vec{X}_M^R\right), \vec{Y}_M^C} \right\} \left(x_{Mk}^R - E\left(x_{Mk}^R\right)\right) \right| \cdot \Delta y_{Ml} \pm (\boldsymbol{T}_{\mathrm{m}})^{-2} \boldsymbol{M}_{\mathrm{m}} \left| \sum_{j=1}^{m} \left\{ \left. \frac{\partial \boldsymbol{T}\left(\vec{X}_T^R, \vec{Y}_T^I\right)}{\partial y_{Tj}^I} \right|_{\vec{X}_T^R, \vec{Y}_T^I = E\left(\vec{X}_T^R\right), \vec{Y}_T^C} \right\} + \sum_{j=1}^{m} \sum_{i=1}^{n} \left\{ \left. \frac{\partial^2 \boldsymbol{T}\left(\vec{X}_T^R, \vec{Y}_T^I\right)}{\partial x_{Ti}^R \partial y_{Tj}^I} \right|_{\vec{X}_T^R, \vec{Y}_T^I = E\left(\vec{X}_T^R\right), \vec{Y}_T^C} \right\} \left(x_{Ti}^R - E\left(x_{Ti}^R\right)\right) \right| \cdot \Delta y_{Tj} \tag{6.66}$$

$$\max_{\delta_{Tj}=+1} / \min_{\delta_{Tj}=-1} \min_{\delta_{Ml}=-1} \Delta \boldsymbol{S} = -(\boldsymbol{T}_{\mathrm{m}})^{-1} \left| \sum_{l=1}^{w} \left\{ \left. \frac{\partial \boldsymbol{M}\left(\vec{X}_M^R, \vec{Y}_M^I\right)}{\partial y_{Ml}^I} \right|_{\vec{X}_M^R, \vec{Y}_M^I = E\left(\vec{X}_M^R\right), \vec{Y}_M^C} \right\} + \sum_{l=1}^{w} \sum_{k=1}^{s} \left\{ \left. \frac{\partial^2 \boldsymbol{M}\left(\vec{X}_M^R, \vec{Y}_M^I\right)}{\partial x_{Mk}^R \partial y_{Ml}^I} \right|_{\vec{X}_M^R, \vec{Y}_M^I = E\left(\vec{X}_M^R\right), \vec{Y}_M^C} \right\} \left(x_{Mk}^R - E\left(x_{Mk}^R\right)\right) \right| \cdot \Delta y_{Ml} \pm (\boldsymbol{T}_{\mathrm{m}})^{-2} \boldsymbol{M}_{\mathrm{m}} \left| \sum_{j=1}^{m} \left\{ \left. \frac{\partial \boldsymbol{T}\left(\vec{X}_T^R, \vec{Y}_T^I\right)}{\partial y_{Tj}^I} \right|_{\vec{X}_T^R, \vec{Y}_T^I = E\left(\vec{X}_T^R\right), \vec{Y}_T^C} \right\} + \sum_{j=1}^{m} \sum_{i=1}^{n} \left\{ \left. \frac{\partial^2 \boldsymbol{T}\left(\vec{X}_T^R, \vec{Y}_T^I\right)}{\partial x_{Ti}^R \partial y_{Tj}^I} \right|_{\vec{X}_T^R, \vec{Y}_T^I = E\left(\vec{X}_T^R\right), \vec{Y}_T^C} \right\} \left(x_{Ti}^R - E\left(x_{Ti}^R\right)\right) \right| \cdot \Delta y_{Tj} \tag{6.67}$$

Here, $|\blacksquare|$ represents the absolute value. $\max\limits_{\delta_{Tj}=+1} / \min\limits_{\delta_{Tj}=-1} \Delta \boldsymbol{S}$ correspond to the maximum and minimum values of $\Delta \boldsymbol{S}$ with respect to δ_{Tj}, respectively. Additionally, $\max\limits_{\delta_{Tj}=+1} / \min\limits_{\delta_{Tj}=-1} \max\limits_{\delta_{Ml}=+1} \Delta \boldsymbol{S}$ and $\max\limits_{\delta_{Tj}=+1} / \min\limits_{\delta_{Tj}=-1} \max\limits_{\delta_{Ml}=-1} \Delta \boldsymbol{S}$ are the maximum and minimum of $\max\limits_{\delta_{Tj}=+1} / \min\limits_{\delta_{Tj}=-1} \Delta \boldsymbol{S}$ with respect to δ_{Ml}, respectively.

Since the IR Δy_{Ml} and Δy_{Tl} derived from the interval vectors are incorporated into the aforementioned formula, the UB $\overline{\varepsilon}^I$ and the LB $\underline{\varepsilon}^I$ of the dynamic response with respect to the interval vectors are not fixed deterministic values but rather intervals themselves.

Consequently, the maximum and minimum values of the UB $\overline{\varepsilon}^I$ in relation to the interval vectors $\vec{\boldsymbol{Y}}_M^I$ and $\vec{\boldsymbol{Y}}_T^I$ can be formulated as follows:

$$\begin{aligned}
\max / \min \left(\overline{\varepsilon}^I\right) &= \boldsymbol{S}_0 + \max_{\delta_{Tj}=+1} / \min_{\delta_{Tj}=-1} \max_{\delta_{Ml}=+1} \Delta \boldsymbol{S} = (\boldsymbol{T}_\mathrm{m})^{-1} \boldsymbol{M}_\mathrm{m} + (\boldsymbol{T}_\mathrm{m})^{-1} \\
&\left\{ \sum_{k=1}^{s} \left\{ \left. \frac{\partial \boldsymbol{M}\left(\vec{\boldsymbol{X}}_M^R, \vec{\boldsymbol{Y}}_M^I\right)}{\partial x_{Mk}^R} \right|_{\vec{\boldsymbol{X}}_M^R, \vec{\boldsymbol{Y}}_M^I = E\left(\vec{\boldsymbol{X}}_M^R\right), \vec{\boldsymbol{Y}}_M^C} \right\} \left(x_{Mk}^R - E\left(x_{Mk}^R\right)\right) \right\} \\
&- (\boldsymbol{T}_\mathrm{m})^{-2} \boldsymbol{M}_\mathrm{m} \left\{ \sum_{i=1}^{n} \left\{ \left. \frac{\partial \boldsymbol{T}\left(\vec{\boldsymbol{X}}_T^R, \vec{\boldsymbol{Y}}_T^I\right)}{\partial x_{Ti}^R} \right|_{\vec{\boldsymbol{X}}_T^R, \vec{\boldsymbol{Y}}_T^I = E\left(\vec{\boldsymbol{X}}_T^R\right), \vec{\boldsymbol{Y}}_T^C} \right\} \left(x_{Ti}^R - E\left(x_{Ti}^R\right)\right) \right\} \\
&+ \left| (\boldsymbol{T}_\mathrm{m})^{-1} \left[\sum_{l=1}^{w} \left\{ \left. \frac{\partial \boldsymbol{M}\left(\vec{\boldsymbol{X}}_M^R, \vec{\boldsymbol{Y}}_M^I\right)}{\partial y_{Ml}^I} \right|_{\vec{\boldsymbol{X}}_M^R, \vec{\boldsymbol{Y}}_M^I = E\left(\vec{\boldsymbol{X}}_M^R\right), \vec{\boldsymbol{Y}}_M^C} \right\} \right. \right. \\
&\left. \left. + \sum_{l=1}^{w} \sum_{k=1}^{s} \left\{ \left. \frac{\partial^2 \boldsymbol{M}\left(\vec{\boldsymbol{X}}_M^R, \vec{\boldsymbol{Y}}_M^I\right)}{\partial x_{Mk}^R \partial y_{Ml}^I} \right|_{\vec{\boldsymbol{X}}_M^R, \vec{\boldsymbol{Y}}_M^I = E\left(\vec{\boldsymbol{X}}_M^R\right), \vec{\boldsymbol{Y}}_M^C} \right\} \left(x_{Mk}^R - E\left(x_{Mk}^R\right)\right) \right] \right| \cdot \Delta y_{Ml} \\
&\pm (\boldsymbol{T}_\mathrm{m})^{-2} \boldsymbol{M}_\mathrm{m} \left| \sum_{j=1}^{m} \left\{ \left. \frac{\partial \boldsymbol{T}\left(\vec{\boldsymbol{X}}_T^R, \vec{\boldsymbol{Y}}_T^I\right)}{\partial y_{Tj}^I} \right|_{\vec{\boldsymbol{X}}_T^R, \vec{\boldsymbol{Y}}_T^I = E\left(\vec{\boldsymbol{X}}_T^R\right), \vec{\boldsymbol{Y}}_T^C} \right\} \right. \\
&\left. + \sum_{j=1}^{m} \sum_{i=1}^{n} \left\{ \left. \frac{\partial^2 \boldsymbol{T}\left(\vec{\boldsymbol{X}}_T^R, \vec{\boldsymbol{Y}}_T^I\right)}{\partial x_{Ti}^R \partial y_{Tj}^I} \right|_{\vec{\boldsymbol{X}}_T^R, \vec{\boldsymbol{Y}}_T^I = E\left(\vec{\boldsymbol{X}}_T^R\right), \vec{\boldsymbol{Y}}_T^C} \right\} \left(x_{Ti}^R - E\left(x_{Ti}^R\right)\right) \right| \cdot \Delta y_{Tj}
\end{aligned} \tag{6.68}$$

Likewise, the maximum and minimum values of the LB $\underline{\varepsilon}^I$ with respect to the interval vectors $\vec{\boldsymbol{Y}}_M^I$ and $\vec{\boldsymbol{Y}}_T^I$ can be expressed as follows

$$
\begin{aligned}
\max/\min\left(\underline{\varepsilon}^{I}\right) = {} & \boldsymbol{S}_0 + \max_{\delta_{Tj}=+1} / \min_{\delta_{Tj}=-1} \min_{\delta_{Ml}=-1} \Delta \boldsymbol{S} = (\boldsymbol{T}_{\mathrm{m}})^{-1} \boldsymbol{M}_{\mathrm{m}} + (\boldsymbol{T}_{\mathrm{m}})^{-1} \\
& \left\{ \sum_{k=1}^{s} \left\{ \left. \frac{\partial \boldsymbol{M}\left(\vec{\boldsymbol{X}}_{M}^{R}, \vec{\boldsymbol{Y}}_{M}^{I}\right)}{\partial x_{Mk}^{R}} \right|_{\vec{X}_M^R, \vec{Y}_M^I = E\left(\vec{X}_M^R\right), \vec{Y}_M^C} \right\} \left(x_{Mk}^{R} - E\left(x_{Mk}^{R}\right)\right) \right\} \\
& - (\boldsymbol{T}_{\mathrm{m}})^{-2} \boldsymbol{M}_{\mathrm{m}} \left\{ \sum_{i=1}^{n} \left\{ \left. \frac{\partial \boldsymbol{T}\left(\vec{\boldsymbol{X}}_{T}^{R}, \vec{\boldsymbol{Y}}_{T}^{I}\right)}{\partial x_{Ti}^{R}} \right|_{\vec{X}_T^R, \vec{Y}_T^I = E\left(\vec{X}_T^R\right), \vec{Y}_T^C} \right\} \left(x_{Ti}^{R} - E\left(x_{Ti}^{R}\right)\right) \right\} \\
& - \left| (\boldsymbol{T}_{\mathrm{m}})^{-1} \left[\sum_{l=1}^{w} \left\{ \left. \frac{\partial \boldsymbol{M}\left(\vec{\boldsymbol{X}}_{M}^{R}, \vec{\boldsymbol{Y}}_{M}^{I}\right)}{\partial y_{Ml}^{I}} \right|_{\vec{X}_M^R, \vec{Y}_M^I = E\left(\vec{X}_M^R\right), \vec{Y}_M^C} \right\} \right. \right. \\
& \left. \left. + \sum_{l=1}^{w} \sum_{k=1}^{s} \left\{ \left. \frac{\partial^2 \boldsymbol{M}\left(\vec{\boldsymbol{X}}_{M}^{R}, \vec{\boldsymbol{Y}}_{M}^{I}\right)}{\partial x_{Mk}^{R} \partial y_{Ml}^{I}} \right|_{\vec{X}_M^R, \vec{Y}_M^I = E\left(\vec{X}_M^R\right), \vec{Y}_M^C} \right\} \left(x_{Mk}^{R} - E\left(x_{Mk}^{R}\right)\right) \right] \right| \cdot \Delta y_{Ml} \\
& \pm (\boldsymbol{T}_{\mathrm{m}})^{-2} \boldsymbol{M}_{\mathrm{m}} \left| \sum_{j=1}^{m} \left\{ \left. \frac{\partial \boldsymbol{T}\left(\vec{\boldsymbol{X}}_{T}^{R}, \vec{\boldsymbol{Y}}_{T}^{I}\right)}{\partial y_{Tj}^{I}} \right|_{\vec{X}_T^R, \vec{Y}_T^I = E\left(\vec{X}_T^R\right), \vec{Y}_T^C} \right\} \right. \\
& \left. + \sum_{j=1}^{m} \sum_{i=1}^{n} \left\{ \left. \frac{\partial^2 \boldsymbol{T}\left(\vec{\boldsymbol{X}}_{T}^{R}, \vec{\boldsymbol{Y}}_{T}^{I}\right)}{\partial x_{Ti}^{R} \partial y_{Tj}^{I}} \right|_{\vec{X}_T^R, \vec{Y}_T^I = E\left(\vec{X}_T^R\right), \vec{Y}_T^C} \right\} \left(x_{Ti}^{R} - E\left(x_{Ti}^{R}\right)\right) \right| \cdot \Delta y_{Tj}
\end{aligned}
\tag{6.69}
$$

6.3.3.2 Mathematical Characteristics of Extrema of Bounds of the LSOAAC Response

For simplicity, all variables in this section are assumed to be mutually independent. Consequently, the covariance between any pair of random variables is zero, which leads to the following conclusion

$$
\operatorname{Cov}\left(x_{Mk}^{R}, x_{Mp}^{R}\right) = D\left(\partial x_{M}^{R}\right), \operatorname{Cov}\left(x_{Ti}^{R}, x_{Tp}^{R}\right) = D\left(\partial x_{T}^{R}\right) \tag{6.70}
$$

Here, Cov and D denote the covariance operator and variance operator, respectively.

Thus, it is readily apparent that both the LB $\underline{\varepsilon}^{I}$ and UB $\overline{\varepsilon}^{I}$ of the LSOAAC response are random functions that depend solely on the random vectors x_{Mk}^{R} and x_{Ti}^{R}. To compute the expectation and variance of the extrema of these bounds for the LSOAAC response, the RIMM outlined in Section 6.3.2.1 can be employed.

The expectation and variance corresponding to both the maximum and minimum values of the UB $\overline{\varepsilon}^{I}$ can be computed as follows:

$$E\left(\max/\min\left(\overline{\varepsilon}^{I}\right)\right) = (\boldsymbol{T}_{\mathrm{m}})^{-1}\,\boldsymbol{M}_{\mathrm{m}} + \left|(\boldsymbol{T}_{\mathrm{m}})^{-1}\sum_{l=1}^{w}\left\{\left\{\left.\frac{\partial \boldsymbol{M}\left(\vec{\boldsymbol{X}}_{M}^{R},\vec{\boldsymbol{Y}}_{M}^{I}\right)}{\partial y_{Ml}^{I}}\right|_{\vec{\boldsymbol{X}}_{M}^{R},\vec{\boldsymbol{Y}}_{M}^{I}=E\left(\vec{\boldsymbol{X}}_{M}^{R}\right),\vec{\boldsymbol{Y}}_{M}^{C}}\right\}\right\}\right|\cdot\Delta y_{Ml} \pm \left|(\boldsymbol{T}_{\mathrm{m}})^{-2}\,\boldsymbol{M}_{\mathrm{m}}\sum_{j=1}^{m}\left\{\left.\frac{\partial \boldsymbol{T}\left(\vec{\boldsymbol{X}}_{T}^{R},\vec{\boldsymbol{Y}}_{T}^{I}\right)}{\partial y_{Tj}^{I}}\right|_{\vec{\boldsymbol{X}}_{T}^{R},\vec{\boldsymbol{Y}}_{T}^{I}=E\left(\vec{\boldsymbol{X}}_{T}^{R}\right),\vec{\boldsymbol{Y}}_{T}^{C}}\right\}\right|\cdot\Delta y_{Tj} \tag{6.71}$$

$$\begin{aligned}
D\left(\max/\min\left(\overline{\varepsilon}^{I}\right)\right) =\; & \left\{(\boldsymbol{T}_{\mathrm{m}})^{-1}\left\{\sum_{k=1}^{s}\left\{\left.\frac{\partial \boldsymbol{M}\left(\vec{\boldsymbol{X}}_{M}^{R},\vec{\boldsymbol{Y}}_{M}^{I}\right)}{\partial x_{Mk}^{R}}\right|_{\vec{\boldsymbol{X}}_{M}^{R},\vec{\boldsymbol{Y}}_{M}^{I}=E\left(\vec{\boldsymbol{X}}_{M}^{R}\right),\vec{\boldsymbol{Y}}_{M}^{C}}\right\}\right\}\right\}^{2} D\left(x_{M}^{R}\right) \\
& + \left\{-(\boldsymbol{T}_{\mathrm{m}})^{-2}\,\boldsymbol{M}_{\mathrm{m}}\left\{\sum_{i=1}^{n}\left\{\left.\frac{\partial \boldsymbol{T}\left(\vec{\boldsymbol{X}}_{T}^{R},\vec{\boldsymbol{Y}}_{T}^{I}\right)}{\partial x_{Ti}^{R}}\right|_{\vec{\boldsymbol{X}}_{T}^{R},\vec{\boldsymbol{Y}}_{T}^{I}=E\left(\vec{\boldsymbol{X}}_{T}^{R}\right),\vec{\boldsymbol{Y}}_{T}^{C}}\right\}\right\}\right\}^{2} D\left(x_{T}^{R}\right) \\
& + \left\{\left|+(\boldsymbol{T}_{\mathrm{m}})^{-1}\left[\sum_{l=1}^{w}\sum_{k=1}^{s}\left\{\left.\frac{\partial^{2} \boldsymbol{M}\left(\vec{\boldsymbol{X}}_{M}^{R},\vec{\boldsymbol{Y}}_{M}^{I}\right)}{\partial x_{Mk}^{R}\,\partial y_{Ml}^{I}}\right|_{\vec{\boldsymbol{X}}_{M}^{R},\vec{\boldsymbol{Y}}_{M}^{I}=E\left(\vec{\boldsymbol{X}}_{M}^{R}\right),\vec{\boldsymbol{Y}}_{M}^{C}}\right\}\right]\cdot\Delta y_{Ml}\right|\right\}^{2} \\
& D\left(x_{M}^{R}\right) \\
& + \left\{\left|\pm(\boldsymbol{T}_{\mathrm{m}})^{-2}\,\boldsymbol{M}_{\mathrm{m}}\left[\sum_{j=1}^{m}\sum_{i=1}^{n}\left\{\left.\frac{\partial^{2} \boldsymbol{T}\left(\vec{\boldsymbol{X}}_{T}^{R},\vec{\boldsymbol{Y}}_{T}^{I}\right)}{\partial x_{Ti}^{R}\,\partial y_{Tj}^{I}}\right|_{\vec{\boldsymbol{X}}_{T}^{R},\vec{\boldsymbol{Y}}_{T}^{I}=E\left(\vec{\boldsymbol{X}}_{T}^{R}\right),\vec{\boldsymbol{Y}}_{T}^{C}}\right\}\cdot\Delta y_{Tl}\right]\right|\right\}^{2} \\
& D\left(x_{T}^{R}\right) \\
& + 2\cdot(\boldsymbol{T}_{\mathrm{m}})^{-1}\left\{\sum_{k=1}^{s}\left\{\left.\frac{\partial \boldsymbol{M}\left(\vec{\boldsymbol{X}}_{M}^{R},\vec{\boldsymbol{Y}}_{M}^{I}\right)}{\partial x_{Mk}^{R}}\right|_{\vec{\boldsymbol{X}}_{M}^{R},\vec{\boldsymbol{Y}}_{M}^{I}=E\left(\vec{\boldsymbol{X}}_{M}^{R}\right),\vec{\boldsymbol{Y}}_{M}^{C}}\right\}\right\}
\end{aligned}$$

$$\cdot\left|(\boldsymbol{T}_{\mathrm{m}})^{-1}\left[\sum_{l=1}^{w}\sum_{k=1}^{s}\left\{\left.\frac{\partial^2\boldsymbol{M}\left(\vec{\boldsymbol{X}}_{\boldsymbol{M}}^{R},\vec{\boldsymbol{Y}}_{\boldsymbol{M}}^{I}\right)}{\partial x_{\boldsymbol{M}k}^{R}\partial y_{\boldsymbol{M}l}^{I}}\right|_{\vec{X}_M^R,\vec{Y}_M^I=E\left(\vec{X}_M^R\right),\vec{Y}_M^C}\right\}\right]\cdot\Delta y_{\boldsymbol{M}l}\right|D\left(x_{\boldsymbol{M}}^{R}\right)$$
$$\pm 2\cdot(\boldsymbol{T}_{\mathrm{m}})^{-2}\boldsymbol{M}_{\mathrm{m}}\left\{\sum_{i=1}^{n}\left\{\left.\frac{\partial\boldsymbol{T}\left(\vec{\boldsymbol{X}}_{\boldsymbol{T}}^{R},\vec{\boldsymbol{Y}}_{\boldsymbol{T}}^{I}\right)}{\partial x_{\boldsymbol{T}i}^{R}}\right|_{\vec{X}_T^R,\vec{Y}_T^I=E\left(\vec{X}_T^R\right),\vec{Y}_T^C}\right\}\right\} \tag{6.72}$$
$$\cdot\left|(\boldsymbol{T}_{\mathrm{m}})^{-2}\boldsymbol{M}_{\mathrm{m}}\left[\sum_{j=1}^{m}\sum_{i=1}^{n}\left\{\left.\frac{\partial^2\boldsymbol{T}\left(\vec{\boldsymbol{X}}_{\boldsymbol{T}}^{R},\vec{\boldsymbol{Y}}_{\boldsymbol{T}}^{I}\right)}{\partial x_{\boldsymbol{T}i}^{R}\partial y_{\boldsymbol{T}j}^{I}}\right|_{\vec{X}_T^R,\vec{Y}_T^I=E\left(\vec{X}_T^R\right),\vec{Y}_T^C}\right\}\cdot\Delta y_{\boldsymbol{T}l}\right]\right|D\left(x_{\boldsymbol{T}}^{R}\right)$$

Likewise, the expectation and variance associated with the maximum and minimum values of the LB $\underline{\varepsilon}^I$ can be computed as follows:

$$E\left(\max/\min\left(\underline{\varepsilon}^I\right)\right)=\boldsymbol{T}_{\mathrm{m}}^{-1}\boldsymbol{M}_{\mathrm{m}}-$$
$$\left|(\boldsymbol{T}_{\mathrm{m}})^{-1}\sum_{l=1}^{w}\left\{\left\{\left.\frac{\partial\boldsymbol{M}\left(\vec{\boldsymbol{X}}_{\boldsymbol{M}}^{R},\vec{\boldsymbol{Y}}_{\boldsymbol{M}}^{I}\right)}{\partial y_{\boldsymbol{M}l}^{I}}\right|_{\vec{X}_M^R,\vec{Y}_M^I=E\left(\vec{X}_M^R\right),\vec{Y}_M^C}\right\}\right\}\right|\cdot\Delta y_{\boldsymbol{M}l}$$
$$\pm\left|(\boldsymbol{T}_{\mathrm{m}})^{-2}\boldsymbol{M}_{\mathrm{m}}\sum_{j=1}^{m}\left\{\left.\frac{\partial\boldsymbol{T}\left(\vec{\boldsymbol{X}}_{\boldsymbol{T}}^{R},\vec{\boldsymbol{Y}}_{\boldsymbol{T}}^{I}\right)}{\partial y_{\boldsymbol{T}j}^{I}}\right|_{\vec{X}_T^R,\vec{Y}_T^I=E\left(\vec{X}_T^R\right),\vec{Y}_T^C}\right\}\right|\cdot\Delta y_{\boldsymbol{T}j} \tag{6.73}$$

$$D\left(\max/\min\left(\underline{\varepsilon}^I\right)\right)=\left((\boldsymbol{T}_{\mathrm{m}})^{-1}\left\{\sum_{k=1}^{s}\left\{\left.\frac{\partial\boldsymbol{M}\left(\vec{\boldsymbol{X}}_{\boldsymbol{M}}^{R},\vec{\boldsymbol{Y}}_{\boldsymbol{M}}^{I}\right)}{\partial x_{\boldsymbol{M}k}^{R}}\right|_{\vec{X}_M^R,\vec{Y}_M^I=E\left(\vec{X}_M^R\right),\vec{Y}_M^C}\right\}\right\}\right)^2 D\left(x_{\boldsymbol{M}}^{R}\right)$$
$$+\left(-(\boldsymbol{T}_{\mathrm{m}})^{-2}\boldsymbol{M}_{\mathrm{m}}\left\{\sum_{i=1}^{n}\left\{\left.\frac{\partial\boldsymbol{T}\left(\vec{\boldsymbol{X}}_{\boldsymbol{T}}^{R},\vec{\boldsymbol{Y}}_{\boldsymbol{T}}^{I}\right)}{\partial x_{\boldsymbol{T}i}^{R}}\right|_{\vec{X}_T^R,\vec{Y}_T^I=E\left(\vec{X}_T^R\right),\vec{Y}_T^C}\right\}\right\}\right)^2 D\left(x_{\boldsymbol{T}}^{R}\right)$$
$$+\left(-\left|(\boldsymbol{T}_{\mathrm{m}})^{-1}\left[\sum_{l=1}^{w}\sum_{k=1}^{s}\left\{\left.\frac{\partial^2\boldsymbol{M}\left(\vec{\boldsymbol{X}}_{\boldsymbol{M}}^{R},\vec{\boldsymbol{Y}}_{\boldsymbol{M}}^{I}\right)}{\partial x_{\boldsymbol{M}k}^{R}\partial y_{\boldsymbol{M}l}^{I}}\right|_{\vec{X}_M^R,\vec{Y}_M^I=E\left(\vec{X}_M^R\right),\vec{Y}_M^C}\right\}\right]\cdot\Delta y_{\boldsymbol{M}l}\right|\right)^2$$
$$D\left(x_{\boldsymbol{M}}^{R}\right)$$

$$+\left(\pm\left|(T_{\mathrm{m}})^{-2} M_{\mathrm{m}}\left[\sum_{j=1}^{m}\sum_{i=1}^{n}\left\{\left.\frac{\partial^2 T\left(\vec{X}_T^R,\vec{Y}_T^I\right)}{\partial x_{Ti}^R \partial y_{Tj}^I}\right|_{\vec{X}_T^R,\vec{Y}_T^I=E\left(\vec{X}_T^R\right),\vec{Y}_T^C}\right\}\cdot \Delta y_{Tl}\right]\right| D\left(x_T^R\right)\right)^2$$

$$-2\cdot(T_{\mathrm{m}})^{-1}\left\{\sum_{k=1}^{s}\left\{\left.\frac{\partial M\left(\vec{X}_M^R,\vec{Y}_M^I\right)}{\partial x_{Mk}^R}\right|_{\vec{X}_M^R,\vec{Y}_M^I=E\left(\vec{X}_M^R\right),\vec{Y}_M^C}\right\}\right\}$$

$$\cdot(T_{\mathrm{m}})^{-1}\left[\sum_{l=1}^{w}\sum_{k=1}^{s}\left\{\left.\frac{\partial^2 M\left(\vec{X}_M^R,\vec{Y}_M^I\right)}{\partial x_{Mk}^R \partial y_{Ml}^I}\right|_{\vec{X}_M^R,\vec{Y}_M^I=E\left(\vec{X}_M^R\right),\vec{Y}_M^C}\right\}\cdot \Delta y_{Ml}\right| D\left(x_M^R\right)$$

$$\pm 2\cdot(T_{\mathrm{m}})^{-2} M_{\mathrm{m}}\left\{\sum_{i=1}^{n}\left\{\left.\frac{\partial T\left(\vec{X}_T^R,\vec{Y}_T^I\right)}{\partial x_{Ti}^R}\right|_{\vec{X}_T^R,\vec{Y}_T^I=E\left(\vec{X}_T^R\right),\vec{Y}_T^C}\right\}\right\}$$

$$\cdot\left|(T_{\mathrm{m}})^{-2} M_{\mathrm{m}}\left[\sum_{j=1}^{m}\sum_{i=1}^{n}\left\{\left.\frac{\partial^2 T\left(\vec{X}_T^R,\vec{Y}_T^I\right)}{\partial x_{Ti}^R \partial y_{Tj}^I}\right|_{\vec{X}_T^R,\vec{Y}_T^I=E\left(\vec{X}_T^R\right),\vec{Y}_T^C}\right\}\cdot \Delta y_{Tl}\right]\right| D\left(x_T^R\right) \tag{6.74}$$

The procedure of the MHUAM for predicting the dynamic response field of the LSOAAC with various random and interval parameters can be summarized as follows:

Step 1. The dynamic response of the LSOAAC involving both random and interval parameters is approximated through a combination of perturbation theory, first-order Taylor series expansion, and first-order Neumann series (as presented in Eq. [6.60]).

Step 2. The intervals corresponding to the LB and UB of the LSOAAC's dynamic response are computed via monotonic analysis, as illustrated in Eqs. (6.68) and (6.69).

Step 3. The expectation and variance of the extrema of the bounds for the LSOAAC response are computed using the random interval moment method, as demonstrated in Eqs. (6.71)–(6.74).

6.3.4 Numerical Examples

In this section, several examples implemented in MATLAB for the LSOAAC are presented to verify the feasibility and effectiveness of the method proposed. The first example is intended to demonstrate the feasibility of solving the problem of the hybrid LSOAAC response field with both random and interval parameters.

The subsequent two examples respectively showcase its capability in addressing the LSOAAC response field problem with purely random parameters and purely interval parameters. For validating the feasibility and efficiency of the proposed method, the HMCM and the IPM are adopted as reference approaches. The deterministic parameters of the LSOAAC are provided in Table 6.3.

6.3.4.1 Hybrid LSOAAC Response Field Problem with Random and Interval Parameters

In this example, the thrust of the luffing cylinder and the gravity of the suspended weight are treated as random parameters following a Gaussian distribution. Meanwhile, the displacement between the lower end of the lifting arm and the luffing cylinder, along with the length of the lifting arm, are regarded as interval parameters. It is assumed that these uncertain parameters are mutually independent, and their specific properties are detailed in Table 6.4.

Based on above description, we have

$$\vec{\boldsymbol{X}}_{\boldsymbol{M}}^{\boldsymbol{R}} = (\vec{X}_{M1}^{R}) = (F), \vec{\boldsymbol{X}}_{\boldsymbol{T}}^{\boldsymbol{R}} = (\vec{X}_{T1}^{R}) = (Q)$$

$$\vec{\boldsymbol{Y}}_{\boldsymbol{M}}^{\boldsymbol{I}} = (\vec{Y}_{M1}^{I}) = (a), \vec{\boldsymbol{Y}}_{\boldsymbol{T}}^{\boldsymbol{I}} = (\vec{Y}_{T1}^{I}) = (L) \tag{6.75}$$

Table 6.3 Deterministic LSOAAC parameters.

Symbol	Description	Value
F_s (N)	Tension of the hoisting rope	5×10^4
G (N)	Gravity of the lifting arm	7×10^4
d (m)	Displacement between the lower end of the lifting arm and the hoisting rope	0.5
L_g (m)	Displacement between the lower end of the lifting arm and its center of gravity	10

Table 6.4 Uncertain LSOAAC parameters.

Uncertain Parameters	Symbol	Description	Expectation or Midpoint Value
Random parameters	F (N)	Thrust of the luffing cylinder	$\mu_F = 4.3 \times 10^5$
	Q (N)	Gravity of the weight in suspension	$\mu_Q = 5 \times 10^4$
Interval parameters	a (m)	Displacement between the lower end of the lifting arm and the luffing cylinder	$a^c = 2.5$
	L (m)	Length of the lifting arm	$L^c = 30$

Based on Eq. (6.60), and in conjunction with the specified conditions, the dynamic response expression for the LSOAAC involving both random and interval parameters can be derived as follows:

$$\begin{aligned}\varepsilon = &\frac{\mu_F a^c + F_s d}{\mu_Q L^c + GL_g} + \frac{\mu_F \cdot \Delta a^I}{\mu_Q L^c + GL_g} - \frac{\mu_F a^c + F_s d}{\left(\mu_Q L^c + GL_g\right)^2} \cdot \mu_Q \cdot \Delta L^I \\ &+ \left[\frac{a^c + \Delta a^I}{\mu_Q L^c + GL_g} - \frac{a^c \cdot \mu_Q \cdot \Delta L^I}{\left(\mu_Q L^c + GL_g\right)^2}\right] \left(x^F - E\left(x^F\right)\right) \\ &+ \left[\frac{\left(\mu_F a^c + F_s d\right) \cdot L^c + \mu_F \cdot L^c \cdot \Delta a^I - \left(\mu_F a^c + F_s d\right) \cdot \mu_Q \cdot \Delta L^I}{\left(\mu_Q L^c + GL_g\right)^2}\right] \left(x^Q - E\left(x^Q\right)\right)\end{aligned} \tag{6.76}$$

Here, we introduce the coefficient of variation V_{x^R} for random variable x^R and the interval change ratio Δy_F for the interval variable y^I as follows:

$$V_{x^R} = \frac{\sigma_{x^R}}{\mu_{x^R}}, \Delta y_F = \frac{\Delta y}{y^c} \tag{6.77}$$

On the basis of Eqs. (6.71) and (6.72), the expectation and variance of the extrema of the UB $\overline{\varepsilon}^I$ computed by the MHUAM can be expressed as follows:

$$E\left(\max/\min\left(\overline{\varepsilon}^I\right)\right) = \frac{\mu_F a^c + F_s d}{\mu_Q L^c + GL_g} + \frac{\mu_F \cdot a^c \cdot \Delta a_F}{\mu_Q L^c + GL_g} \pm \frac{\mu_F a^c + F_s d}{\left(\mu_Q L^c + GL_g\right)^2} \cdot \mu_Q \cdot L^c \cdot L^F \tag{6.78}$$

$$\begin{aligned}D\left(\max/\min\left(\overline{\varepsilon}^I\right)\right) = &\left(\frac{a^c}{\mu_Q L^c + GL_g}\right)^2 \cdot \left(\mu_F \cdot V_F\right)^2 \\ &+ \left(\frac{a^c \cdot \Delta a_F}{\mu_Q L^c + GL_g}\right)^2 \cdot \left(\mu_F \cdot V_F\right)^2 \\ &+ \left(-\frac{a^c \cdot \mu_Q \cdot L^c \cdot \Delta L_F}{\left(\mu_Q L^c + GL_g\right)^2}\right)^2 \cdot \left(\mu_F \cdot V_F\right)^2 \\ &+ \left(\frac{\left(\mu_F a^c + F_s d\right)}{\left(\mu_Q L^c + GL_g\right)^2} \cdot L^c\right)^2 \cdot \left(\mu_Q \cdot V_Q\right)^2 \\ &+ \left(\frac{\mu_F \cdot L^c \cdot a^c \cdot \Delta a_F}{\left(\mu_Q L^c + GL_g\right)^2}\right)^2 \cdot \left(\mu_Q \cdot V_Q\right)^2\end{aligned}$$

$$
\begin{aligned}
&+\left(-\frac{(\mu_F a^c + F_s d)\cdot \mu_Q \cdot L^c \cdot \Delta L_F}{(\mu_Q L^c + GL_g)^2}\right)^2 \cdot (\mu_Q \cdot V_Q)^2 \\
&+2\cdot \frac{a^c}{\mu_Q L^c + GL_g}\cdot \frac{a^c \cdot \Delta a_F}{\mu_Q L^c + GL_g}\cdot (\mu_F \cdot V_F)^2 \\
&\pm 2\left(\frac{a^c}{\mu_Q L^c + GL_g}\cdot \frac{a^c \cdot \mu_Q \cdot L^c \cdot \Delta L_F}{(\mu_Q L^c + GL_g)^2}\right. \\
&\left.+\frac{a^c \cdot \Delta a_F}{\mu_Q L^c + GL_g}\cdot \frac{a^c \cdot \mu_Q \cdot L^c \cdot \Delta L_F}{(\mu_Q L^c + GL_g)^2}\right)\cdot (\mu_F \cdot V_F)^2 \\
&+2\cdot \frac{(\mu_F a^c + F_s d)}{(\mu_Q L^c + GL_g)^2}\cdot L^c \cdot \frac{\mu_F \cdot L^c \cdot a^c \cdot \Delta a_F}{(\mu_Q L^c + GL_g)^2}\cdot (\mu_Q \cdot V_Q)^2 \\
&\pm 2\left(\frac{(\mu_F a^c + F_s d)}{(\mu_Q L^c + GL_g)^2}\cdot L^c \cdot \frac{(\mu_F a^c + F_s d)\cdot \mu_Q \cdot L^c \cdot \Delta L_F}{(\mu_Q L^c + GL_g)^2}\right. \\
&\left.+\frac{\mu_F \cdot L^c \cdot a^c \cdot \Delta a_F}{(\mu_Q L^c + GL_g)^2}\cdot \frac{(\mu_F a^c + F_s d)\cdot \mu_Q \cdot L^c \cdot \Delta L_F}{(\mu_Q L^c + GL_g)^2}\right)\cdot (\mu_Q \cdot V_Q)^2
\end{aligned}
\tag{6.79}
$$

Likewise, relying on Eqs. (6.73) and (6.74), the expectation and variance of the extrema of the LB $\underline{\varepsilon}^I$ computed by the MHUAM can be expressed as follows:

$$
E\left(\max/\min\left(\underline{\varepsilon}^I\right)\right) = \frac{\mu_F a^c + F_s d}{\mu_Q L^c + GL_g} - \frac{\mu_F \cdot a^c \cdot \Delta a_F}{\mu_Q L^c + GL_g} \pm \frac{\mu_F a^c + F_s d}{(\mu_Q L^c + GL_g)^2}\cdot \mu_Q \cdot L^c \cdot L^F
\tag{6.80}
$$

$$
\begin{aligned}
D\left(\max/\min\left(\underline{\varepsilon}^I\right)\right) = &+\left(\frac{a^c \cdot \Delta a_F}{\mu_Q L^c + GL_g}\right)^2 \cdot (\mu_F \cdot V_F)^2 + \left(\frac{a^c \cdot \Delta a_F}{\mu_Q L^c + GL_g}\right)^2 \cdot (\mu_F \cdot V_F)^2 \\
&+\left(-\frac{a^c \cdot \mu_Q \cdot L^c \cdot \Delta L_F}{(\mu_Q L^c + GL_g)^2}\right)^2 \cdot (\mu_F \cdot V_F)^2 \\
&+\left(\frac{(\mu_F a^c + F_s d)}{(\mu_Q L^c + GL_g)^2}\cdot L^c\right)^2 \cdot (\mu_Q \cdot V_Q)^2 \\
&+\left(\frac{\mu_F \cdot L^c \cdot a^c \cdot \Delta a_F}{(\mu_Q L^c + GL_g)^2}\right)^2 \cdot (\mu_Q \cdot V_Q)^2 \\
&+\left(-\frac{(\mu_F a^c + F_s d)\cdot \mu_Q \cdot L^c \cdot \Delta L_F}{(\mu_Q L^c + GL_g)^2}\right)^2 \cdot (\mu_Q \cdot V_Q)^2 \\
&-2\cdot \frac{a^c}{\mu_Q L^c + GL_g}\cdot \frac{a^c \cdot \Delta a_F}{\mu_Q L^c + GL_g}\cdot (\mu_F \cdot V_F)^2
\end{aligned}
$$

$$
\begin{aligned}
&\pm 2\left(\frac{a^c}{\mu_Q L^c + GL_g}\cdot\frac{a^c\cdot\mu_Q\cdot L^c\cdot\Delta L_F}{(\mu_Q L^c + GL_g)^2}+\frac{a^c\cdot\Delta a_F}{\mu_Q L^c + GL_g}\cdot\frac{a^c\cdot\mu_Q\cdot L^c\cdot\Delta L_F}{(\mu_Q L^c + GL_g)^2}\right)\cdot(\mu_F\cdot V_F)^2\\
&-2\cdot\frac{(\mu_F a^c + F_s d)}{(\mu_Q L^c + GL_g)^2}\cdot L^c\cdot\frac{\mu_F\cdot L^c\cdot a^c\cdot\Delta a_F}{(\mu_Q L^c + GL_g)^2}\cdot(\mu_Q\cdot V_Q)^2\\
&\pm 2\left(\frac{(\mu_F a^c + F_s d)}{(\mu_Q L^c + GL_g)^2}\cdot L^c\cdot\frac{(\mu_F a^c + F_s d)\cdot\mu_Q\cdot L^c\cdot\Delta L_F}{(\mu_Q L^c + GL_g)^2}\right.\\
&\left.+\frac{\mu_F\cdot L^c\cdot a^c\cdot\Delta a_F}{(\mu_Q L^c + GL_g)^2}\cdot\frac{(\mu_F a^c + F_s d)\cdot\mu_Q\cdot L^c\cdot\Delta L_F}{(\mu_Q L^c + GL_g)^2}\right)\cdot(\mu_Q\cdot V_Q)^2
\end{aligned}
\tag{6.81}
$$

where V_F and V_Q represent the coefficients of variation of the random parameters F and Q, respectively. Meanwhile, Δa_F and ΔL_F denote the interval change ratios of the interval parameters a and L, respectively.

Based on the above analysis, we define that the values of the coefficient of variation for random variables and the interval change ratio for interval variables both range from 0 to 0.1. Let DD represent the dispersal degree of uncertain variables; in this case, DD will also vary from 0 to 0.1.

To examine the differences in expectation and variance resulting from different uncertain parameters of the LSOAAC under various hybrid uncertainty models, as well as the distinct effects of different uncertain parameters on the LSOAAC response field, we adopt the hybrid uncertainty model $\mathrm{DD1} = V_F = V_Q = \Delta a_F = \Delta L_F$ as a reference. Simulations for the LSOAAC response field problem using the MHUAM were conducted in MATLAB R2014a on a computer equipped with a 2.5 GHz Intel(R) Core(TM) i7-4710MQ CPU. The expectations and variances of the extrema of the bounds of the LSOAAC response, calculated by the MHUAM under different hybrid uncertainty models, are plotted in Figures 6.15–6.18 and Figures 6.19–6.22, respectively.

Figures 6.15 and 6.19 show the expectation and variance of the extrema of the bounds of the LSOAAC response calculated by the MHUAM under two scenarios: one including V_F, and the other excluding V_F. Specifically, the two hybrid uncertainty models considered are $\mathrm{DD1} = V_F = V_Q = \Delta a_F = \Delta L_F$ and $\mathrm{DD2} = V_Q = \Delta a_F = \Delta L_F, V_F = 0$. It is evident that changes in V_F have no influence on either the expectation or the variance of the extrema of the bounds of the LSOAAC response. This indicates that the thrust of the luffing cylinder F does not affect the LSOAAC response field. Furthermore, it can be observed that all eight curves exhibit an increasing trend as DD increases, demonstrating that the fluctuation of uncertain parameters impacts the expectation and variance of the extrema of the bounds of the LSOAAC response.

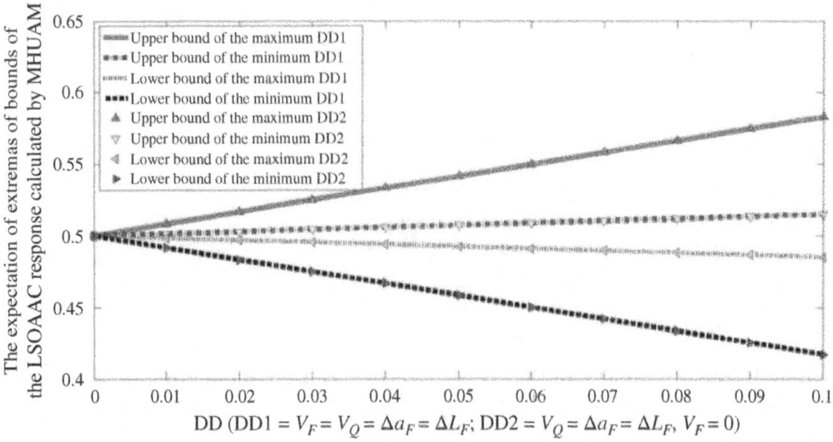

Figure 6.15 The expectation of extrema of bounds of the LSOAAC response calculated by MHUAM with/without V_F.

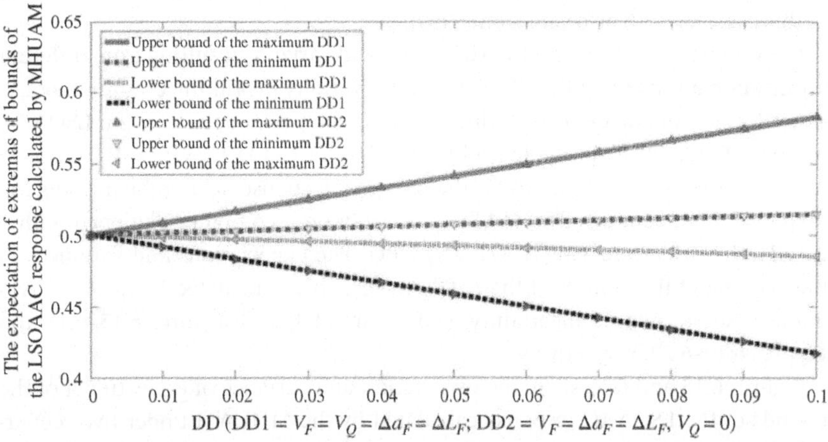

Figure 6.16 The expectation of extrema of bounds of the LSOAAC response calculated by MHUAM with/without V_Q.

Figures 6.16 and 6.18 illustrate the expectation and variance of the extrema of the bounds of the LSOAAC response calculated by the MHUAM under two conditions: one with V_Q, included and the other with V_Q excluded. Specifically, the two hybrid uncertainty models in question are DD1 $= V_F = V_Q = \Delta a_F = \Delta L_F$ and DD2 $= V_F = \Delta a_F = \Delta L_F, V_Q = 0$. It is clear that changes in V_Q affect the variance of the extrema of the bounds of the LSOAAC response but have no impact on the expectation of these extrema. This implies that the gravity of the suspended weight

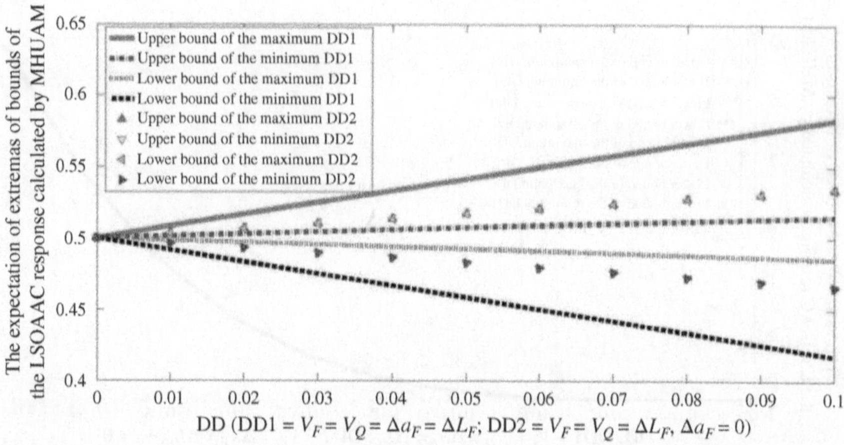

Figure 6.17 The expectation of extrema of bounds of the LSOAAC response calculated by MHUAM with/without Δa_F.

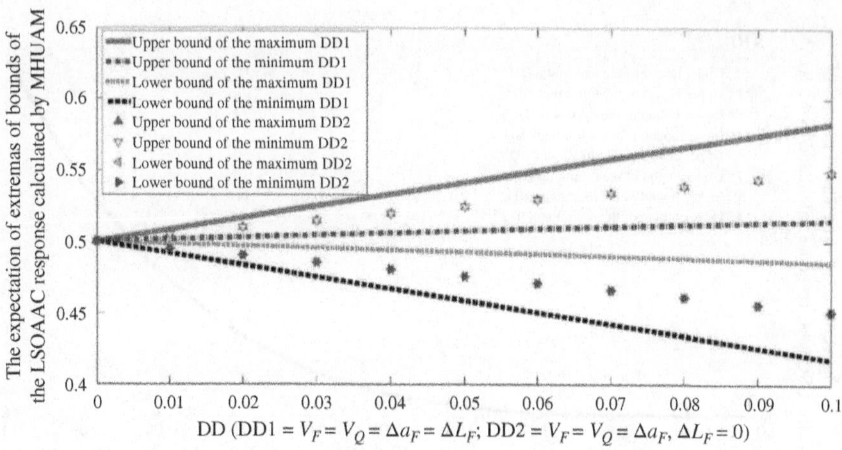

Figure 6.18 The expectation of extrema of bounds of the LSOAAC response calculated by MHUAM with/without ΔL_F.

Q exerts a significant influence on the variance of the extrema of the bounds of the LSOAAC response, while having no effect on their expectation.

Figures 6.17 and 6.21 show the expectation and variance of the extrema of the bounds of the LSOAAC response calculated by the MHUAM under two scenarios: one including Δa_F, and the other excluding Δa_F. Specifically, the two hybrid uncertainty models considered are $\text{DD1} = V_F = V_Q = \Delta a_F = \Delta L_F$ and

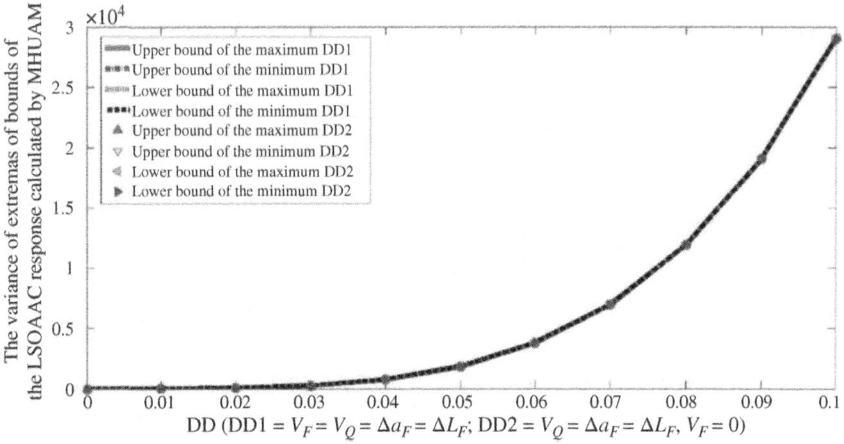

Figure 6.19 The variance of bounds of the LSOAAC response calculated by MHUAM with/without V_F.

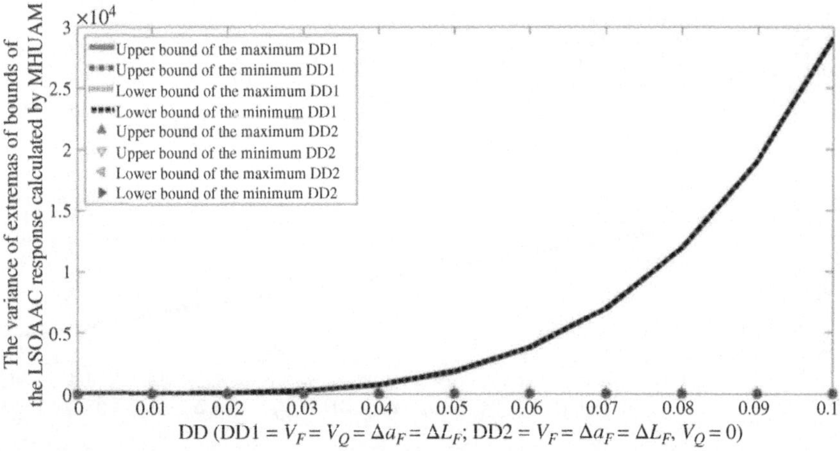

Figure 6.20 The variance of bounds of the LSOAAC response calculated by MHUAM with/without V_Q.

$DD2 = V_F = V_Q = \Delta L_F, \Delta a_F = 0$. It is evident that changes in Δa_F affect the expectation of the extrema of the bounds of the LSOAAC response but have no impact on the variance of these extrema. This indicates that the displacement a between the lower end of the lifting arm and the luffing cylinder exerts a significant influence on the expectation of the extrema of the bounds of the LSOAAC response, while having no effect on their variance.

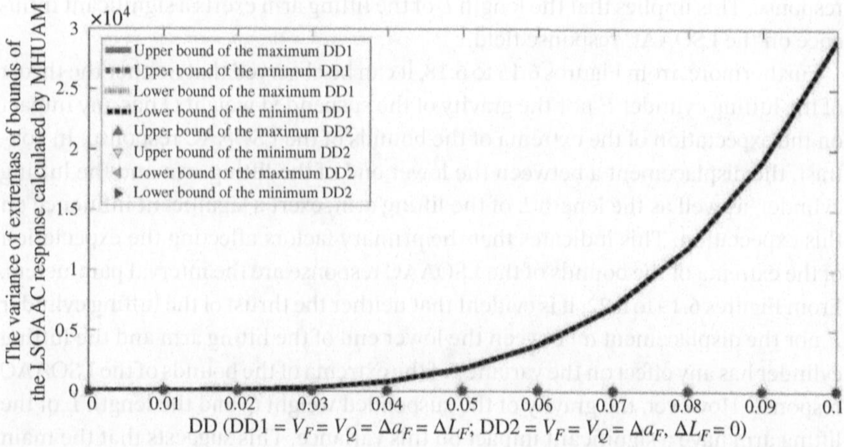

Figure 6.21 The variance of bounds of the LSOAAC response calculated by MHUAM with/without Δa_F.

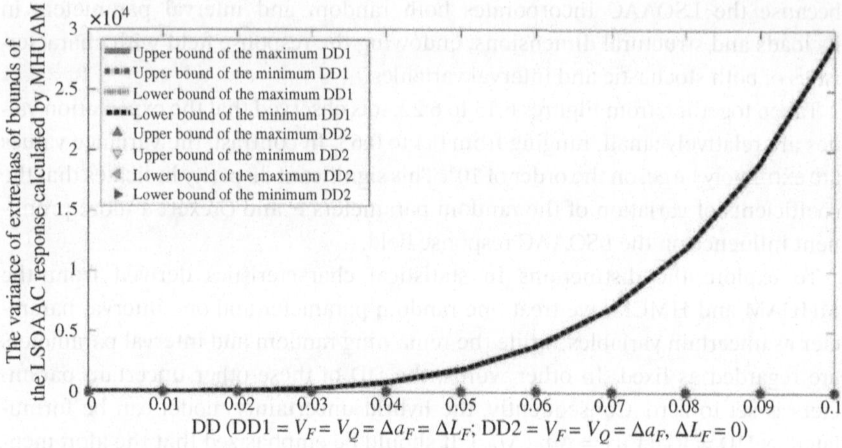

Figure 6.22 The variance of bounds of the LSOAAC response calculated by MHUAM with/without ΔL_F.

Figures 6.18 and 6.22 illustrate the expectation and variance of the extrema of the bounds of the LSOAAC response calculated by the MHUAM under two conditions: one with ΔL_F, included and the other with ΔL_F excluded. Specifically, the two hybrid uncertainty models in question are $\text{DD1} = V_F = V_Q = \Delta a_F = \Delta L_F$ and $\text{DD2} = V_F = V_Q = \Delta a_F, \Delta L_F = 0$. It is clear that changes in ΔL_F affect both the expectation and the variance of the extrema of the bounds of the LSOAAC response. This implies that the length L of the lifting arm exerts a significant influence on the LSOAAC response field.

Furthermore, from Figures 6.15 to 6.18, it can be observed that neither the thrust of the luffing cylinder F nor the gravity of the suspended weight Q has any impact on the expectation of the extrema of the bounds of the LSOAAC response. In contrast, the displacement a between the lower end of the lifting arm and the luffing cylinder, as well as the length L of the lifting arm, exert a significant influence on this expectation. This indicates that the primary factors affecting the expectation of the extrema of the bounds of the LSOAAC response are the interval parameters. From Figures 6.19 to 6.22, it is evident that neither the thrust of the luffing cylinder F nor the displacement a between the lower end of the lifting arm and the luffing cylinder has any effect on the variance of the extrema of the bounds of the LSOAAC response. However, the gravity of the suspended weight Q and the length L of the lifting arm have a significant impact on this variance. This suggests that the main influences on the variance of the extrema of the bounds of the LSOAAC response stem from both random and interval parameters. Furthermore, as DD increases, the LSOAAC response field increases correspondingly. This phenomenon arises because the LSOAAC incorporates both random and interval parameters in its loads and structural dimensions, endowing the response field with characteristics of both stochastic and interval variables.

Taken together, from Figures 6.15 to 6.22, it is observed that the expectation values are relatively small, ranging from 0.4 to 0.65. In contrast, the variance values are extremely large, on the order of 10^4. This significant disparity indicates that the coefficients of variation of the random parameters F and Q exert a more prominent influence on the LSOAAC response field.

To explore the distinctions in statistical characteristics derived from the MHUAM and HMCM, we treat one random parameter and one interval parameter as uncertain variables, while the remaining random and interval parameters are regarded as fixed. In other words, the DD of these other uncertain parameters is set to zero. Consequently, the hybrid uncertainty model can be formulated as $\text{DD} = V_F\ (V_Q) = \Delta a_F(\Delta L_F)$. It should be emphasized that the aforementioned approach is applicable to analyzing other hybrid uncertainty models, with comparable results. For the sake of simplifying the discussion, this scenario is taken as a typical case. The HMCM, involving 10^8 samples, is employed as a reference solution and executed in three steps, as outlined in the flowchart shown in Figure 6.23. First, a value is chosen within the range of the interval variable. Second, 10^4 simulations are conducted for the random variable to derive a set of expectation and variance for the response. Third, 10^4 values are selected for the interval variable, covering its range from the LB to the UB at a fixed interval. This yields a large number of expectations and variances, from which the extreme values can be easily determined.

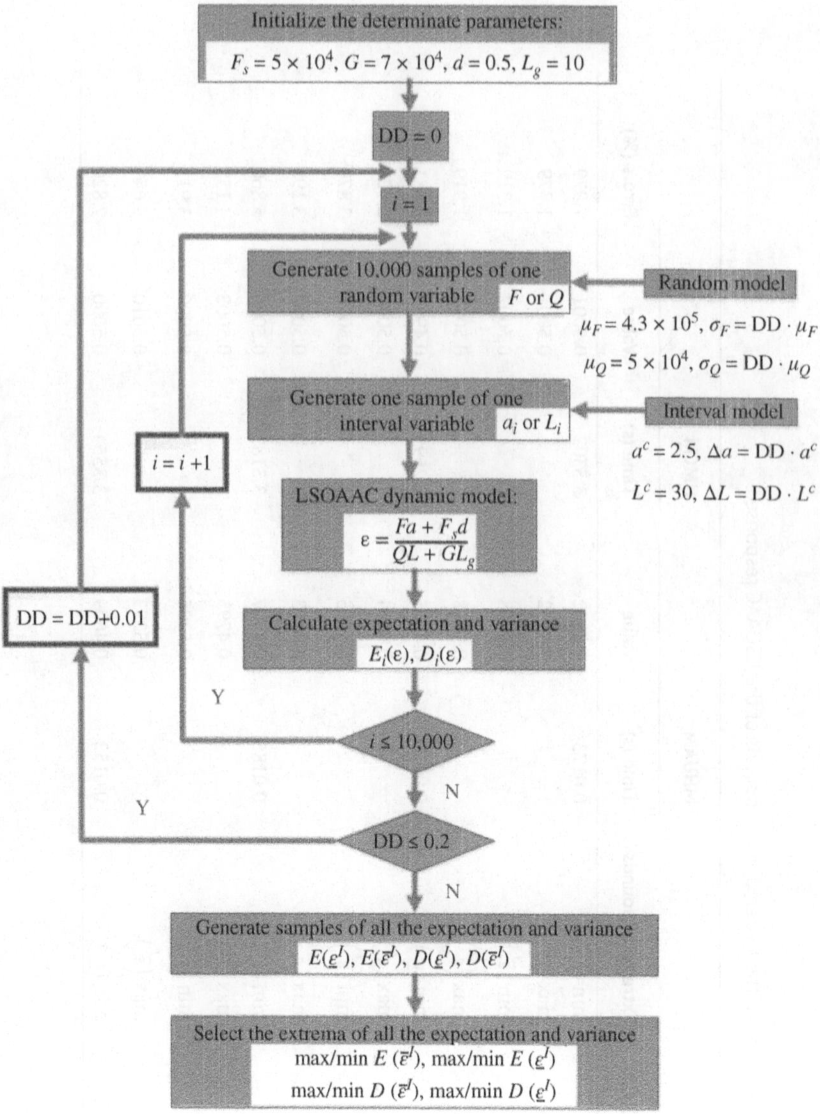

Figure 6.23 Flowchart of generating extrema of the LSOAAC response calculated by HMCM.

To more explicitly examine the effectiveness of the MHUAM in predicting the dynamic response field of the LSOAAC, the relative errors between the expectations and variances obtained via the MHUAM and those from the HMCM are presented in Tables 6.5 and 6.6, respectively.

Table 6.5 The expectation of the extrema of the bounds of the LSOAAC response.

		MHUAM		HMCM		
$DD = V_F(V_Q) = a_F(L_F)$	Extrema of bounds	Time (s)	Value	Time (s)	Value	Errors (%)
0.02	$\min\left(\underline{\varepsilon}^I\right)$	0.0173	0.4902	3.3302	0.5001	1.979
	$\max\left(\underline{\varepsilon}^I\right)$		0.4932		0.5001	1.379
	$\min\left(\overline{\varepsilon}^I\right)$		0.5068		0.5002	1.319
	$\max\left(\overline{\varepsilon}^I\right)$		0.5098		0.5002	1.919
0.04	$\min\left(\underline{\varepsilon}^I\right)$	0.0129	0.4805	3.2695	0.4999	3.881
	$\max\left(\underline{\varepsilon}^I\right)$		0.4864		0.5000	2.720
	$\min\left(\overline{\varepsilon}^I\right)$		0.5195		0.5001	3.879
	$\max\left(\overline{\varepsilon}^I\right)$		0.5163		0.5003	3.198
0.06	$\min\left(\underline{\varepsilon}^I\right)$	0.0185	0.4707	3.3183	0.5000	5.860
	$\max\left(\underline{\varepsilon}^I\right)$		0.4795		0.5003	4.158
	$\min\left(\overline{\varepsilon}^I\right)$		0.5205		0.5009	3.913
	$\max\left(\overline{\varepsilon}^I\right)$		0.5293		0.5010	5.649
0.08	$\min\left(\underline{\varepsilon}^I\right)$	0.0153	0.4609	3.3554	0.5000	7.820

Table 6.6 The variance of the extrema of the bounds of the LSOAAC response.

		MHUAM		HMCM		
$DD = V_F(V_Q) = a_F(L_F)$	Extrema of bounds	Time (s)	Value	Time (s)	Value	Errors (%)
0.02	$\min\left(\underline{\varepsilon}^I\right)$	0.0173	4.8948E−05	3.3302	5.1147E−05	4.30
	$\max\left(\underline{\varepsilon}^I\right)$		5.2697E−05		5.4936E−05	4.08
	$\min\left(\overline{\varepsilon}^I\right)$		9.5878E−05		1.0048E−04	4.58
	$\max\left(\overline{\varepsilon}^I\right)$		9.6345E−05		1.0087E−04	4.49
0.04	$\min\left(\underline{\varepsilon}^I\right)$	0.0129	1.8738E−04	3.2695	2.0376E−04	8.04
	$\max\left(\underline{\varepsilon}^I\right)$		1.9875E−04		2.1576E−04	7.88
	$\min\left(\overline{\varepsilon}^I\right)$		3.6423E−04		3.9658E−04	8.16
	$\max\left(\overline{\varepsilon}^I\right)$		3.8234E−04		4.1758E−04	8.44
0.06	$\min\left(\underline{\varepsilon}^I\right)$	0.0185	4.0014E−04	3.3183	4.5926E−04	12.87
	$\max\left(\underline{\varepsilon}^I\right)$		4.4564E−04		5.0548E−04	11.84
	$\min\left(\overline{\varepsilon}^I\right)$		7.8702E−04		8.9328E−04	11.90
	$\max\left(\overline{\varepsilon}^I\right)$		8.1949E−04		9.3594E−04	12.44

(Continued)

Table 6.6 (Continued)

		MHUAM		HMCM		
0.08	min $(\underline{\varepsilon}^{I})$	0.0153	8.3060E−04	3.3554	9.7004E−04	14.37
	max $(\underline{\varepsilon}^{I})$		8.5517E−04		9.9627E−04	14.16
	min $(\overline{\varepsilon}^{I})$		0.0011		0.0013	15.38
	max $(\overline{\varepsilon}^{I})$		0.0012		0.0014	14.29
	min $(\underline{\varepsilon}^{I})$	0.0162	0.0013	3.3214	0.0016	18.75
	max $(\underline{\varepsilon}^{I})$		0.0014		0.0017	17.65
	min $(\overline{\varepsilon}^{I})$		0.0019		0.0025	24.00
	max $(\overline{\varepsilon}^{I})$		0.0020		0.0026	23.08

From Tables 6.5 and 6.6, it can be observed that when the DD is 0.04 or less, the relative errors of the expectation and variance obtained by the MHUAM are close to the reference values from the HMCM. This indicates that the results calculated by the MHUAM are entirely acceptable when the DD of uncertain parameters is not high. As the DD increases from 0.04 to 0.1, the LSOAAC response field generated by the MHUAM deviates significantly from that produced by the HMCM. This suggests that small variations in load parameters and structural parameters may result in large deviations in the LSOAAC response field. It also implies that the MHUAM, which is based on the RIPM, first-order Taylor series expansion, and first-order Neumann series, is more suitable for predicting the LSOAAC response field with narrow uncertainties. In other words, the impact of neglecting higher-order terms in the Taylor series expansion and Neumann series is unpredictable and uncontrollable. Computational cost is another criterion for evaluating the performance of numerical methods. From the results in the tables, it is evident that the MHUAM is much more efficient and significantly reduces the execution time compared to the HMCM.

To conduct a deeper analysis of how the random characteristics of the luffing cylinder thrust and the suspended weight's gravity influence the dynamic response field of the LSOAAC with both random and interval parameters, we set the scenario as follows: the DD of one random parameter ranges from 0 to 0.2, while the DD of the other uncertain parameters is fixed at 0.1. Accordingly, the hybrid uncertainty models are defined as DD1 $= V_Q, V_F = \Delta a_F = \Delta L_F = 0.1$, and DD2 $= V_F, V_Q = \Delta a_F = \Delta L_F = 0.1$. Figure 6.24 illustrates the variances of the minimum value of the LB of the LSOAAC response calculated by the MHUAM under the hybrid uncertainty models DD1 and DD2. Figure 6.25 presents the variances of the minimum values of both the LB and UB of the LSOAAC response under the hybrid uncertainty model DD2, as computed by the MHUAM. Additionally, Figure 6.26 shows the variances of the minimum and maximum values of the LB of the LSOAAC response under the hybrid uncertainty model DD2, derived from the MHUAM calculations.

From Figure 6.24, it is evident that when the DD is less than 0.1, the impact of randomness in the luffing cylinder thrust F is greater than that of randomness in the suspended weight's gravity Q. Conversely, when the DD exceeds 0.1, the randomness in Q exerts a more pronounced effect. It should be noted that the MHUAM can be applied to analyze other variances of the extrema of the LSOAAC response bounds, with comparable results. For the sake of simplifying the discussion, this scenario is taken as a representative example. From Figures 6.25 and 6.26, it can be observed that as the coefficient of variation V_F increases, the interval width between the variances of the minimum value of the LB and the UB of the LSOAAC response shows an upward trend. In contrast, the interval width between the variances of the minimum and maximum values of the LB of the LSOAAC

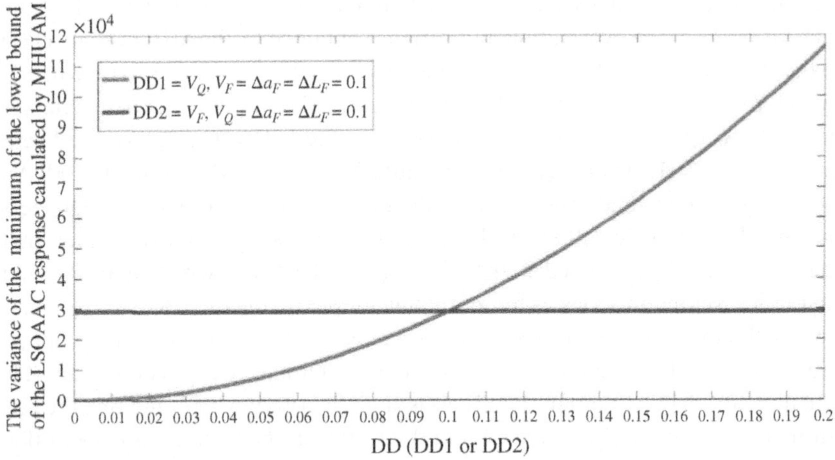

Figure 6.24 The variance of the minimum of the LB of the LSOAAC response calculated by MHUAM.

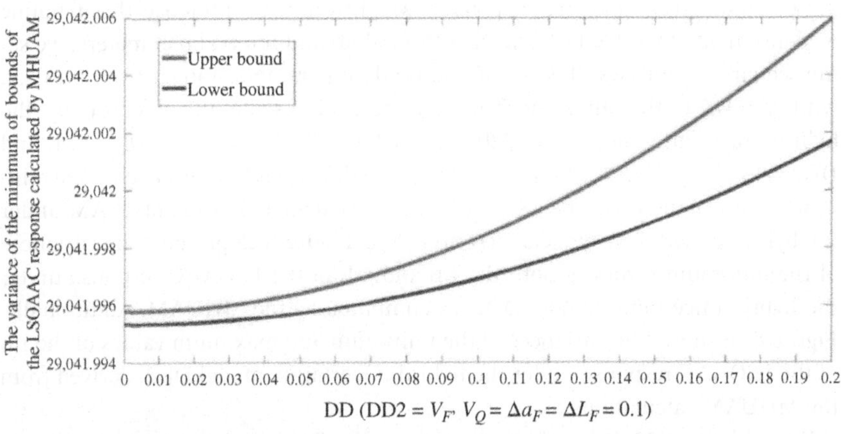

Figure 6.25 The variance of the minimum of the bounds of the LSOAAC response calculated by MHUAM.

response remains constant as V_F increases. Similarly, the MHUAM can be used to analyze another hybrid uncertainty model (DD $= V_Q, V_F = \Delta a_F = \Delta L_F = 0.1$), yielding identical results. To avoid redundancy, these results are not elaborated.

To conduct an in-depth analysis of how the interval characteristics of the displacement a (between the lower end of the lifting arm and the luffing cylinder) and the lifting arm length L affect the dynamic response field of the LSOAAC with both random and interval parameters, we use the hybrid uncertainty model DD $= V_F = V_Q = \Delta a_F = \Delta L_F$ as a reference. By the definition of uniform distribution, random variables can be modeled as interval variables. As a result, the

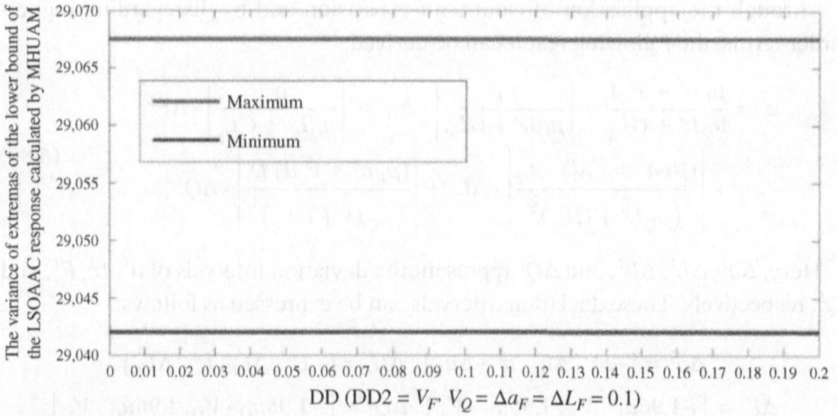

Figure 6.26 The variance of extrema of the LB of the LSOAAC response calculated by MHUAM.

dynamic response field of the LSOAAC under the hybrid uncertainty model is equivalent to that under a pure interval model. A comparison will thus be drawn between the dynamic response field of the LSOAAC under the hybrid uncertainty model and that under the pure interval model.

For random parameters (the thrust of the luffing cylinder F and the gravity of the suspended weight Q) following a uniform distribution, according to the characteristics of the uniform distribution curve, when the area covered by a certain interval reaches 95%, the LB and UB of standard normal random parameters are known to be (−1.96, +1.96). On this basis, the LB and UB of the aforementioned random parameters can be derived as follows:

$$\begin{aligned} \underline{F} &= \mu_F - 1.96\sigma_F, \overline{F} = \mu_F + 1.96\sigma_F \\ \underline{Q} &= \mu_Q - 1.96\sigma_Q, \overline{Q} = \mu_Q + 1.96\sigma_Q \end{aligned} \tag{6.82}$$

Building on Eq. (6.49), the input moment M and output moment T of the LSOAAC can be derived as follows:

$$\begin{aligned} M &= \mu_F a^c + F_s d + \mu_F \Delta a^I + a^c \Delta F^I \\ T &= \mu_Q L^c + G L_g + \mu_Q \Delta L^I + L^c \Delta Q^I \end{aligned} \tag{6.83}$$

With reference to Eq. (6.57), the interval dynamic response equilibrium equation of the LSOAAC can be re-expressed as follows:

$$\varepsilon' = \left(\mu_Q L^c + G L_g + \mu_Q \Delta L^I + L^c \Delta Q^I\right)^{-1} \left(\mu_F a^c + F_s d + \mu_F \Delta a^I + a^c \Delta F^I\right) \tag{6.84}$$

where ε' is the interval LSOAAC response.

Through the application of Neumann expansion and by disregarding higher-order terms, the following result can be derived:

$$\begin{aligned}\varepsilon' = &\frac{\mu_F a^c + F_s d}{\mu_Q L^c + G L_g} + \left|\frac{\mu_F}{\mu_Q L^c + G L_g}\right| \cdot \Delta a^I + \left|\frac{a^c}{\mu_Q L^c + G L_g}\right| \cdot \Delta F^I \\ &+ \left|\frac{(\mu_F a^c + F_s d) \cdot \mu_Q}{(\mu_Q L^c + G L_g)^2}\right| \cdot \Delta L^I + \left|\frac{(\mu_F a^c + F_s d)\, L^c}{(\mu_Q L^c + G L_g)^2}\right| \cdot \Delta Q^I\end{aligned} \tag{6.85}$$

Here, Δa^I, ΔL^I, ΔF^I, and ΔQ^I represent the deviation intervals of a^I, L^I, F^I, and Q^I, respectively. These deviation intervals can be expressed as follows:

$$\begin{aligned}\Delta a^I = [-a^c \cdot \Delta a_F, a^c \cdot \Delta a_F]\,, \Delta L^I = [-L^c \cdot \Delta L_F, L^c \cdot \Delta L_F] \\ \Delta F^I = [-1.96\mu_F \cdot V_F, 1.96\mu_F \cdot V_F]\,, \Delta Q^I = [-1.96\mu_Q \cdot V_Q, 1.96\mu_Q \cdot V_Q]\end{aligned} \tag{6.86}$$

In accordance with the IPM [37], the LB and UB of the dynamic response of the LSOAAC under the pure interval model can be expressed as follows:

$$\begin{aligned}\underline{\varepsilon'} = &\frac{\mu_F a^c + F_s d}{\mu_Q L^c + G L_g} + \left|\frac{\mu_F}{\mu_Q L^c + G L_g}\right| \cdot a^c \cdot \Delta a_F + \left|\frac{a^c}{\mu_Q L^c + G L_g}\right| \cdot 1.96\mu_F \cdot V_F \\ &+ \left|\frac{(\mu_F a^c + F_s d) \cdot \mu_Q}{(\mu_Q L^c + G L_g)^2}\right| \cdot L^c \cdot \Delta L_F + \left|\frac{(\mu_F a^c + F_s d)\, L^c}{(\mu_Q L^c + G L_g)^2}\right| \cdot 1.96\mu_Q \cdot V_Q\end{aligned} \tag{6.87}$$

$$\begin{aligned}\overline{\varepsilon'} = &\frac{\mu_F a^c + F_s d}{\mu_Q L^c + G L_g} - \left|\frac{\mu_F}{\mu_Q L^c + G L_g}\right| \cdot a^c \cdot \Delta a_F - \left|\frac{a^c}{\mu_Q L^c + G L_g}\right| \cdot 1.96\mu_F \cdot V_F \\ &- \left|\frac{(\mu_F a^c + F_s d) \cdot \mu_Q}{(\mu_Q L^c + G L_g)^2}\right| \cdot L^c \cdot \Delta L_F - \left|\frac{(\mu_F a^c + F_s d)\, L^c}{(\mu_Q L^c + G L_g)^2}\right| \cdot 1.96\mu_Q \cdot V_Q\end{aligned} \tag{6.88}$$

Based on the above analysis, the expectations of the bounds of the LSOAAC response field calculated using the IPM and MHUAM are presented in Figure 6.27. In this figure, the LB and UB obtained from the MHUAM and IPM follow the order: MHUAM (LB), MHUAM (UB), IPM (LB), IPM (UB).

It is clear that the UB of the LSOAAC response field derived from the MHUAM is smaller than that calculated via the IPM, while the LB obtained by the MHUAM is larger than that from the IPM. From this, we can draw the following conclusions: if all random parameters of the LSOAAC are treated as interval variables, when the DD is small, the bounds of the LSOAAC response field obtained using the IPM

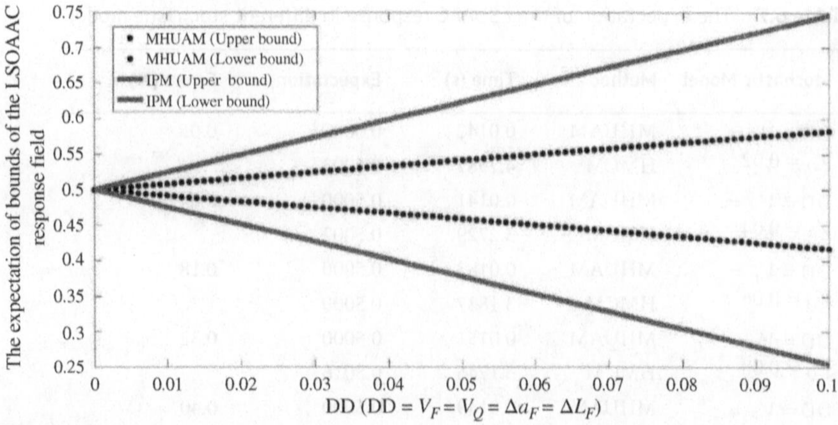

Figure 6.27 The expectation of bounds of the LSOAAC response field obtained by IPM and MHUAM.

may be conservative. When the DD is large, as the DD increases, the interval width of the LSOAAC response field calculated by the IPM expands more rapidly than that from the MHUAM. Consequently, the IPM may lead to either an unsafe or an overly conservative estimation of the LSOAAC response field.

6.3.4.2 LSOAAC Response Field Problem with Pure Random Parameters

In this example, the luffing cylinder thrust F and the suspended weight's gravity Q are also treated as random parameters following a Gaussian distribution, with $\mu_F = 4.3 \times 10^5$ N, $\mu_Q = 5 \times 10^4$ N. These two random parameters are assumed to be mutually independent, so the stochastic model can be defined as DD $= V_F = V_Q$. In the various stochastic models listed in Tables 6.7 and 6.8, the other two uncertain parameters—the displacement a between the lower end of the lifting arm and the luffing cylinder, and the lifting arm length L—are taken as constants, specifically $a = 2.5$ m, $L = 30$ m. The expectations and variances calculated by both the MHUAM and the HMCM are presented in Tables 6.7 and 6.8, respectively, where the HMCM with 10^8 samples serves as the reference solution.

From Tables 6.7 and 6.8, it can be observed that varying DDs in loads exert distinct impacts on the LSOAAC response field. In comparison with the HMCM, the results generated by the MHUAM exhibit acceptable accuracy while effectively reducing the substantial computational cost associated with the HMCM. This example validates the feasibility of applying the MHUAM to address the problem of the LSOAAC response field involving purely random parameters.

Table 6.7 The expectation of the LSOAAC response in different stochastic model.

Stochastic Model	Method	Time (s)	Expectation	Errors (%)
$DD = V_F = V_Q = 0.02$	MHUAM	0.0142	0.5000	0.06
	HMCM	4.7987	0.5003	
$DD = V_F = V_Q = 0.04$	MHUAM	0.0141	0.5000	0.04
	HMCM	3.2729	0.5002	
$DD = V_F = V_Q = 0.06$	MHUAM	0.0183	0.5000	0.18
	HMCM	3.1837	0.5009	
$DD = V_F = V_Q = 0.08$	MHUAM	0.0151	0.5000	0.32
	HMCM	3.1746	0.5016	
$DD = V_F = V_Q = 0.10$	MHUAM	0.0149	0.5000	0.40
	HMCM	3.1673	0.5020	

Table 6.8 The variance of the LSOAAC response in different stochastic model.

Stochastic Model	Method	Time (s)	Variance	Errors (%)
$DD = V_F = V_Q = 0.02$	MHUAM	0.0142	1.4199E−04	2.84
	HMCM	4.7987	1.3807E−04	
$DD = V_F = V_Q = 0.04$	MHUAM	0.0141	5.6798E−04	0.91
	HMCM	3.2729	5.6287E−04	
$DD = V_F = V_Q = 0.06$	MHUAM	0.0183	0.0012	7.69
	HMCM	3.1837	0.0013	
$DD = V_F = V_Q = 0.08$	MHUAM	0.0151	0.0023	4.17
	HMCM	3.1746	0.0024	
$DD = V_F = V_Q = 0.10$	MHUAM	0.0149	0.0035	2.78
	HMCM	3.1673	0.0036	

To further explore how the random characteristics of the luffing cylinder thrust F and the suspended weight's gravity Q, affect the dynamic response field of the LSOAAC with purely random parameters, two additional stochastic models are adopted as references: $DD1 = V_F, V_Q = 0$; $DD2 = V_Q, V_F = 0$. Figure 6.28 presents the variances of the LSOAAC response under these different stochastic models, as calculated by the MHUAM.

It is evident that the variance of the LSOAAC response field calculated by the MHUAM exhibits distinct variation trends across different stochastic models. Moreover, the impact of randomness in the luffing cylinder thrust F is greater than that of randomness in the suspended weight's gravity Q.

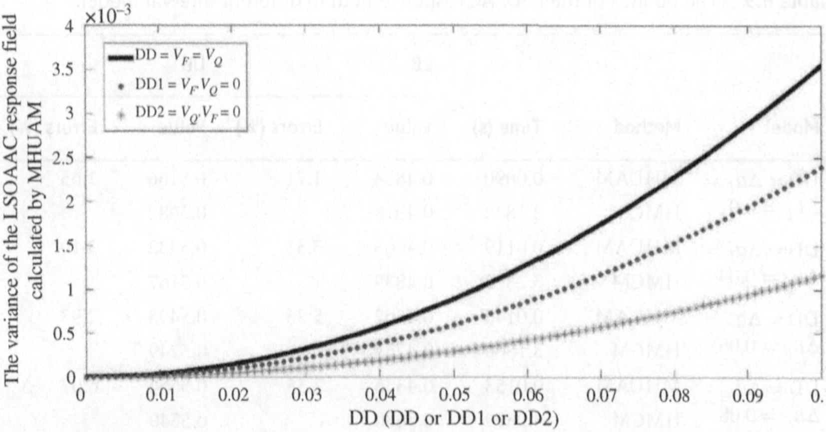

Figure 6.28 The variance of the LSOAAC response field in different stochastic model calculated by MHUAM.

6.3.4.3 LSOAAC Response Field Problem with Pure Interval Parameters

In this example, the displacement a between the lower end of the lifting arm and the luffing cylinder, and the length L of the lifting arm are treated as interval parameters, with $a^c = 2.5$ m and $L^c = 30$ m. These two interval parameters are assumed to be independent of each other, so the interval model can be defined as $\mathrm{DD} = \Delta a_F = \Delta L_F$. In the various interval models listed in Table 6.9, the other two uncertain parameters—the thrust F of the luffing cylinder and the gravity Q of the suspended weight—are taken as constants, specifically $F = 4.3 \times 10^5$ N and $Q = 5 \times 10^4$ N. The bounds of the LSOAAC response field calculated by both the MHUAM and the HMCM are presented in Table 6.9, where the HMCM with 10^8 samples serve as the reference solution. In the relevant figure, the LB and UB of the LSOAAC response are denoted as LB and UB, respectively.

From Table 6.9, it can be seen that varying DDs in structural parameters have distinct impacts on the LSOAAC response field. Compared with the HMCM, the results obtained by the MHUAM show acceptable accuracy while effectively reducing the enormous computational cost of the HMCM. This example confirms the feasibility of applying the MHUAM to solve the problem of the LSOAAC response field with pure interval parameters. To further explore how the uncertain characteristics of the displacement a (between the lower end of the lifting arm and the luffing cylinder) and the lifting arm length L, affect the dynamic response field of the LSOAAC with pure interval parameters, two additional interval models are used as references: $\mathrm{DD1} = \Delta a_F, \Delta L_F = 0$; $\mathrm{DD2} = \Delta L_F, \Delta a_F = 0$. The bounds of the LSOAAC response field under these different interval models, calculated by the MHUAM, are plotted in Figure 6.29.

Table 6.9 The bounds of the LSOAAC response field in different interval model.

Model	Method	Time (s)	LB		UB	
			Value	Errors (%)	Value	Errors (%)
$DD = \Delta a_F = \Delta L_F = 0.02$	MHUAM	0.0090	0.4834	1.71	0.5166	1.65
	HMCM	3.1842	0.4918		0.5082	
$DD = \Delta a_F = \Delta L_F = 0.04$	MHUAM	0.0119	0.4668	3.53	0.5332	3.19
	HMCM	3.2322	0.4839		0.5167	
$DD = \Delta a_F = \Delta L_F = 0.06$	MHUAM	0.0146	0.4502	5.36	0.5498	2.93
	HMCM	3.1806	0.4757		0.5249	
$DD = \Delta a_F = \Delta L_F = 0.08$	MHUAM	0.0153	0.4336	7.35	0.5664	6.07
	HMCM	3.2699	0.4680		0.5340	
$DD = \Delta a_F = \Delta L_F = 0.10$	MHUAM	0.0165	0.4170	9.43	0.5830	7.43
	HMCM	3.1485	0.4604		0.5427	

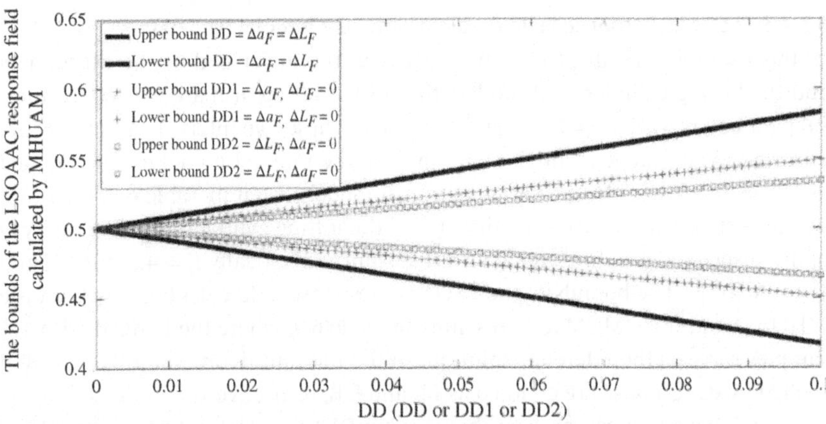

Figure 6.29 The bounds of the LSOAAC response field in different interval model calculated by MHUAM.

It is evident that the bounds of the LSOAAC response field calculated by the MHUAM exhibit distinct variation trends across different interval models. Furthermore, the impact of uncertainty in the displacement a (between the lower end of the lifting arm and the luffing cylinder) is greater than that of uncertainty in the lifting arm length L.

6.3.5 Conclusion

By utilizing perturbation theory and the random interval moment method, this section puts forward a revised hybrid uncertainty analysis approach to forecast the dynamic response field issue in the LSOAAC that involves both random and interval parameters. All uncertainties regarding structural dimensions, input forces, and external loads are taken into full consideration, which enhances the objectivity of the hybrid uncertain dynamic response equilibrium equation for the LSOAAC. Uncertain parameters with a specific probability distribution are modeled as random variables, while those defined merely by LB and UB are treated as interval variables. Within the proposed method, the dynamic response expression of the LSOAAC is approximated through the integration of random interval perturbation, which is rooted in stochastic perturbation analysis. With the help of the random interval moment method and monotonic analysis technology, the mathematical properties of the extrema of the bounds of the dynamic response are successfully predicted.

In contrast to HMCM, numerical outcomes from several examples concerning the hybrid LSOAAC response field have confirmed that MHUAM is feasible and efficient when addressing narrow uncertainties. When the random variables within the LSOAAC featuring hybrid uncertain parameters are treated as interval variables, the results indicate that MHUAM can yield a safer and less conservative assessment of the hybrid LSOAAC response field in comparison to IPM. Regarding the influence on the expectation of the hybrid LSOAAC response field, the primary effect originates from all interval parameters (specifically, the displacement between the lower end of the lifting arm and the luffing cylinder, and the length of the lifting arm). As for the variance of the hybrid LSOAAC response field, the main contributors are the output-related random and interval parameters (i.e., the gravity of the suspended weight and the length of the lifting arm). A study into how the randomness of random parameters impacts the hybrid LSOAAC response field shows that the luffing cylinder thrust and the gravity of the suspended weight have varying effects on the hybrid LSOAAC response field under different DDs. When compared with HMCM, the proposed method also proves efficient in solving problems related to the pure random and pure interval LSOAAC response fields. Furthermore, the distinct impacts of different uncertain parameters on the pure LSOAAC response field have been explored in depth. Nevertheless, it should be pointed out that the proposed method, which is based on first-order Taylor series expansion and first-order Neumann series while ignoring higher-order terms, is not applicable to hybrid LSOAAC response field problems where variables have large variation ranges. Although taking higher-order series expansions into account can enhance the accuracy of the method, it will require more computational resources. Extending the application of this method to cooperative cable parallel manipulators for multiple mobile cranes with random and interval parameters represents a valuable area for future research.

6.4 Conclusion

This chapter completes dynamic analysis of rigid–flexible coupling hoisting robots with uncertainty. First, static uncertainty analysis with fuzzy parameters was accomplished, followed by dynamic uncertainty analysis with hybrid random and interval parameters, providing a reference for future uncertainty analysis in the rigid–flexible coupled crane dynamics.

References

1 Merlet J P, Daney D. A portable, modular parallel wire crane for rescue operations. IEEE International Conference on Robotics and Automation, Alaska, USA, 2010.
2 Leban F, Gonzalez J, Parker G. Inverse kinematic control of a dual crane system experiencing base motion. IEEE Transactions on Control Systems Technology, 2015, 23: 331–339.
3 Scalera L, Gallina P, Seriani S, et al. Cable-based robotic crane (CBRC): design and implementation of overhead traveling cranes based on variable radius drums. IEEE Transactions on Robotics, 2018, 34(2): 1–8.
4 Lu B, Fang Y C, Sun N. Modeling and nonlinear coordination control for an underactuated dual overhead crane system. Automatica, 2018, 91: 244–255.
5 Hussein H, Santos J C, Izard J B, et al. Smallest maximum cable tension determination for cable-driven parallel robots. IEEE Transactions on Robotics, 2021, 99: 1–20.
6 Sun N, Fang Y C, Chen H. Adaptive antiswing control for cranes in the presence of rail length constraints and uncertainties. Nonlinear Dynamics, 2015, 81 (1–2): 41–51.
7 Zi B, Zhou B. A modified hybrid uncertain analysis method for dynamic response field of the LSOAAC with random and interval parameters. Journal of Sound and Vibration, 2016, 374: 111–137.
8 Eldred M S, Swiler L, Tang G. Mixed aleatory-epistemic uncertainty quantification with stochastic expansions and optimization-based interval estimation. Reliability Engineering and System Safety, 2011, 96(9): 1092–1113.
9 Elishakoff I, Yao J T P. Probabilistic methods in the theory of structures. ASME Journal of Applied Mechanics, 1984, 51(2): 451–452.
10 Moens D, Hanss M. Non-probabilistic finite element analysis for parametric uncertainty treatment in applied mechanics: recent advances. Finite Elements in Analysis and Design, 2011, 47(1): 4–16.

11 Neumaier A. Interval Methods for Systems of Equations. Cambridge, UK: Cambridge University Press, 1990.

12 Qiu Z P, Elishakoff I. Anti-optimization of structures with large uncertain-but-non-random parameters via interval analysis. Computer Methods Applied Mechanics Engineering, 1998, 152: 361–372.

13 Sofi A, Muscolino G, Elishakoff I. Natural frequencies of structures with interval parameters. Journal of Sound and Vibration, 2015, 347: 79–95.

14 Xia B Z, Yu D J. Modified sub-interval perturbation finite element method for 2D acoustic field prediction with large uncertain-but-bounded parameters. Journal of Sound and Vibration, 2012, 331: 3774–3790.

15 Shafer G. A Mathematical Theory of Evidence. Princeton: Princeton University Press, 1976.

16 Lü H, Shangguan W B, Yu D. An imprecise probability approach for squeal instability analysis based on evidence theory. Journal of Sound and Vibration, 2016, 387: 96–113.

17 Yin S W, Yu D J, Yin H, et al. A new evidence-theory-based method for response analysis of acoustic system with epistemic uncertainty by using Jacobi expansion. Computer Methods in Applied Mechanics and Engineering, 2017, 322: 419–440.

18 Cao L X, Liu J, Jiang C, et al. Evidence-based structural uncertainty quantification by dimension reduction decomposition and marginal interval analysis. Journal of Mechanical Design, 2020, 142(5): 051701.

19 Ben-Haim Y, Elishakoff I. Convex Models of Uncertainty in Applied Mechanics. Amsterdam: Elsevier Science, 1990.

20 Xia B Z, Yu D J. Response analysis of acoustic field with convex parameters. Journal of Vibration and Acoustics, 2014, 136: 041017.

21 Zadeh L A. Fuzzy sets. Information and Control, 1965, 8(3): 338–353.

22 Rao S S, Sawyer J P. A fuzzy element approach for the analysis of imprecisely defined system. AIAA Journal, 1995, 33(12): 2364–2370.

23 Chen L, Rao S S. Fuzzy finite-element approach for the vibration analysis of imprecisely-defined systems. Finite Elements in Analysis and Design, 1997, 27(1): 69–83.

24 Massa F, Tison T, Lallemand B. A fuzzy procedure for the static design of imprecise structures. Computer Methods in Applied Mechanics and Engineering, 2006, 195 (9–12): 925–941.

25 Wang C, Qiu Z P. Uncertain temperature field prediction of heat conduction problem with fuzzy parameters. International Journal of Heat and Mass Transfer, 2015, 91: 725–733.

26 Balu A S, Rao B N. High dimensional model representation based formulations for fuzzy finite element analysis of structures. Finite Elements in Analysis and Design, 2012, 50: 217–230.

27 Yin H, Yu D J, Yin S W. Fuzzy interval finite element/statistical energy analysis for mid-frequency analysis of built-up systems with mixed fuzzy and interval parameters. Journal of Sound and Vibration, 2016, 380: 192–212.

28 Wang C, Qiu Z P. Subinterval perturbation methods for uncertain temperature field prediction with large fuzzy parameters. International Journal of Thermal Sciences, 2016, 100: 381–390.

29 Gao W, Song C, Tin-Loi F. Probabilistic interval analysis for structures with uncertainty. Structural Safety, 2010, 32: 191–199.

30 Gao W, Wu D, Song C, et al. Hybrid probabilistic interval analysis of bar structures with uncertainty using a mixed perturbation Monte-Carlo method. Finite Elements in Analysis and Design, 2011, 47: 643–652.

31 Xia B, Lü H, Yu D, et al. Reliability-based design optimization of structural systems under hybrid probabilistic and interval model. Computers and Structures, 2015, 160: 126–134.

32 Zhou B, Zi B, Qian S. Dynamics-based nonsingular interval model and luffing angular response field analysis of the DACS with narrowly bounded uncertainty. Nonlinear Dynamics, 2017, 90: 2599–2626.

33 Zi B, Qian S, Ding H, et al. Design and analysis of cooperative cable parallel manipulators for multiple mobile cranes. International Journal of Advanced Robotic Systems, 2012, 9: 1–9.

34 Wang C, Qiu Z. Hybrid uncertain analysis for steady-state heat conduction with random and interval parameters. International Journal of Heat and Mass Transfer, 2015, 80: 319–328.

35 Wang C, Qiu Z. Hybrid uncertain analysis for temperature field prediction with random, fuzzy and interval parameters. International Journal of Thermal Sciences, 2015, 98: 124–134.

36 Xia B, Yin S, Liu J. A new random interval method for response analysis of structural–acoustic system with interval random variables. Applied Acoustics, 2015, 99: 31–42.

37 Wang C, Qiu Z. An interval perturbation method for exterior acoustic field prediction with uncertain-but-bounded parameters. Journal of Fluids and Structures, 2014, 49: 441–449.

7

Trajectory Planning and Tracking Control of Rigid–Flexible Coupling Hoisting Robots

7.1 Preamble

A specific kind of rigid–flexible coupling hoisting robot has become known as a cable-suspended parallel robot (CSPRs), during which flexible cables are used in place of rigid linkages to drive the end-effector. CSPRs in which gravity acts as a cable tension and all cable attachment points are above the end-effector. This is accomplished by reversing all of the CSPRs taut cables and using the end-effector's gravity as a downward push. Wide workspace, reduced inertia, load capacity, high precision and accelerated motions, ease of construction and disassembly, reduced cost and consumption of energy, portability, and reconfiguration are just a few of the many alluring advantages that CSPRs provide over standard parallel robots. In recent years, CSPRs have been extensively utilized in a variety of fields due to their appealing merits, such as hoisting heavy loads [1, 2], large radio telescopes [3, 4], fast manipulation [5], wind tunnel testing [6], camera robots [7], 3D printers [8–10], rehabilitation robots [11, 12], construction of buildings [13, 14], and haptic devices [15]. However, one inherent drawback of CSPRs is that cables can only drag the end-effector—they cannot push it [16, 17].

Some scholars have looked at the CSPR's trajectory planning. A dynamic trajectory planning technology for six-degree-of-freedom (DOF) CSPRs was introduced by Jiang et al. [18]. An analogous passive mechanical system to the CSPR was demonstrated. It was suggested to use a trajectory design that goes beyond the robot's stationary workspace. In addition to establishing the mathematical model, the tilt-and-torsion angle convention can be used to confine the trajectory's rotational component. The dynamic differential equations governing the trajectory's translation component can become linear under specific circumstances. Thus, the natural frequency of a linear system of equivalent passive spring with constant stiffness was gained, and a general solution for periodic and natural trajectory was produced. This method was adopted to the trajectory of pure translation and more sophisticated movement including the pose change. Dion-Gauvin et al. [19] proposed a dynamic trajectory planning technology for 3-DOF CSPRs. The idea of the trajectory path originated from a hypocycloid curve, which was inserted in the plane determined by the acceleration vector at the start and end points. The presented movement can guarantee acceleration continuity at each end point and zero instantaneous velocity. The positive tension of cable can be ensured by appropriately selecting the number of arcs for the hypocycloid. Utilizing this trajectory, contiguous target sites that might be outside the robot's static workspace can be successively joined. In comparison with earlier techniques, this technology can generate an extensive selection of target spots that are approachable.

The existing research on trajectory planning fixes the height and angle of the cable mast of CSPRs; however, this chapter takes into account the variations in the under-constrained cable-suspended parallel robot's (UCPR's) cable mast's angle and height. The UCPR's point-to-point trajectory is planned using three algebraic techniques. Three translational DOFs can be achieved via the UCPR end-effector with three cables. By adjusting the cable mast's angle via a gear drive along its circular base, the likelihood of cables colliding with obstructions is decreased. A hydraulic cylinder can change the cable mast's height, increasing the end-effector's reaching workspace. The provided trajectory, which consists of arcs and lines, is a closed curve with six end-to-end connected segments. In order to guarantee that the acceleration is continuous, the connecting points of trajectory segments are designed to have zero instantaneous velocity. The robot's impact and vibration can be efficiently avoided by using the planned acceleration curves to ensure a smooth transition at the connecting sites.

Numerous studies have used fuzzy PID controllers for robot trajectory tracking control in recent years [20], and it is undeniable that these controllers have good control effects. A fuzzy PID controller enhances the control effect of a PID controller on nonlinear factors; however, the universe, its membership function, and fuzzy rules for a fuzzy controller must be chosen and designed with experiences and specialist expertise. Additionally, if the design is unjustified, the control effect may be poor. Therefore, some experts suggest optimizing the fuzzy PID controller's design settings to increase the controller's logic and trajectory tracking control impact. Since it is challenging to characterize the optimization problem of a fuzzy controller using explicit or implicit mathematical functions, a swarm intelligence optimization algorithm that is independent of mathematical functions can resolve the practical complex optimization problem. In order to improve the robot's trajectory tracking ability, Richa et al. employed a genetic algorithm to optimize the PID controller parameters of the planar two-link robot. The goal function was the weight of the integral change square controller output and the overall square error [21].

A fuzzy PID control strategy based on an adaptive whale optimization algorithm (FPID-AWOA) is presented to increase the control effect of CDPRs in order to address the issues of high precision trajectory tracking control of CDPRs in the current control schemes. First, this study provides fuzzy rules relevant to CDPRs based on the configuration parameters and fuzzy rules that are currently in use. The scale factor and quantization factor of the fuzzy PID controller are then optimized using an optimization approach that is based on the fuzzy PID controller. This study introduces the weight function and differential variation interference factor to enhance the optimization performance of the whale optimization algorithm, taking into account its limitations in terms of accuracy and efficiency. After that, the fuzzy PID controller is updated to use the recently created whale optimization method. The suggested FPID-AWOA can successfully increase the tracking accuracy of the CDPR trajectory, according to numerical calculations and experiments conducted under a spiral trajectory.

Cable-driven robots have garnered a lot of attention both domestically and internationally due to their many benefits, and they have emerged as a new robotics research hotspot. The cable-driven parallel robot, also referred to as a CDPR, is the sort of parallel robot that garners the most interest. Instead of stiff linkages like in conventional rigid-link parallel robots, the end-effector in CDPRs is suspended by several flexible cables [22]. In order to achieve movement and positioning of the cable's end within the space range or to make the end produce the anticipated pulling force to finish the relevant activity, the fixed position drive device drives and releases the cable as needed. The cable drive unit exhibits significant physical and geometric nonlinearity.

It features some viscous damping in addition to a high frictional resistance. It is a dynamic, intricate system [23]. Zou [24] suggested an initiative force servo control method for the force control of a cable-driven training robot in terms of control and monitoring techniques for cable-driven robots. To get rid of extra force, Xing [25] used speed feedforward adjustment. The adaptive fuzzy control strategy that was suggested to achieve the cable parallel robot's monitoring function has been tested by Xia [26].

This part uses the well-known cable force control platform as the study object, examines the cable drive unit's force control mechanism, and creates PID controller parameters to enhance the cable drive unit's output force characteristics. An experimental platform monitoring system is constructed by combining the aforementioned control techniques. In addition to monitoring the cable's condition and the experimental platform's operational parameters, this system can optimize the cable drive unit's output force. It significantly increases the effectiveness of monitoring and offers a workable theoretical and practical reference for the system's future implementation on the cable-driven robot entities.

This section's remaining content is arranged as follows. Section 7.2 introduces the trajectory planning for rigid–flexible coupling hoisting robots. Section 7.3 introduces the fuzzy control for rigid–flexible coupling hoisting robots. Section 7.4 introduces force control for rigid–flexible coupling hoisting robots. Finally, some conclusions are summarized in Section 7.5.

7.2 Trajectory Planning for RFCHRs

In this section, a configuration for a UCPR with an angle- and height-variable cable mast is proposed. The UCPR's end-effector, driven by three cables, can perform motions with three translational DOFs. The modeling of the robot's inverse kinematics and dynamics, considering the mast's geometry, has been completed. A motion trajectory for the end-effector, comprising six segments, is formulated. For planning this trajectory, three algebraic approaches were applied: the fifth-degree polynomial, the cycloid trajectory, and the double-S velocity curve. These algebraic methods ensure that velocity is zero at the transition points between segments and that acceleration curves are continuous and smooth. The findings suggest that the most desirable trajectory is achieved using the fifth-degree polynomial method. Based on this optimal planning, both numerical simulations and experiments were performed for the specified trajectory. At points of sudden velocity shifts, a comparative analysis was conducted on the length and tension variation curves for the three cables with and without planning. Throughout the experimental phase, the UCPR's end-effector was tracked in real time using an OptiTrack motion capture system. This process served to validate the overall effectiveness and feasibility of employing fifth-degree polynomial planning.

7.2.1 Inverse Kinematic Modeling

The architecture of the three-cable UCPR is presented in the diagram in Figure 7.1. Driven by gears, each cable mast can pivot along the circular base, a design that minimizes the probability of cables interfering with obstacles. At the top of each mast, a pulley's height can be adjusted by a hydraulic cylinder, thereby expanding the operational workspace of the end-effector. The modeling proceeds under the common assumption that all cables are inelastic and have negligible mass. Solving the inverse kinematics for this UCPR therefore entails computing the cable lengths required to achieve a specific end-effector pose.

The kinematic model of the UCPR is formulated using two essential reference frames. A global, stationary frame $OXYZ$ is established with its origin O at the geometric center of the circular base and its Z-axis aligned with

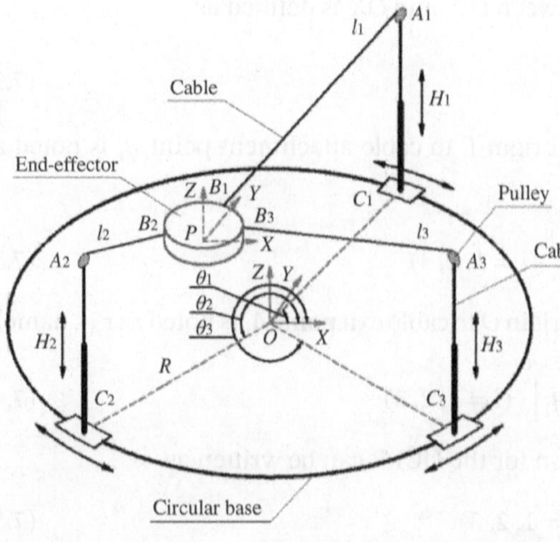

Figure 7.1 Structural diagram of the UCPR with three cables.

the vertical direction. A local, moving frame $PXYZ$ is rigidly attached to the end-effector, with its origin P located at the centroid. A_i $(i = 1, 2, 3)$ denotes cable exit point, and B_i $(i = 1, 2, 3)$ represents cable attachment point between cable and the end-effector. R stands for the radius of the circular base of the robot. H_i $(i = 1, 2, 3)$ denotes the height of cable mast. ΔH_i $(i = 1, 2, 3)$ denotes the amount of change in the height of cable mast. θ_i $(i = 1, 2, 3)$ represents the angle between OC_i $(i = 1, 2, 3)$ and OX. $\Delta\theta_i$ $(i = 1, 2, 3)$ denotes the amount of change in the angle of cable mast. h stands for the thickness of the end-effector. The position vector of the origin P in the frame $OXYZ$ is expressed as $\mathbf{r}_P = \begin{bmatrix} x_P & y_P & z_P \end{bmatrix}^T$. The generalized coordinates considering the pose of the end-effector as well as the angles and heights of the cable masts in the frame $OXYZ$ can be expressed as $\mathbf{q} = \begin{bmatrix} x & y & z & \Delta\theta_1 & \Delta\theta_2 & \Delta\theta_3 & \Delta H_1 & \Delta H_2 & \Delta H_3 \end{bmatrix}^T$. Since the UCPR's end-effector possesses only three translational degrees of freedom, the orientation between the two frames is constant, and the rotation matrix from $PXYZ$ to $OXYZ$ is therefore defined as

$$ {}^O_P\mathbf{R} = \begin{bmatrix} 1 & 0 & 0 \\ 0 & 1 & 0 \\ 0 & 0 & 1 \end{bmatrix} \tag{7.1} $$

The angle θ_i of cable mast between OC_i and OX is defined as

$$\theta_i = \frac{2\pi (i-1)}{3}\ (i = 1, 2, 3) \tag{7.2}$$

The position vector from the origin P to cable attachment point B_i is noted as $\mathbf{r}_{B_i}$, namely

$$\mathbf{r}_{B_i} = \begin{bmatrix} r\cos\theta_i & r\sin\theta_i & h \end{bmatrix}^T\ (i = 1, 2, 3) \tag{7.3}$$

The position vector from the origin O to cable exit point A_i is noted as $\mathbf{r}_{A_i}$, namely

$$\mathbf{r}_{A_i} = \begin{bmatrix} R\cos\theta_i & R\sin\theta_i & H_i \end{bmatrix}^T\ (i = 1, 2, 3) \tag{7.4}$$

The inverse kinematic equation for the UCPR can be written as

$$\mathbf{l}_i = \mathbf{r}_{A_i} - {}^{O}_{P}\mathbf{R} \cdot \mathbf{r}_{B_i} - \mathbf{r}_P\ (i = 1, 2, 3) \tag{7.5}$$

where $\mathbf{l}_i$ denotes the length vector of the ith cable.

The length scalar of the ith cable can be denoted as

$$l_i = \sqrt{\mathbf{l}_i^T \cdot \mathbf{l}_i}\ (i = 1, 2, 3) \tag{7.6}$$

Differentiating Eq. (7.5) with respect to time establishes the kinematic relationship between the generalized velocities and the cable velocities, which is expressed as

$$\dot{\mathbf{l}} = -\mathbf{J}\dot{\mathbf{q}} \tag{7.7}$$

where $\dot{\mathbf{q}} = \begin{bmatrix} \dot{x} & \dot{y} & \dot{z} & \Delta\dot{\theta}_1 & \Delta\dot{\theta}_2 & \Delta\dot{\theta}_3 & \Delta\dot{H}_1 & \Delta\dot{H}_2 & \Delta\dot{H}_3 \end{bmatrix}^T$ denotes the velocity vector of the generated coordinates, $\dot{\mathbf{l}} = \begin{bmatrix} \dot{l}_1 & \dot{l}_2 & \dot{l}_3 \end{bmatrix}^T$ is the velocity vector of cables, and $\mathbf{J}$ stands for the Jacobian matrix which can be expressed as

$$\mathbf{J} = \begin{bmatrix} \frac{\partial l_1}{\partial x} & \frac{\partial l_1}{\partial y} & \frac{\partial l_1}{\partial z} & \frac{\partial l_1}{\partial \Delta\theta_1} & \frac{\partial l_1}{\partial \Delta\theta_2} & \frac{\partial l_1}{\partial \Delta\theta_3} & \frac{\partial l_1}{\partial \Delta H_1} & \frac{\partial l_1}{\partial \Delta H_2} & \frac{\partial l_1}{\partial \Delta H_3} \\ \frac{\partial l_2}{\partial x} & \frac{\partial l_2}{\partial y} & \frac{\partial l_2}{\partial z} & \frac{\partial l_2}{\partial \Delta\theta_1} & \frac{\partial l_2}{\partial \Delta\theta_2} & \frac{\partial l_2}{\partial \Delta\theta_3} & \frac{\partial l_2}{\partial \Delta H_1} & \frac{\partial l_2}{\partial \Delta H_2} & \frac{\partial l_2}{\partial \Delta H_3} \\ \frac{\partial l_3}{\partial x} & \frac{\partial l_3}{\partial y} & \frac{\partial l_3}{\partial z} & \frac{\partial l_3}{\partial \Delta\theta_1} & \frac{\partial l_3}{\partial \Delta\theta_2} & \frac{\partial l_3}{\partial \Delta\theta_3} & \frac{\partial l_3}{\partial \Delta H_1} & \frac{\partial l_3}{\partial \Delta H_2} & \frac{\partial l_3}{\partial \Delta H_3} \end{bmatrix} \tag{7.8}$$

7.2.2 Dynamic Modeling

Figure 7.2 illustrates the force analysis of the UCPR. The model considers three primary actuation inputs: T_i $(i = 1, 2, 3)$ representing the tension in each cable; T_{θ_i} $(i = 1, 2, 3)$ represents the torque acting on the mast about the Z-axis.

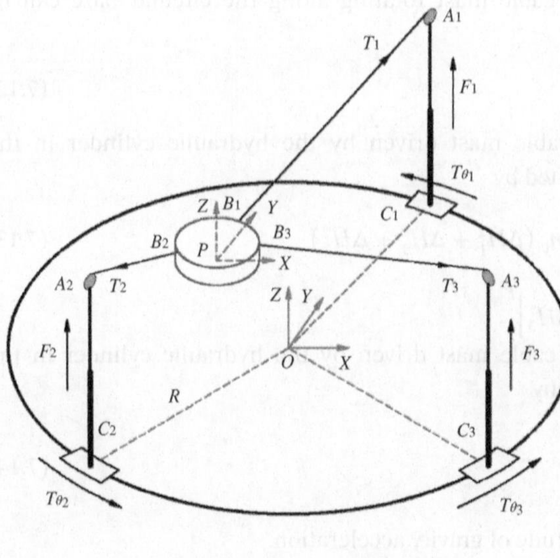

Figure 7.2 Force analysis diagram of the UCPR.

F_i $(i = 1, 2, 3)$ is the upward force applied by the mast to the pulley. The dynamic equations of motion are derived using the Lagrange method, which provides an explicit mapping between forces and motion. The foundation for this method is the system's Lagrangian, which is formulated by subtracting the total potential energy from the total kinetic energy.

The kinetic energy of the end-effector can be written as

$$K_1 = \frac{1}{2} m_e \dot{\mathbf{q}}_e^T \dot{\mathbf{q}}_e = \frac{1}{2} m \left(\dot{x}^2 + \dot{y}^2 + \dot{z}^2\right) \tag{7.9}$$

where m_e denotes the mass of the end-effector, and $\dot{\mathbf{q}}_e = \begin{bmatrix} \dot{x} & \dot{y} & \dot{z} \end{bmatrix}^T$.

The potential energy of the end-effector can be expressed as

$$U_1 = m_e g z \tag{7.10}$$

where g denotes the magnitude of gravity acceleration.

The kinetic energy of cable mast rotating along the circular base can be depicted as

$$K_2 = \frac{1}{2} I_\theta \dot{\mathbf{q}}_\theta^T \dot{\mathbf{q}}_\theta = \frac{1}{2} I_\theta \left(\Delta\dot{\theta}_1^2 + \Delta\dot{\theta}_2^2 + \Delta\dot{\theta}_3^2\right) \tag{7.11}$$

where $I_\theta = m_c R^2$ denotes the inertia moment of cable mast along the circular base, m_c represents the mass of cable mast, and $\dot{\mathbf{q}}_\theta = \begin{bmatrix} \Delta\dot{\theta}_1 & \Delta\dot{\theta}_2 & \Delta\dot{\theta}_3 \end{bmatrix}^T$.

The potential energy of cable mast rotating along the circular base can be described as

$$U_2 = 0 \tag{7.12}$$

The kinetic energy of cable mast driven by the hydraulic cylinder in the Z-direction can be represented by

$$K_3 = \frac{1}{2} m_c \dot{\mathbf{q}}_H^T \dot{\mathbf{q}}_H^T = \frac{1}{2} m_c \left(\Delta \dot{H}_1^2 + \Delta \dot{H}_2^2 + \Delta \dot{H}_3^2\right) \tag{7.13}$$

where $\dot{\mathbf{q}}_H = \begin{bmatrix} \Delta \dot{H}_1 & \Delta \dot{H}_2 & \Delta \dot{H}_3 \end{bmatrix}^T$.

The potential energy of cable mast driven by the hydraulic cylinder in the Z-direction can be denoted by

$$U_3 = \sum_{i=1}^{3} m_c g \Delta H_i \tag{7.14}$$

where g denotes the magnitude of gravity acceleration.

The Lagrange formulation of the UCPR can be described by

$$\begin{aligned} L &= K_1 + K_2 + K_3 - U_1 - U_2 - U_3 \\ &= \frac{1}{2} m_e \dot{\mathbf{q}}_e^T \dot{\mathbf{q}}_e + \frac{1}{2} I_\Theta \dot{\mathbf{q}}_\Theta^T \dot{\mathbf{q}}_\Theta + \frac{1}{2} m_c \dot{\mathbf{q}}_H^T \dot{\mathbf{q}}_H^T \\ &\quad - m_e g z - \sum_{i=1}^{3} m_c g \Delta H_i \end{aligned} \tag{7.15}$$

The motion equation of the UCPR can be established using the Lagrange formulation as follows:

$$\frac{d}{dt}\left(\frac{\partial L}{\partial \dot{\mathbf{q}}}\right) - \frac{\partial L}{\partial \mathbf{q}} = \mathbf{Q} \tag{7.16}$$

where $\mathbf{Q} = \begin{bmatrix} T_1 & T_2 & T_3 & T_{\Theta_1} & T_{\Theta_2} & T_{\Theta_3} & F_1 & F_2 & F_3 \end{bmatrix}^T$ represents the generalized force of the UCPR.

Substituting Eq. (7.15) into Eq. (7.16) leads to

$$\begin{cases} \frac{d}{dt}\left(\frac{\partial L}{\partial \dot{x}}\right) = m_e \ddot{x}, \ \frac{d}{dt}\left(\frac{\partial L}{\partial \dot{y}}\right) = m_e \ddot{y}, \ \frac{d}{dt}\left(\frac{\partial L}{\partial \dot{z}}\right) = m_e \ddot{z}, \\ \frac{d}{dt}\left(\frac{\partial L}{\partial \Delta \dot{\Theta}_1}\right) = I_\Theta \Delta \ddot{\Theta}_1, \ \frac{d}{dt}\left(\frac{\partial L}{\partial \Delta \dot{\Theta}_2}\right) = I_\Theta \Delta \ddot{\Theta}_2, \ \frac{d}{dt}\left(\frac{\partial L}{\partial \Delta \dot{\Theta}_3}\right) = I_\Theta \Delta \ddot{\Theta}_3, \\ \frac{d}{dt}\left(\frac{\partial L}{\partial \Delta \dot{H}_1}\right) = m_c \Delta \ddot{H}_1, \ \frac{d}{dt}\left(\frac{\partial L}{\partial \Delta \dot{H}_2}\right) = m_c \Delta \ddot{H}_2, \ \frac{d}{dt}\left(\frac{\partial L}{\partial \Delta \dot{H}_3}\right) = m_c \Delta \ddot{H}_3 \end{cases} \tag{7.17}$$

The partial derivative of the Lagrange formulation with regard to the generalized coordinates can be derived as

$$\begin{cases} \dfrac{\partial L}{\partial x} = 0, \dfrac{\partial L}{\partial y} = 0, \dfrac{\partial L}{\partial z} = -m_e g, \\ \dfrac{\partial L}{\partial \Delta\theta_1} = 0, \dfrac{\partial L}{\partial \Delta\theta_2} = 0, \dfrac{\partial L}{\partial \Delta\theta_3} = 0, \\ \dfrac{\partial L}{\partial \Delta H_1} = -m_c g, \dfrac{\partial L}{\partial \Delta H_1} = -m_c g, \dfrac{\partial L}{\partial \Delta H_1} = -m_c g \end{cases} \tag{7.18}$$

The dynamic model of the UCPR can be obtained by

$$\frac{d}{dt}\left(\frac{\partial L}{\partial \dot{\mathbf{q}}}\right) - \frac{\partial L}{\partial \mathbf{q}} = \begin{bmatrix} m_e & 0 & 0 & 0 & 0 & 0 & 0 & 0 & 0 \\ 0 & m_e & 0 & 0 & 0 & 0 & 0 & 0 & 0 \\ 0 & 0 & m_e & 0 & 0 & 0 & 0 & 0 & 0 \\ 0 & 0 & 0 & I_\theta & 0 & 0 & 0 & 0 & 0 \\ 0 & 0 & 0 & 0 & I_\theta & 0 & 0 & 0 & 0 \\ 0 & 0 & 0 & 0 & 0 & I_\theta & 0 & 0 & 0 \\ 0 & 0 & 0 & 0 & 0 & 0 & m_c & 0 & 0 \\ 0 & 0 & 0 & 0 & 0 & 0 & 0 & m_c & 0 \\ & 0 & 0 & 0 & 0 & 0 & 0 & 0 & m_c \end{bmatrix} \begin{bmatrix} \ddot{x} \\ \ddot{y} \\ \ddot{z} \\ \Delta\ddot{\theta}_1 \\ \Delta\ddot{\theta}_2 \\ \Delta\ddot{\theta}_3 \\ \Delta\ddot{H}_1 \\ \Delta\ddot{H}_2 \\ \Delta\ddot{H}_3 \end{bmatrix} + \begin{bmatrix} 0 \\ 0 \\ m_e g \\ 0 \\ 0 \\ 0 \\ m_c g \\ m_c g \\ m_c g \end{bmatrix}$$
$$= \mathbf{M}\ddot{\mathbf{q}} + \mathbf{G} = -\mathbf{J}^T \cdot \mathbf{T} \tag{7.19}$$

where $\mathbf{T} = \begin{bmatrix} T_1 & T_1 & T_3 \end{bmatrix}^T$ represents the tension vector of three cables.

7.2.3 Point-to-point Trajectory Planning

Figure 7.3 depicts the prescribed trajectory for the UCPR's end-effector, which forms a closed loop composed of both linear and arc segments. Assume that the sequence of motion for the end-effector is as follows: $P0 \rightarrow P1 \rightarrow P2 \rightarrow P3 \rightarrow P4 \rightarrow P5 \rightarrow P0$. The Cartesian coordinates for these waypoints are defined as: $P0\,(0, 0, 300)$, $P1(100, 0, 300)$, $P2\,(100, 0, 200)$, $P3\,(0, -100, 200)$, $P4\,(-100, 0, 200)$, and $P5\,(-100, 0, 300)$. Consequently, the overall trajectory is mathematically described by six distinct segments, which are formulated as

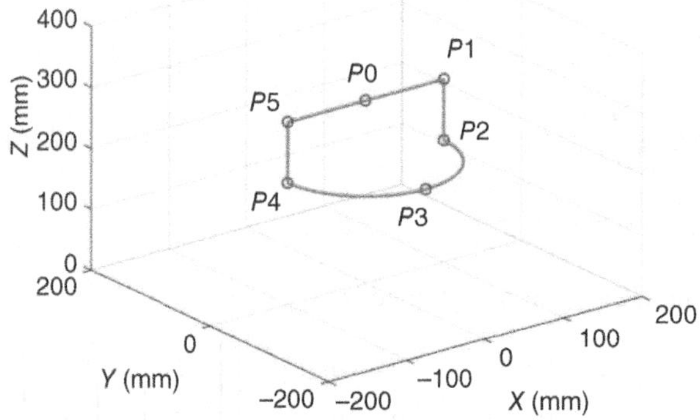

Figure 7.3 Given trajectory of the end-effector.

$$x = \begin{cases} 10 \cdot t\,(0 \le t \le 10), & P0 \to P1, \\ 100\,(10 < t \le 20), & P1 \to P2, \\ 100 \cdot \cos\left(\frac{\pi}{40}(t-20)\right)\,(20 < t \le 60), & P2 \to P3 \to P4, \\ -100\,(60 < t \le 70), & P4 \to P5, \\ 10 \cdot t - 800\,(70 < t \le 80), & P5 \to P0 \end{cases} \tag{7.20}$$

$$y = \begin{cases} 0\,(0 \le t \le 10), & P0 \to P1, \\ 0\,(10 < t \le 20), & P1 \to P2, \\ -100 \cdot \sin\left(\frac{\pi}{40}(t-20)\right)\,(20 < t \le 60), & P2 \to P3 \to P4, \\ 0\,(60 < t \le 70), & P4 \to P5, \\ 0\,(70 < t \le 80), & P5 \to P0 \end{cases} \tag{7.21}$$

$$z = \begin{cases} 300\,(0 \le t \le 10), & P0 \to P1, \\ -10 \cdot t + 400\,(10 < t \le 20), & P1 \to P2, \\ 200\,(20 < t \le 60), & P2 \to P3 \to P4, \\ 10 \cdot t - 400\,(60 < t \le 70), & P4 \to P5, \\ 300\,(70 < t \le 80), & P5 \to P0 \end{cases} \tag{7.22}$$

where t denotes the motion time of the end-effector along the given trajectory.

Based on this time constraint, the trajectory is then planned using three distinct algebraic methods: the fifth-degree polynomial, the cycloid trajectory, and the double-S velocity curve. A comparative analysis is subsequently conducted between the planned and unplanned motion profiles along the X-, Y-, and Z-axes to evaluate the performance of each planning approach.

7.2.3.1 Fifth-degree Polynomial Method

For the trajectory planned with the fifth-degree polynomial method, the corresponding position, velocity, and acceleration profiles are formulated as follows:

$$\begin{aligned} p\left(t\right) = p_0 + a_1\left(t - t_0\right) + a_2\left(t - t_0\right)^2 + a_3\left(t - t_0\right)^3 \\ + a_4\left(t - t_0\right)^4 + a_5\left(t - t_0\right)^5 \end{aligned} \tag{7.23}$$

$$\begin{aligned} \dot{p}\left(t\right) = a_1 + 2a_2\left(t - t_0\right) + 3a_3\left(t - t_0\right)^2 \\ + 4a_4\left(t - t_0\right)^3 + 5a_5\left(t - t_0\right)^4 \end{aligned} \tag{7.24}$$

$$\begin{aligned} \ddot{p}\left(t\right) = 2a_2 + 6a_3\left(t - t_0\right) + 12a_4\left(t - t_0\right)^2 \\ + 20a_5\left(t - t_0\right)^3 \end{aligned} \tag{7.25}$$

where t_0 denotes the time of the initial moment of each trajectory segment; t_1 stands for the time of the final moment of each trajectory segment; $p\left(t_0\right) = p_0$ represents the initial position of each trajectory segment; $p\left(t_1\right) = p_1$ denotes the final position of each trajectory segment; $\dot{p}\left(t_0\right) = v_0$ stands for the initial velocity of each trajectory segment; $\dot{p}\left(t_1\right) = v_1$ denotes the final velocity of each trajectory segment; $\ddot{p}\left(t_0\right) = a_0$ represents the initial acceleration of each trajectory segment; $\ddot{p}\left(t_1\right) = a_1$ denotes the final acceleration of each trajectory segment. In this case, by defining $\Delta T = t_1 - t_0$ and $\Delta p = p_1 - p_0$, the coefficients of the polynomial can be derived by

$$\begin{cases} a_0 = p_0, \\ a_1 = v_0, \\ a_2 = \dfrac{1}{2}a_0, \\ a_3 = \dfrac{1}{2\Delta T^3}\left[20\Delta p - \left(8v_1 + 12v_0\right)\Delta T - \left(3a_0 - a_1\right)\Delta T^2\right], \\ a_4 = \dfrac{1}{2\Delta T^4}\left[-30\Delta p + \left(14v_1 + 16v_0\right)\Delta T + \left(3a_0 - 2a_1\right)\Delta T^2\right], \\ a_5 = \dfrac{1}{2\Delta T^5}\left[12\Delta p - 6\left(v_1 + v_0\right)\Delta T + \left(a_1 - a_0\right)\Delta T^2\right] \end{cases} \tag{7.26}$$

Figures 7.4–7.6 illustrate the trajectories along the X-, Y-, and Z-axes, respectively, after applying the fifth-degree polynomial planning method. Within these figures, the planned motion profiles for position, velocity, and acceleration are represented. For comparison, the corresponding unplanned curves for position, velocity, and acceleration are depicted as dotted lines.

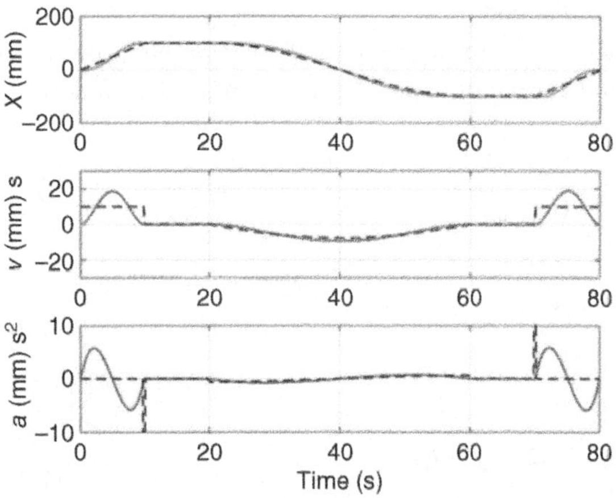

Figure 7.4 Position, velocity, and acceleration curves of fifth-degree polynomial planning in the X-direction.

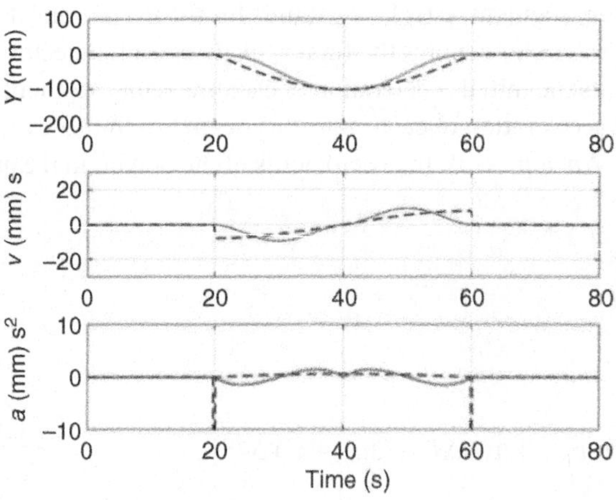

Figure 7.5 Position, velocity, and acceleration curves of fifth-degree polynomial planning in the Y-direction.

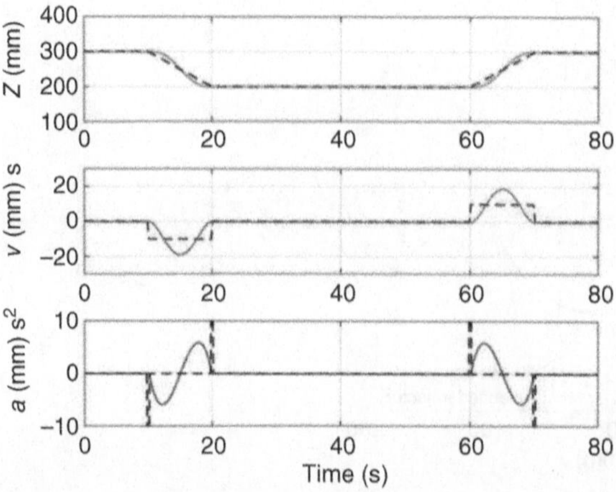

Figure 7.6 Position, velocity, and acceleration curves of fifth-degree polynomial planning in the Z-direction.

Analysis of the motion in the X-direction reveals that the planned position profile closely tracks the original, unplanned path. The primary advantage of the planning method becomes evident at points $P1$ and $P5$. In the unplanned trajectory, these points are characterized by instantaneous velocity changes, which theoretically result in infinite acceleration. Conversely, the fifth-degree polynomial planning method introduces a smooth, parabolic transition for the velocity at these junctures. Consequently, the acceleration follows a continuous sinusoidal profile, which is crucial for preventing the mechanical shock and vibration associated with abrupt motion changes.

Regarding the Y-direction, the planned position profile perfectly coincides with the unplanned path along the linear segments $P0 \rightarrow P1 \rightarrow P2$ and $P4 \rightarrow P5 \rightarrow P0$. A slight lag is introduced in the planned trajectory during the curved segment $P2 \rightarrow P3$, which is then compensated for as the planned path moves ahead of the reference in the final segment $P3 \rightarrow P4$. Consistent with the findings for the X-axis, the unplanned motion exhibits abrupt velocity changes at transition points $P2$ and $P4$, leading to theoretical infinite accelerations. The fifth-degree polynomial planning successfully resolves this by ensuring smooth parabolic velocity and sinusoidal acceleration transitions, thereby maintaining the system's dynamic stability.

In the Z-direction, the planned position curve is basically consistent with the unplanned position curve. The unplanned velocity at points $P1$, $P2$, $P4$, and $P5$ suddenly changes, and the acceleration becomes infinite. After fifth-degree polynomial planning, the velocity at points $P1$, $P2$, $P4$, and $P5$ achieves a parabolic smooth transition, and the acceleration realizes a sinusoid smooth transition.

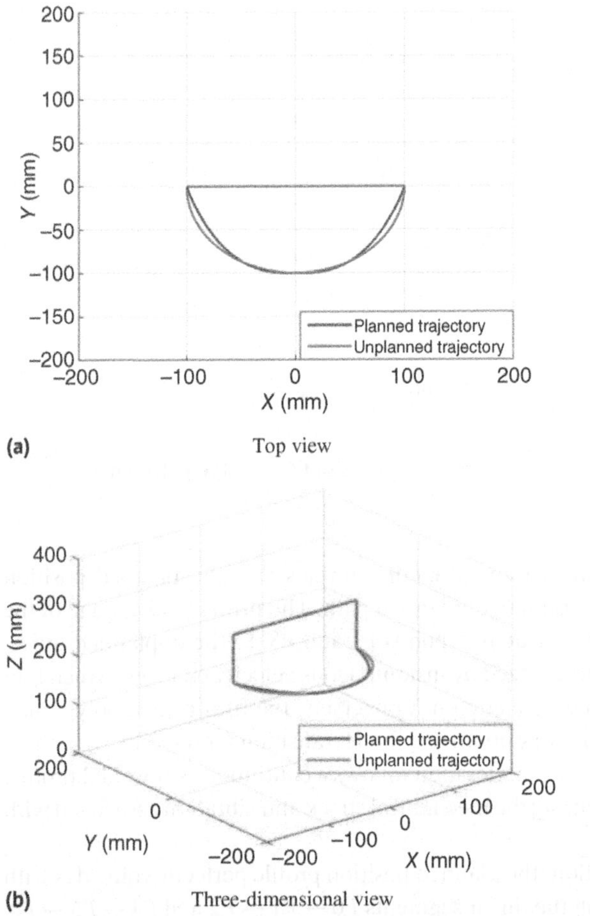

Figure 7.7 Planned and unplanned trajectories for fifth-degree polynomial.

Planned and unplanned trajectories for fifth-degree polynomial is illustrated in Figure 7.7, in which the planned trajectories are completely consistent with the unplanned trajectories in the segments of $P0 \rightarrow P1$, $P1 \rightarrow P2$, $P4 \rightarrow P5$, and $P5 \rightarrow P0$, as well as the planned trajectories is basically consistent with the unplanned trajectories in the segments of $P2 \rightarrow P3 \rightarrow P4$.

7.2.3.2 Cycloid Trajectory Method

A continuous acceleration curve can be obtained by using a cycloid trajectory method, which the equations of position, velocity, and acceleration can be expressed by

$$
\begin{aligned}
f(t) &= (f_1 - f_0)\left(\frac{t-t_0}{t_1-t_0} - \frac{1}{2\pi}\sin\frac{2\pi(t-t_0)}{t_1-t_0}\right) + f_0 \\
&= \Delta f\left(\frac{t-t_0}{\Delta T} - \frac{1}{2\pi}\sin\frac{2\pi(t-t_0)}{\Delta T}\right) + f_0
\end{aligned}
\tag{7.27}
$$

$$
\dot{f}(t) = \frac{\Delta f}{\Delta T}\left(1 - \cos\frac{2\pi(t-t_0)}{\Delta T}\right) \tag{7.28}
$$

$$
\ddot{f}(t) = \frac{2\pi\Delta f}{\Delta T^2}\sin\frac{2\pi(t-t_0)}{\Delta T} \tag{7.29}
$$

where t_0 denotes the time of the initial moment of each trajectory segment, t_1 represents the time of the final moment of each trajectory segment; $f(t_0) = f_0$ represents the initial position of each trajectory segment, $f(t_1) = f_1$ denotes the final position of each trajectory segment; and defining $\Delta T = t_1 - t_0$ and $\Delta p = p_1 - p_0$.

The trajectories in the X-, Y-, and Z-directions after cycloid trajectory planning are shown in Figures 7.8–7.10. The solid lines respectively represent the planned position, velocity and acceleration curves, while the dashed lines represent the unplanned position, velocity and acceleration curves.

Similar to the polynomial method, cycloid trajectory planning for the X-direction produces a position profile that closely follows the reference path. This method also successfully addresses the reference trajectory's key deficiency: the velocity discontinuities at points $P1$ and $P5$, which would otherwise induce infinite accelerations. The cycloid algorithm replaces these abrupt changes with

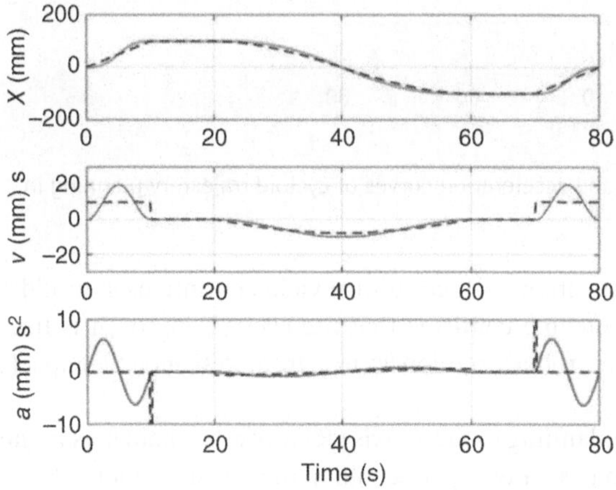

Figure 7.8 Position, velocity, and acceleration curves of cycloid trajectory planning in the X-direction.

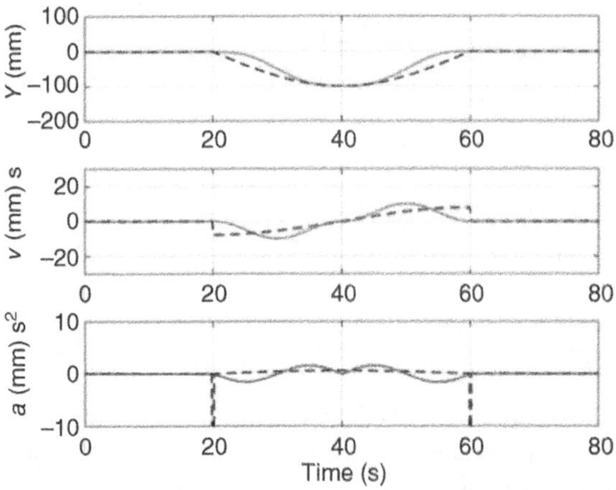

Figure 7.9 Position, velocity, and acceleration curves of cycloid trajectory planning in the Y-direction.

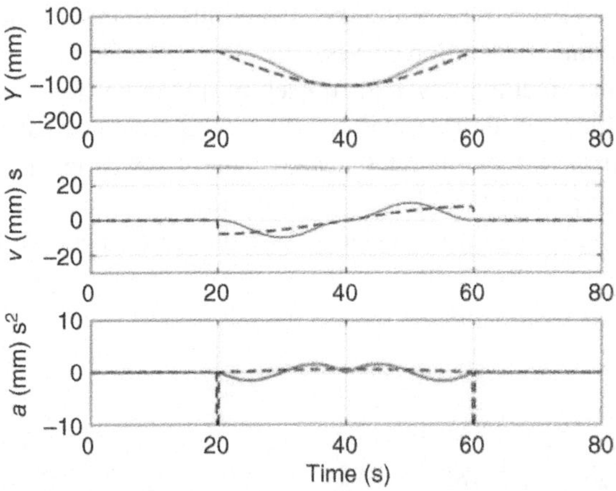

Figure 7.10 Position, velocity, and acceleration curves of cycloid trajectory planning in the Z-direction.

smooth parabolic velocity transitions, which in turn yield continuous sinusoidal acceleration profiles. This outcome confirms that the cycloid approach is likewise effective at ensuring the robot's motion is free from detrimental impacts and vibrations.

The application of cycloid planning to the Y-axis yields results analogous to the polynomial method. Positional fidelity is perfect along the linear segments $P0 \rightarrow P1 \rightarrow P2$ and $P4 \rightarrow P5 \rightarrow P0$. As expected, a minor lag is observed in the curved section $P2 \rightarrow P3$, which is subsequently compensated for by a leading trajectory in the final segment $P3 \rightarrow P4$ to meet the end-point conditions. Crucially, this method also rectifies the kinematic discontinuities at points $P2$ and $P4$ by replacing the instantaneous velocity shifts of the reference path with smooth parabolic transitions, leading to continuous sinusoidal acceleration profiles.

The analysis of the Z-direction motion further demonstrates the effectiveness of the cycloid planning method. As with the other axes, the planned position profile adheres closely to the reference path. This axis is notable for having four points of kinematic discontinuity in the unplanned trajectory ($P1$, $P2$, $P4$, and $P5$), each representing an instantaneous velocity change and a corresponding acceleration singularity. The cycloid planner effectively resolves these by imposing smooth, parabolic velocity blends and sinusoidal acceleration profiles at all four locations, ensuring a dynamically stable motion.

Figure 7.11 presents a comparison between the planned cycloid trajectory and the original unplanned path. The two trajectories are shown to coincide perfectly along segments $P0 \rightarrow P1$, $P1 \rightarrow P2$, $P4 \rightarrow P5$, and $P5 \rightarrow P0$. Furthermore, within

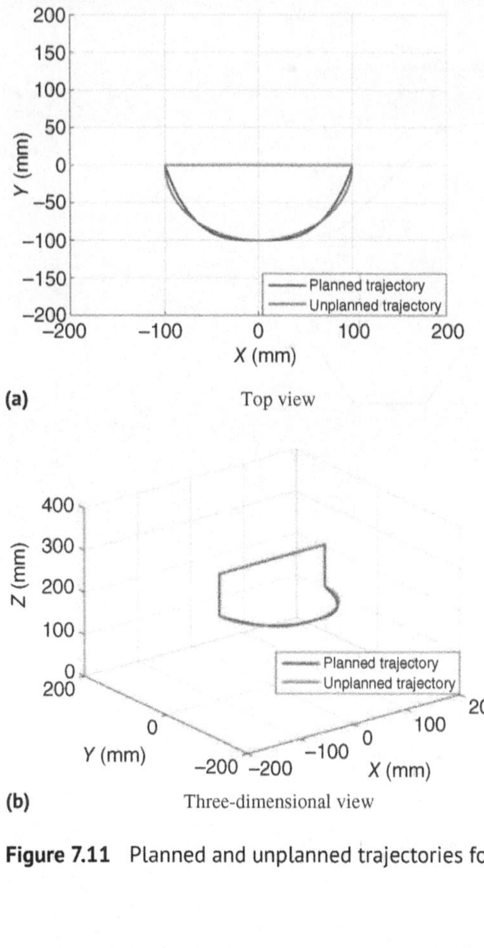

Figure 7.11 Planned and unplanned trajectories for cycloid trajectory.

segment $P2 \rightarrow P3 \rightarrow P4$, the planned path continues to track the reference trajectory with high fidelity.

7.2.3.3 Double-S Velocity Curve Method

To ensure continuous acceleration and thereby prevent harmful vibrations, a motion profile with a higher degree of smoothness is required. An effective approach, illustrated in Figure 7.12, is to construct a piecewise linear acceleration curve. The resulting velocity profile generated by this method is characterized by constant-velocity phases seamlessly joined by parabolic transitions.

The performance of the double-S velocity curve planning is visualized for the X-, Y-, and Z-axes in Figures 7.13–7.15. Each figure compares the planned trajectory (solid lines) against the unplanned reference (dotted lines).

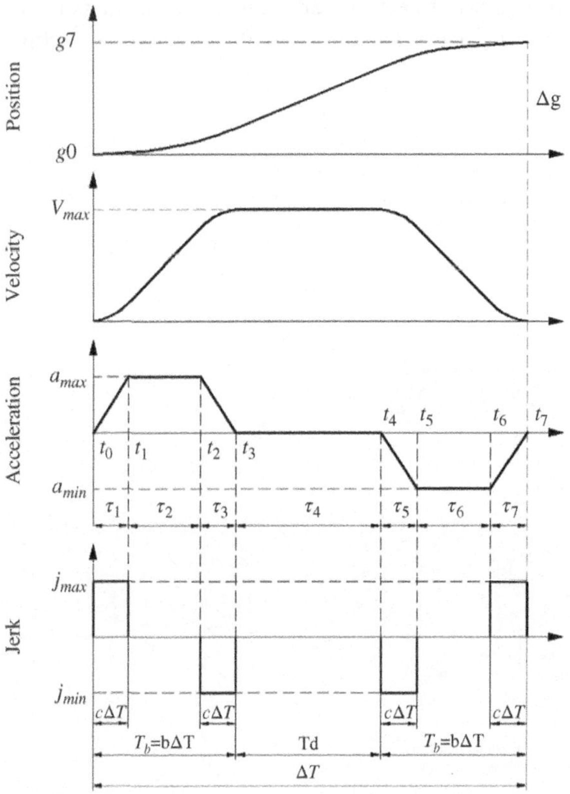

Figure 7.12 Typical curves of position, velocity, acceleration, and jerk for double-S trajectory.

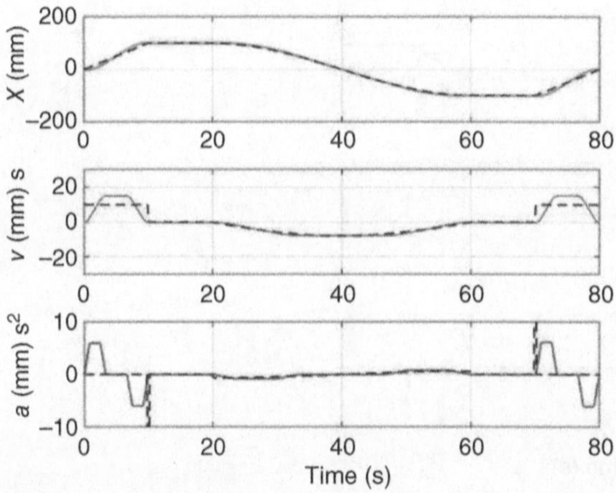

Figure 7.13 Position, velocity, and acceleration curves of double-S velocity curve planning in the X-direction.

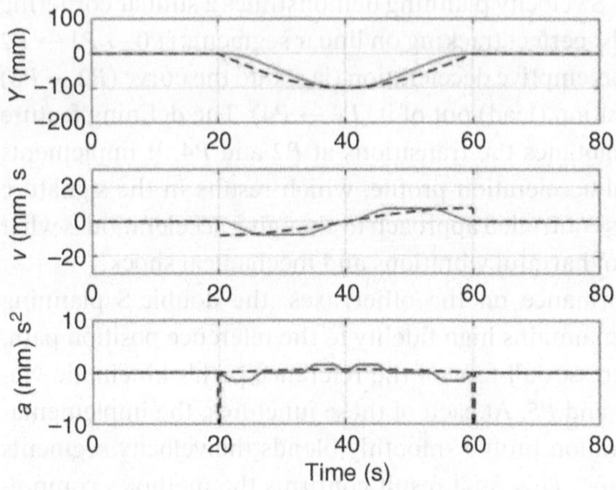

Figure 7.14 Position, velocity, and acceleration curves of double-S velocity curve planning in the Y-direction.

Applying the double-S velocity curve method to the X-axis, the planned position remains highly consistent with the reference path. The method's unique contribution is evident at points $P1$ and $P5$, where it resolves the reference path's velocity discontinuities. Instead of a simple parabolic blend, this technique generates its signature S-shaped velocity profile for the transitions. This S-curve is a direct result of a trapezoidal acceleration profile, which ensures continuity and prevents the infinite acceleration spikes. Consequently, this jerk-limited approach also effectively suppresses mechanical shock and vibration.

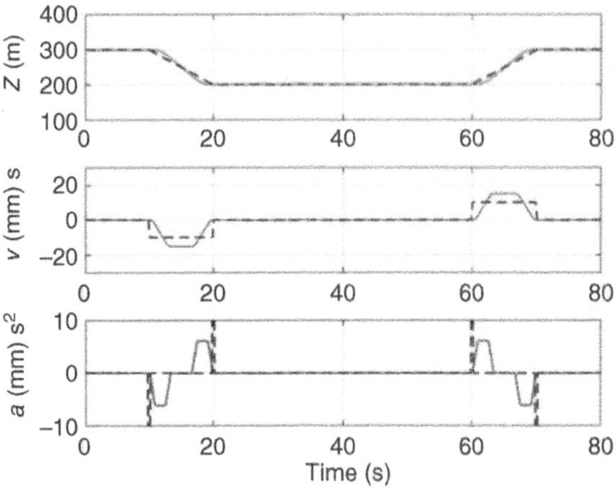

Figure 7.15 Position, velocity, and acceleration curves of double-S velocity curve planning in the Z-direction.

For the Y-axis, the double-S velocity planning demonstrates a similar cornering strategy to the other methods: perfect tracking on linear segments ($P0 \rightarrow P1 \rightarrow P2$ and $P4 \rightarrow P5 \rightarrow P0$) with a preemptive deceleration (lag) into the curve ($P2 \rightarrow P3$) and a compensatory acceleration (lead) out of it ($P3 \rightarrow P4$). The defining feature of this method is how it negotiates the transitions at $P2$ and $P4$. It implements its characteristic trapezoidal acceleration profile, which results in the signature S-shaped velocity curve. This controlled approach to changing acceleration is what guarantees the elimination of harmful vibrations and mechanical shock.

Consistent with its performance on the other axes, the double-S planning method for the Z-direction maintains high fidelity to the reference position path. The algorithm robustly addresses all four of the reference path's kinematic violations at points $P1$, $P2$, $P4$, and $P5$. At each of these junctures, the implementation of a trapezoidal acceleration profile smoothly blends the velocity segments into the characteristic S-shape. This final result confirms the method's comprehensive ability to generate a dynamically smooth and vibration-free motion across the entire workspace.

The trajectory resulting from the double-S velocity curve planning is depicted in Figure 7.16. While the path shows perfect correspondence with the unplanned trajectory on the straight segments ($P0 \rightarrow P1$, $P1 \rightarrow P2$, $P4 \rightarrow P5$, and $P5 \rightarrow P0$), it introduces a more significant deviation during the curved portion ($P2 \rightarrow P3 \rightarrow P4$) than was observed with either the polynomial or cycloid methods.

The comparative analysis reveals that the fifth-degree polynomial method offers the most advantageous performance. While both the polynomial and cycloid methods produce sinusoidal acceleration profiles, the polynomial approach is superior due to its lower peak velocities and accelerations, leading to better tracking accuracy. The double-S method, though capable of achieving lower peak

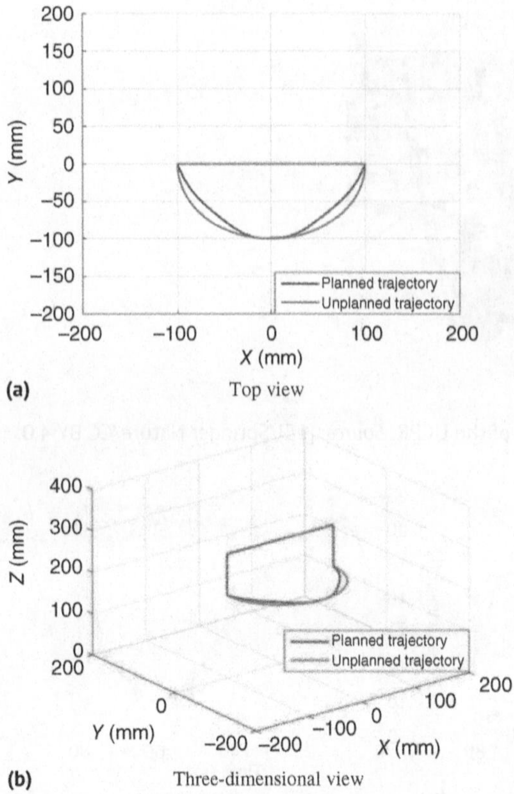

Figure 7.16 Planned and unplanned trajectories for double-S velocity curve.

velocities than the polynomial, is ultimately less desirable. It introduces higher peak accelerations and exhibits a larger tracking error. More importantly, its trapezoidal acceleration profile contains discontinuities in jerk (sharp corners), which can lead to mechanical shock. The smooth, sinusoidal acceleration from the polynomial method avoids this issue entirely. Consequently, considering tracking fidelity, acceleration magnitude, and jerk continuity, the fifth-degree polynomial planning is determined to be the optimal choice [27].

7.2.4 Numerical Simulation and Experiments

The experiments are completed on the basis of fifth-degree polynomial planning. The experiment prototype of the UCPR is illustrated in Figure 7.17, three cable masts are evenly distributed on the circular base, that is, $\theta_1 = 0$, $\theta_2 = \frac{2\pi}{3}$, $\theta_3 = \frac{4\pi}{3}$, and $\Delta\theta_1 = \Delta\theta_2 = \Delta\theta_3 = 0$, as well as three cable masts have the same height, that is, $H_1 = H_2 = H_3 = 660$ mm and $\Delta H_1 = \Delta H_2 = \Delta H_3 = 0$. The end-effector is

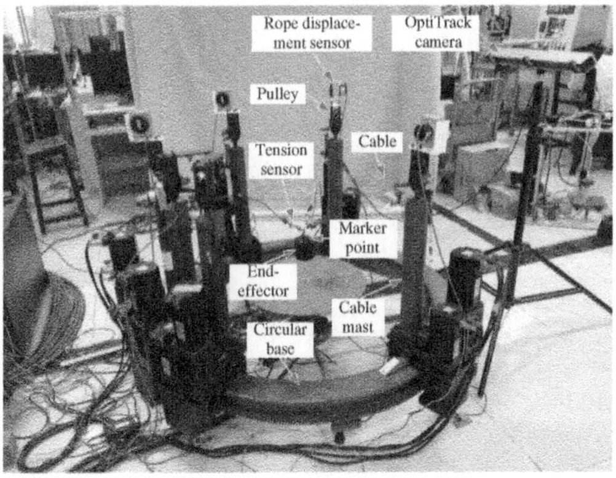

Figure 7.17 Experiment prototype of the UCPR. *Source*: [15]/Springer Nature/CC BY 4.0.

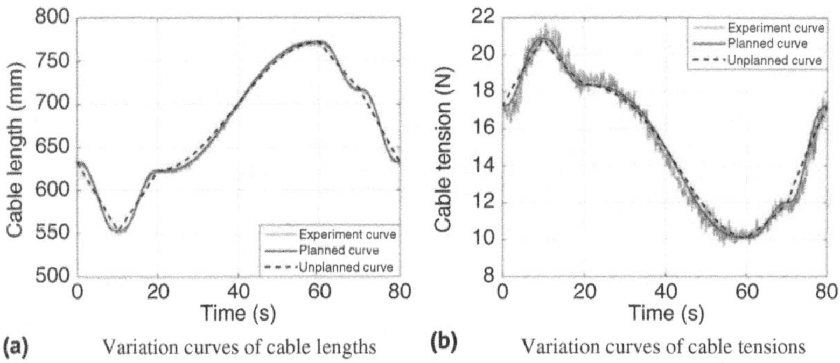

Figure 7.18 Variation curves of lengths and tensions of cable 1.

considered as a point-mass, which the mass is $m_e = 3$ kg. The mass of cable mast is $m_c = 20$ kg.

The variation curves of lengths of cables 1, 2, and 3 are demonstrated in Figures 7.18a, 7.19a, and 7.20a, and the variation curves of tensions of cables 1, 2, and 3 are illustrated in Figures 7.18b, 7.19b, and 7.20b. The fine solid line denotes the experiment curve, the heavy solid line represents the planned curve after fifth-degree polynomial planning, and the dotted line stands for the unplanned curve.

The figures clearly illustrate the benefit of trajectory planning at points of high velocity change. The unplanned trajectory results in sharp inflections in the cable length and tension profiles, which are indicative of discontinuities that would induce mechanical shock and vibration during acceleration and deceleration. In contrast, the profiles generated by the fifth-degree polynomial plan exhibit smooth, continuous transitions, effectively mitigating these detrimental dynamic

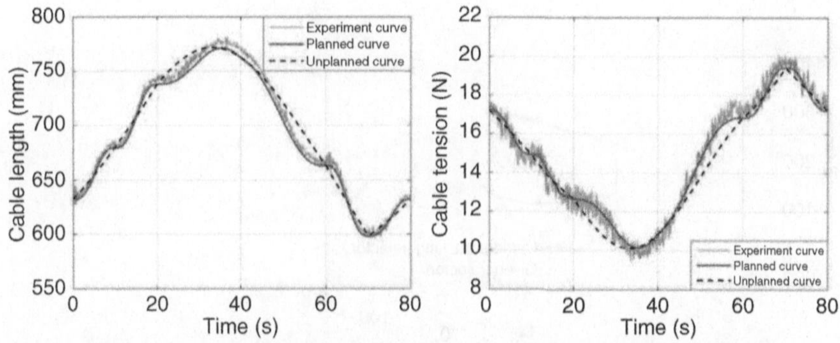

(a) Variation curves of cable lengths (b) Variation curves of cable tensions

Figure 7.19 Variation curves of lengths and tensions of cable 2.

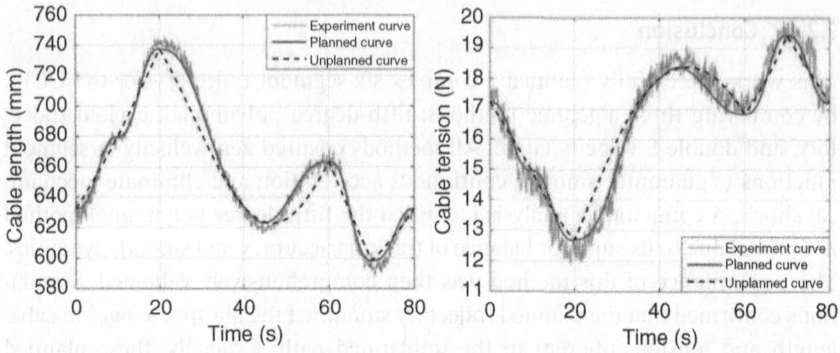

(a) Variation curves of cable lengths (b) Variation curves of cable tensions

Figure 7.20 Variation curves of lengths and tensions of cable 3.

effects. Most importantly, the experimental data for both cable length and tension show strong agreement with the simulated planned curves. This close correlation between the theoretical plan and the physical implementation serves as a robust validation of the effectiveness and feasibility of the fifth-degree polynomial planning method.

To independently verify the robot's tracking performance, an OptiTrack motion capture system was employed for real-time tracking of the end-effector. Figure 7.21 presents a comparison between the experimentally measured trajectory and the commanded reference path. In the figure, the path captured by the camera system is depicted as a heavy solid line, while the given trajectory is represented by a fine solid line. The two paths show a very high degree of correlation, confirming that the end-effector accurately followed the intended trajectory. This strong agreement also substantiates the high measurement precision of the OptiTrack system, reinforcing the validity of the experimental findings.

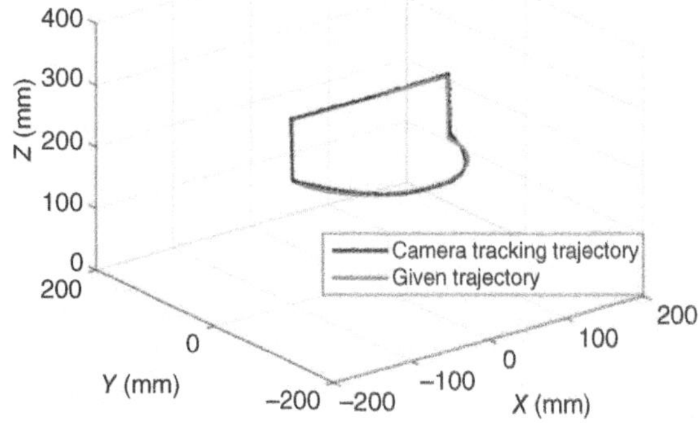

Figure 7.21 Camera tracking trajectory and given trajectory.

7.2.5 Conclusion

This work successfully planned a complex six-segment trajectory for the UCPR by comparing three algebraic methods: fifth-degree polynomial, cycloid trajectory, and double-S velocity curve. All methods ensured zero velocity at segment junctions to guarantee smooth, continuous acceleration and eliminate mechanical shock. A comparative analysis identified the fifth-degree polynomial method as optimal due to its superior balance of tracking accuracy and smooth dynamics. The performance of this method was then comprehensively validated. Simulations confirmed that the planned trajectory smoothed the abrupt changes in cable length and tension inherent in the unplanned path. Crucially, these planned profiles were shown to be in strong agreement with experimental data. Final validation using an OptiTrack system confirmed that the robot's end-effector accurately followed the commanded trajectory, affirming the effectiveness and practical feasibility of the proposed planning approach.

7.3 Fuzzy Control for RFCHRs

This section suggests FPID-AWOA for tracking the trajectory of a rigid–flexible coupling hoisting robot in a large load environment of Chapter 2. The FPID-AWOA has a higher RFCHR trajectory tracking accuracy than the traditional fuzzy PID control strategy, the fuzzy PID control strategy based on whale optimization algorithm (FPID-WOA), and the fuzzy PID control strategy based on particle swarm optimization (FPID-PSOA). Meanwhile, new fuzzy rules are created, and a new fuzzy PID controller is designed, before the AWOA is introduced to optimize quantization and scale factors of the fuzzy PID controller to obtain the optimal solution.

7.3.1 Fuzzy Trajectory Tracking Control

This subsection addresses the issue of the dynamic equation's inability to be represented in detail using a numerical approach. The fuzzy PID controller is then subjected to an adaptive whale optimization technique.

7.3.1.1 Numerical Solutions to Dynamic Model Dynamic Modeling

Equation (2.43) clearly shows that the acceleration $\ddot{\Theta}_t$ at time t is represented as

$$\ddot{\Theta}_t = \boldsymbol{M}^{-1}(\Theta_t)\left[\tau_t - \boldsymbol{C}\left(\dot{\Theta}_t, \Theta_t\right)\dot{\Theta}_t - \boldsymbol{G}(\Theta_t)\right] \tag{7.30}$$

where Θ_t, $\dot{\Theta}_t$, and $\ddot{\Theta}_t$ represent angle, angular velocity, and angular acceleration vectors of joints at time t; and τ_t represents the joint torque at time t. However, values of $\boldsymbol{M}(\Theta)$, $\boldsymbol{C}\left(\dot{\Theta}, \Theta\right)\dot{\Theta}$, and $\boldsymbol{G}(\Theta)$ at time t can be numerically calculated. Equation (2.43) can therefore be extended in a different way,

$$\begin{bmatrix}\tau_1\\ \tau_2\\ \tau_3\end{bmatrix} = \begin{bmatrix}\boldsymbol{M}_{11} & \boldsymbol{M}_{12} & \boldsymbol{M}_{13}\\ \boldsymbol{M}_{21} & \boldsymbol{M}_{22} & \boldsymbol{M}_{23}\\ \boldsymbol{M}_{31} & \boldsymbol{M}_{32} & \boldsymbol{M}_{33}\end{bmatrix}\begin{bmatrix}\ddot{\theta}_1\\ \ddot{\theta}_2\\ \ddot{\theta}_3\end{bmatrix} + \begin{bmatrix}\boldsymbol{C}_{11} & \boldsymbol{C}_{12} & \boldsymbol{C}_{13}\\ \boldsymbol{C}_{21} & \boldsymbol{C}_{22} & \boldsymbol{C}_{23}\\ \boldsymbol{C}_{31} & \boldsymbol{C}_{32} & \boldsymbol{C}_{33}\end{bmatrix}\begin{bmatrix}\dot{\theta}_1\\ \dot{\theta}_2\\ \dot{\theta}_3\end{bmatrix} + \begin{bmatrix}\boldsymbol{G}_1\\ \boldsymbol{G}_2\\ \boldsymbol{G}_3\end{bmatrix} \tag{7.31}$$

The elements in $\boldsymbol{M}(\Theta)$ are all functions of Θ and independent of $\dot{\Theta}$, $\ddot{\Theta}$, and g, based on Eq. (7.31); all of the elements in $\boldsymbol{C}\left(\dot{\Theta}, \Theta\right)$ are functions of $\dot{\Theta}$ and Θ and independent of $\ddot{\Theta}$ and g; and all of the elements in $\boldsymbol{G}(\Theta)$ are functions of Θ and g and independent of $\dot{\Theta}$ and $\ddot{\Theta}$. Thus, one sets $\ddot{\Theta} = \begin{bmatrix}1\\0\\0\end{bmatrix}$, $\dot{\Theta} = \begin{bmatrix}0\\0\\0\end{bmatrix}$, $\Theta = \begin{bmatrix}\theta_1\\ \theta_2\\ \theta_3\end{bmatrix}$, $g = 0$, and based on Eq. (7.31) yields

$$\begin{bmatrix}\tau_1\\ \tau_2\\ \tau_3\end{bmatrix} = \begin{bmatrix}\boldsymbol{M}_{11} & \boldsymbol{M}_{12} & \boldsymbol{M}_{13}\\ \boldsymbol{M}_{21} & \boldsymbol{M}_{22} & \boldsymbol{M}_{23}\\ \boldsymbol{M}_{31} & \boldsymbol{M}_{32} & \boldsymbol{M}_{33}\end{bmatrix}\begin{bmatrix}1\\0\\0\end{bmatrix} + \begin{bmatrix}0\\0\\0\end{bmatrix} + \begin{bmatrix}0\\0\\0\end{bmatrix} = \begin{bmatrix}\boldsymbol{M}_{11}\\ \boldsymbol{M}_{21}\\ \boldsymbol{M}_{31}\end{bmatrix} \tag{7.32}$$

Similarly, one sets $\ddot{\Theta} = \begin{bmatrix}0\\1\\0\end{bmatrix}$, and based on Eq. (7.31), one has $\begin{bmatrix}\tau_1\\ \tau_2\\ \tau_3\end{bmatrix} = \begin{bmatrix}\boldsymbol{M}_{12}\\ \boldsymbol{M}_{22}\\ \boldsymbol{M}_{32}\end{bmatrix}$; one sets $\ddot{\Theta} = \begin{bmatrix}0\\0\\1\end{bmatrix}$, and based on Eq. (7.32), one has $\begin{bmatrix}\tau_1\\ \tau_2\\ \tau_3\end{bmatrix} = \begin{bmatrix}\boldsymbol{M}_{13}\\ \boldsymbol{M}_{23}\\ \boldsymbol{M}_{33}\end{bmatrix}$. Similarly, one sets

$\ddot{\mathbf{\Theta}} = \begin{bmatrix} 0 \\ 0 \\ 0 \end{bmatrix}$, $\dot{\mathbf{\Theta}} = \begin{bmatrix} \dot{\theta}_1 \\ \dot{\theta}_2 \\ \dot{\theta}_3 \end{bmatrix}$, $\mathbf{\Theta} = \begin{bmatrix} \theta_1 \\ \theta_2 \\ \theta_3 \end{bmatrix}$, and g = 9.8, and based on Eq. (7.31), one has

$$\begin{bmatrix} \tau_1 \\ \tau_2 \\ \tau_3 \end{bmatrix} = \begin{bmatrix} 0 \\ 0 \\ 0 \end{bmatrix} + \begin{bmatrix} C_{11} & C_{12} & C_{13} \\ C_{21} & C_{22} & C_{23} \\ C_{31} & C_{32} & C_{33} \end{bmatrix} \begin{bmatrix} \dot{\theta}_1 \\ \dot{\theta}_2 \\ \dot{\theta}_3 \end{bmatrix} + \begin{bmatrix} G_1 \\ G_2 \\ G_3 \end{bmatrix} \tag{7.33}$$

Based on Eqs. (7.31)–(7.33), the values of $\boldsymbol{M}(\mathbf{\Theta}_t)$ and $\boldsymbol{C}\left(\dot{\mathbf{\Theta}}_t, \mathbf{\Theta}_t\right)\dot{\mathbf{\Theta}}_t + \boldsymbol{G}(\mathbf{\Theta}_t)$ in Eq. (7.30) can be calculated.

7.3.1.2 Fuzzy PID Controller

Each joint angle's error e_j and error rate e_{c_j} are taken into account as input variables in the fuzzy PID controller, and one has

$$e_j = \theta_{d_j} - \theta_j, \; j = 1, 2, 3 \tag{7.34}$$

$$e_{c_j} = \dot{\theta}_{d_j} - \dot{\theta}_j \tag{7.35}$$

where $\dot{\theta}_{d_j}$ is the ideal joint angle; θ_j is the actual joint angle; $\dot{\theta}_{d_j}$ is the ideal angular velocity vector; and $\dot{\theta}_j$ is the actual angular velocity vector. The fuzzy PID controller uses the parameters ΔK_p, ΔK_i, and ΔK_d to accomplish adaptive modification. The expression for the output torque $\tau_j(t)$ is

$$\begin{aligned} \tau_j(t) &= \left(K_{po} + \Delta K_p\right) e_j + \left(K_{io} + \Delta K_i\right) \int_0^t e_j dt + \left(K_{do} + \Delta K_d\right) \frac{de_j}{dt} \\ &= \left(K_{po} e_j + K_{io} \int_0^t e_j dt + K_{do} \frac{de_j}{dt}\right) + \left(\Delta K_p e_j + \Delta K_i \int_0^t e_j dt + \Delta K_d \frac{de_j}{dt}\right) \end{aligned} \tag{7.36}$$

where K_{po}, K_{io}, and K_{do} are the initial values of PID in manually selected fuzzy PID controller.

It ought to be noted that both the control accuracy and the control cost increase with the number of fuzzy subsets [28, 29]. The error e_j, the error rate e_{c_j}, and the output variables ΔK_p, ΔK_i, and ΔK_d are therefore among the input variables with seven fuzzy subsets. Figure 7.22 displays the membership function curves for the input variables, such as error e_j and error rate e_{c_j}, and the output variables, such as ΔK_p, ΔK_i, and ΔK_d. The fuzzy subsets of e_j and e_{c_j} are NB: negative bigness, NM: negative medium, NS: negative small, ZO: zero, PS: positive small, PM: positive medium, and PB: positive bigness. Both e_j and e_{c_j} have ranges of variation

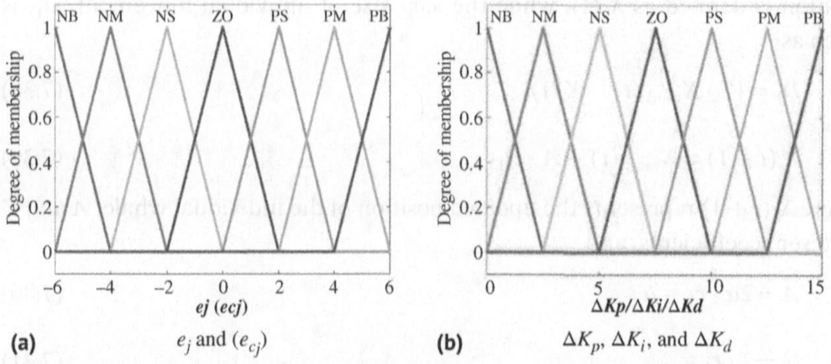

Figure 7.22 Curves of membership functions of the input and output variables.

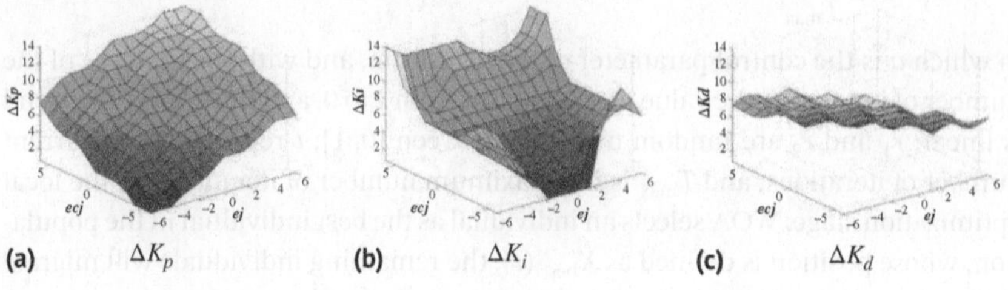

Figure 7.23 Fuzzy regular surface diagrams.

of [−6, 6]. The fuzzy subsets of ΔK_p, ΔK_i, and ΔK_d are also NB: negative bigness, NM: negative medium, NS: negative small, ZO: zero, PS: positive small, PM: positive medium, and PB: positive bigness. In the meantime, the ranges of variation of ΔK_p, ΔK_i, and ΔK_d are [0, 15]. Figure 7.23 displays the fuzzy regular surface diagrams of ΔK_p, ΔK_i, and ΔK_d. The fuzzy PID controller's ΔK_p, ΔK_i, and ΔK_d can be computed using the centroid technique [30] and Figures 7.22 and 7.23,

$$\mu_j = \frac{\sum_{j=0}^{n} \mu_c\left(z_j\right) \cdot z_j}{\sum_{j=0}^{n} \mu_c\left(z_j\right)} \tag{7.37}$$

where μ_j represents the exact values of the output variables of the fuzzy controller after defuzzification, z_j represents the values of fuzzy control quantity in the fuzzy field, and $\mu_c\left(z_j\right)$ represents the membership values of z_j.

7.3.1.3 Fuzzy Control Strategy Based on an Adaptive Whale Optimization Algorithm

During the global search stage, the WOA randomly selects an individual as the lucky individual in the population, whose position is defined as $X_{\text{rand}}\ (t)$, and other individuals move to the chosen lucky individual. The individual whale's current position is defined as $X(t)$, while the step size of individual movement, D_1, is given as

$$D_1 = |C \cdot X_{\text{rand}}(t) - X(t)| \tag{7.38}$$

$$X(t+1) = X_{\text{rand}}(t) - A \cdot D_1 \tag{7.39}$$

where $X(t+1)$ represents the updated position of the individual whale; A and C represent coefficients, and

$$A = 2a \cdot r_1 - a \tag{7.40}$$

$$C = 2 \cdot r_2 \tag{7.41}$$

$$a = 2 - \frac{2t}{T_{\max}} \tag{7.42}$$

in which a is the control parameter of the algorithm, and with the increase of the number of iterations, the value of a decreases from 2 to 0, and the decreasing trend is linear; r_1 and r_2 are random numbers between [0, 1]; t represents the current number of iterations; and $T_{\max}$ is the maximum number of iterations. In the local optimization stage, WOA selects an individual as the best individual in the population, whose position is defined as $X_{\text{best}}(t)$, the remaining individuals will migrate to $X_{\text{best}}(t)$ and the individual migration step is expressed as

$$D_2 = |C \cdot X_{\text{best}}(t) - X(t)| \tag{7.43}$$

The distance D' from the $X_{\text{best}}(t)$ to the $X(t)$ is expressed as

$$D' = |X_{\text{best}}(t) - X(t)| \tag{7.44}$$

Based on Eqs. (7.43) and (7.44), the $X(t+1)$ is expressed as

$$X(t+1) = \begin{cases} X_{\text{best}}(t) - A \cdot D_2, \\ D' \cdot e^{bl} \cdot \cos(2\pi l) + X_{\text{best}}(t) \end{cases} \tag{7.45}$$

where b is a constant coefficient; and l is a random number between [−1, 1]. Based on Eqs. (7.39) and (7.45), when the probability factor $p \geq 0.5$, the $X(t+1)$ is expressed as

$$X(t+1) = D' \cdot e^{bl} \cdot \cos(2\pi l) + X_{\text{best}}(t) \tag{7.46}$$

when the probability factor $p < 0.5$, the $X(t+1)$ is expressed as

$$X(t+1) = \begin{cases} X_{\text{rand}}(t) - A \cdot D_1, & |A| \geq 1, \\ X_{\text{best}}(t) - A \cdot D_2, & |A| < 1 \end{cases} \tag{7.47}$$

WOA incorporates a weight function and a differential mutation interference factor because to its sluggish convergence rate, limited solving precision, and tendency to fall into local optimality. The weight function ω is introduced to govern the updating position of whales,

$$\omega = \omega_{\max} - (\omega_{\max} - \omega_{\min})\left(\frac{t}{T_{\max}}\right)^{\frac{1}{t}} \tag{7.48}$$

where the range of ω is $[\omega_{\min}, \omega_{\max}]$. Based on Eqs. (7.46) and (7.47), and by incorporating weight, one has

$$X(t+1) = \omega \cdot D' \cdot e^{bl} \cdot \cos(2\pi l) + X_{\text{best}}(t) \tag{7.49}$$

$$X(t+1) = X_{\text{rand}}(t) - \omega \cdot A \cdot D_1 \tag{7.50}$$

In addition, the differential mutation interference factor γ is introduced, and

$$\gamma = \delta \cdot (X_{\text{best}}(t) - X(t)) \tag{7.51}$$

where δ is an adaptive mutation factor, and

$$\delta = \delta_0 \cdot 2^{\varepsilon} \tag{7.52}$$

$$\varepsilon = e^{1-\frac{T_{\max}}{T_{\max}+1-t}} \tag{7.53}$$

where the mutation parameter is δ_0. In the beginning, $\delta = 2\delta_0$. WOA may more easily exit the local optimal solution and prevent premature termination as the number of iterations increases because the value of δ progressively approaches δ_0. By using the differential variation interference factor γ and using Eq. (7.47), one has

$$X(t+1) = X_{\text{best}}(t) - A \cdot D_2 + \gamma \tag{7.54}$$

When compared to other optimization algorithms, AWOA offers the following benefits: high solution accuracy, quick discovery of optimal solutions, ease of use, fewer parameters to be changed, improved convergence speed, global search capability, and the ability to jump out of the local optimum. It also works effectively for a variety of optimization problems. Thus, the fuzzy PID controller via AWOA is optimized in this chapter. Figure 7.24 displays the control diagram for the suggested FPID-AWOA. Kinematics converts the ideal trajectory X_d in Cartesian space into the ideal joint angle in joint space. The designed AWOA uses the error between the ideal and actual joint angles as well as the error change rate as inputs to optimize the fuzzy parameters by obtaining the optimal PID parameters. The robot is then controlled to achieve high precision trajectory tracking.

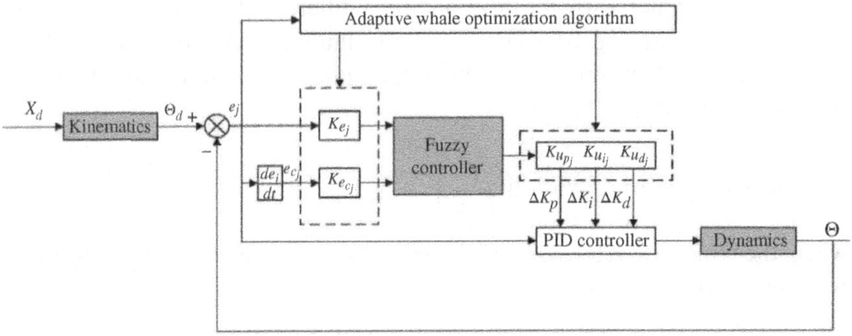

Figure 7.24 The control diagram of the proposed FPID-AWOA.

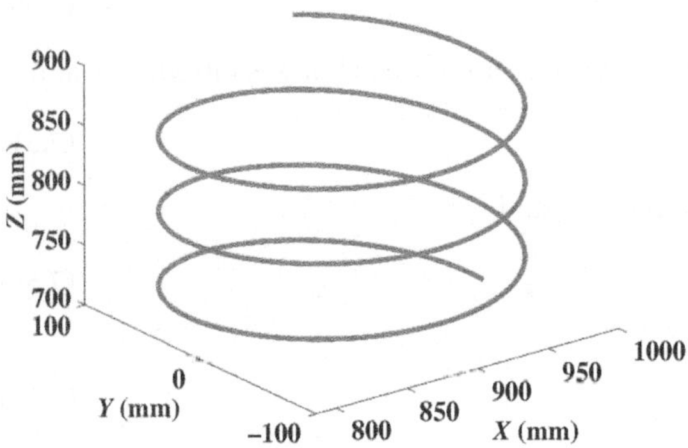

Figure 7.25 Space spiral track.

7.3.2 Numerical Simulations

This part compares and analyzes the trajectory tracking control performances of FPID, FPID-WOA, FPID-PSOA, and FPID-AWOA in order to observe the control performance of the suggested FPID-AWOA. The system simulation test is also conducted.

Assuming that the end-effector's trajectory at point B_4 in space is a cylindrical spiral and taking into account the maximum position the robot can reach as well as the experimental effect, as illustrated in Figure 7.25, the end-effector's trajectory equation is stated as

$$\begin{cases} x = 100 \cdot \cos(t) + 886, \\ y = 100 \cdot \sin(t), \\ z = 10t + 700 \end{cases} \tag{7.55}$$

where t is a certain point in a period.

$K_{po} = 500$, $K_{io} = 10$, and $K_{do} = 10$ are the fuzzy PID controller's initial values. WOA, PSOA, and AWOA are used to optimize the K_{e_j}, K_{ec_j}, K_{up_j}, K_{ui_j}, and K_{ud_j}. We choose $\int_0^{\infty} e(t)^2 dt$ as the fitness function. The simulation length is 20 sec, the population size is 10, the number of iterations is 100, and the range of values for ω is $[0.6, 0.9]$.

Figures 7.26–7.28 display position tracking and the errors of θ_1, θ_2, and θ_3 in the joint space, whereas Figure 7.29 displays the trajectory tracking errors in the Cartesian space. It is evident from Figures 7.26 to 7.29 that the smooth trajectory tracking can be accomplished by the four distinct control techniques. But when it comes to tracking accuracy and stability, it's clear that FPID-AWOA excels. On the one hand, FPID-AWOA has the smallest overshoot, fluctuating only once before stabilizing, as seen in Figures 7.26–7.28b. In comparison to FPID-AWOA, FPID, FPID-WOA, and FPID-PSOA have larger overshoots and fluctuation times that are more than twice as long.

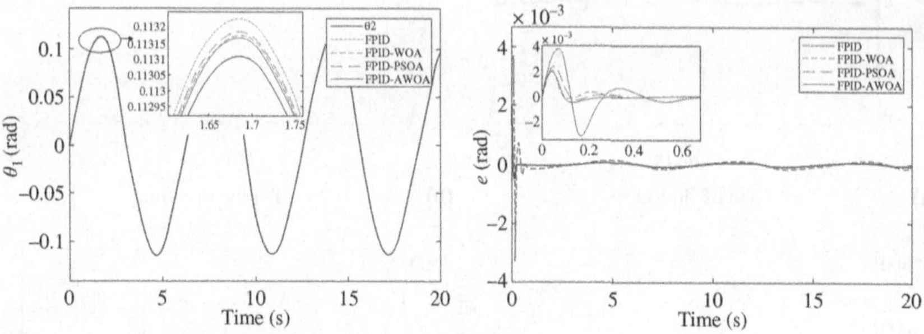

(a) Position tracking (b) The error

Figure 7.26 Position tracking and the error of θ_1.

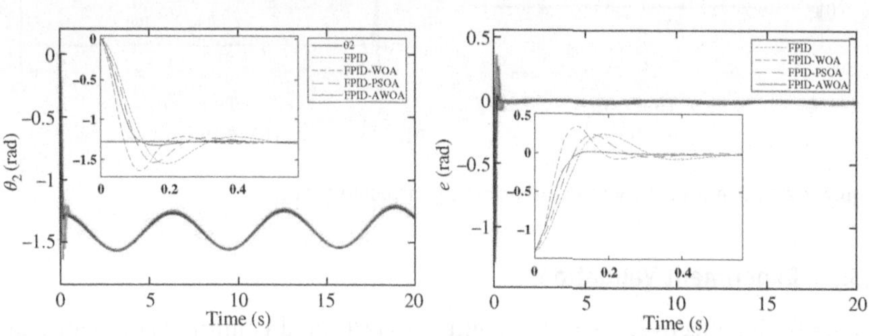

(a) Position tracking (b) The error

Figure 7.27 Position tracking and the error of θ_2.

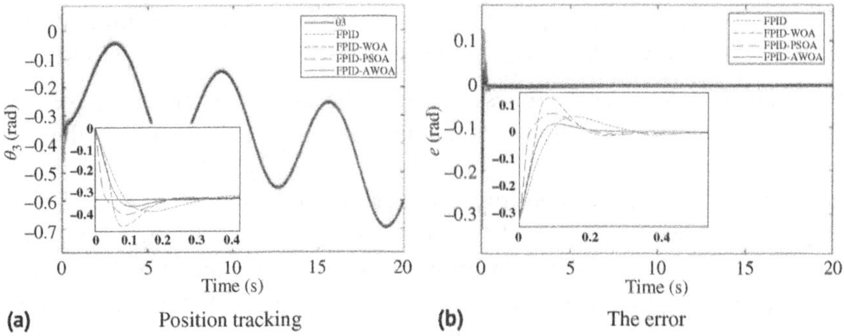

Figure 7.28 Position tracking and the error of θ_3.

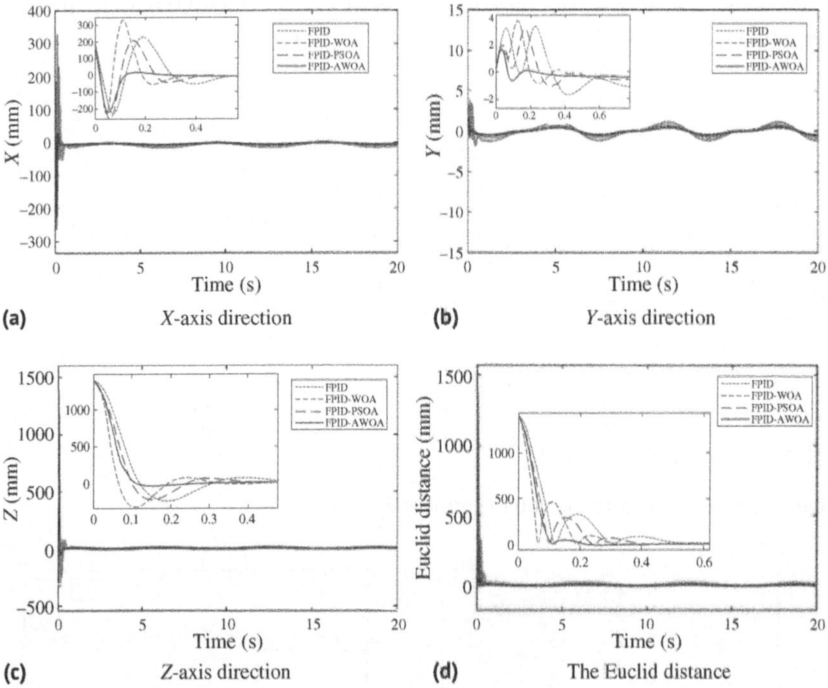

Figure 7.29 The trajectory tracking errors in Cartesian space.

7.3.3 Experiment Validation

To verify the effectiveness of the control algorithm, a control experiment platform for CDPR was constructed. The platform used for the control experiment is shown in [31]. Delta PLC is responsible for controlling CDPR. Delta software is responsible for writing the control program. The NOKOV multi-camera visual capture system records the trajectory of CDPR. The final motion trajectory of CDPR is the trajectory of the visual calibration target captured by the camera.

The verification trajectory is the spiral trajectory in Figure 7.25, which is similar to the simulations. It is confirmed that FPID-AWOA outperforms FPID-WOA in terms of stability, response speed, and tracking accuracy based on Eqs. (7.48) and (7.51) as well as the simulation findings above. Only the FPID, FPID-PSOA, and FPID-WOA trajectory tracking experiments are chosen for comparison in order to keep the experiment simple. Position tracking and spiral trajectory errors in the experiment are displayed in Figures 7.30–7.32 when the spiral trajectory is followed. In the entire tracking process, FPID-AWOA is the closest to the ideal trajectory, and the error fluctuation is negligible, as shown in Figures 7.30–7.32. High tracking accuracy is indicated by the trajectory approach, and great stability

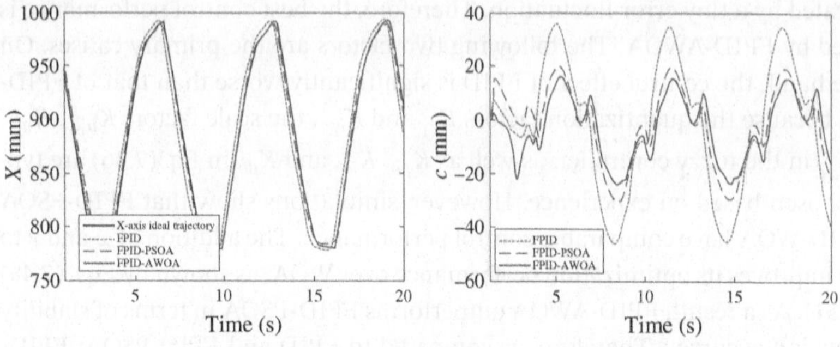

(a) Position tracking (b) The error

Figure 7.30 Position tracking and the error in the X-axis direction in the spiral trajectory in the experiment.

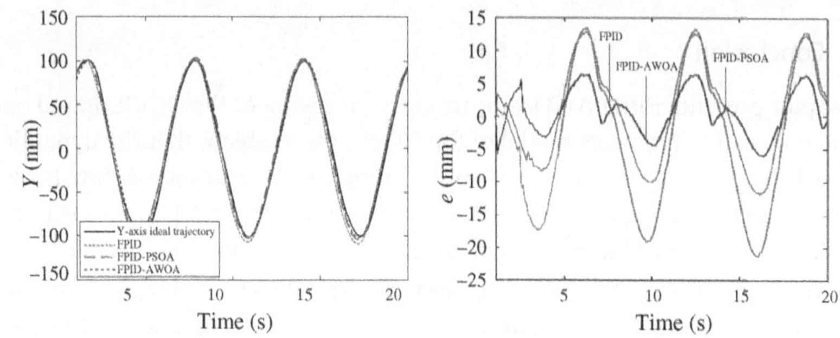

(a) Position tracking (b) The error

Figure 7.31 Position tracking and the error in the Y-axis direction in the spiral trajectory in the experiment.

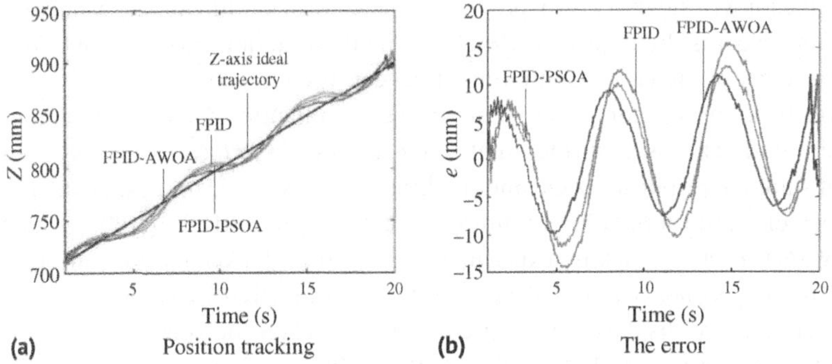

Figure 7.32 Position tracking and the error in the Z-axis direction in the spiral trajectory in the experiment.

is indicated by a tiny error fluctuation. Therefore, the best control performance is achieved by FPID-AWOA. The following two factors are the primary causes. On the one hand, the control effect of FPID is significantly worse than that of FPID-AWOA because the quantization factors K_{e_j} and K_{ec_j}, the scale factors $K_{u_{pj}}$, $K_{u_{ij}}$, and $K_{u_{dj}}$ in the fuzzy controller, as well as K_{po}, K_{io}, and K_{do} in Eq. (7.36) are typically chosen based on experience. However, simulations show that FPID-PSOA and FPID-WOA have comparable control performance. The addition of ω and γ to AWOA improves its optimization performance over WOA, as shown by Eqs. (7.48) and (7.51). As a result, FPID-AWOA outperforms FPID-PSOA in terms of stability and tracking accuracy. Therefore, as compared to FPID and FPID-PSOA, FPID-AWOA efficiently minimizes all faults and attains the highest control accuracy. Naturally, the experiment's findings align with the numerical models' conclusions as well.

7.3.4 Conclusion

This chapter presents FPID-AWOA for trajectory tracking of the CDPR. Based on the kinematics and dynamics model of the CDPR, the problem that the dynamic equation is too complex to express is solved numerically, and then a new fuzzy rule is proposed and a new fuzzy PID controller is designed. The quantization factors K_{e_j} and K_{ec_j} as well as the scale factors $K_{u_{pj}}$, $K_{u_{ij}}$, and $K_{u_{dj}}$ in the fuzzy PID controller are then optimized using AWOA. It should be noted that in order to enhance the optimization performance of WOA, this chapter enhances AWOA by adding weight and differential mutation interference factor based on traditional WOA. Numerical simulations show that FPID-AWOA may considerably increase the control accuracy of CDPR in the Cartesian space or joint space while reducing all errors when compared to other control techniques.

7.4 Force Control for RFCHRs

An experimental platform focused on cable driving force control was constructed with the goal of examining the structural features and operational procedures of the cable-driven robot. The platform uses a double cable driven system, meaning that two motors drive one cable, which moves in a reciprocating motion. A PID controller based on the idea of structural invariance is devised because the cable system is very nonlinear and readily disrupted. Through the OPC server, the experimental platform master control PLC completes the data interaction with the monitoring software. It then uses RS232 communication to achieve driving force control of the experimental platform. At last, the application is able to monitor the platform state in real time. According to the experimental results, the cable's driving force control is a huge success. The accuracy and real-time performance of the monitoring interface also meet expectations.

7.4.1 Construction of Experimental Platform

The cable force control experiment platform's structure is shown in [32]. Six fixed pulleys, three stepper motor linear modules, two DC torque motors (with encoder), tow cable drums, two tension sensors, two displacement sensors, and two cable connectors are among its primary components. The experimental platform's electrical components are also presented in [32]. The fundamental idea is that one torque motor drags the cable end at a steady speed (as load) while another torque motor creates driving force on the cable (as driving force). During operation, the torque motor functions as a drive or load in the second half of the stretching cycle and as a load or drive in the first half. A reciprocating pair of pulling is finished in one stretching cycle. The cable's real-time tension is measured by the tension sensor attached to the fixed pulley, and its elongation is measured by the displacement sensor. The encoder measures the cable speed.

7.4.2 Cable Driving Force Control Method

The torque motor powers the drum to enable the cable's retracting and unwinding movement when the cable drive unit is operating. We only look at the cable's pulling force when examining its driving force because it can only be stretched and not compressed when in operation. The driving force of the cable is influenced by a number of variables, including the nonlinearity of the cable itself and the spinning and friction of the motor rotor and pulley in the system. Both Coulomb and static friction will have a significant effect on the real physical system [33]. This section investigates how the cable driving force is controlled for a few of these factors.

7.4.2.1 The Establishment of Mathematical Models

The drive torque of the system is the motor's output torque (in half-cycle, as drive force), refer to the mechanism model in [32]. During the transmission process, the motor rotor, cable drum, fixed pulley, and other parts will lose their driving force. For example, the moments of inertia produced during start/stop T_{JA}, T_{JD}, and T_{JA_i}; the viscous friction torques T_{BA}, T_{BD}, and T_{BP_i}; and the static friction torques of the motor rotor, cable drum, and fixed pulleys T_{fA}, T_{fD}, and T_{fP_i}. For the overall system,

$$T_f = T_{fA} + T_{fD} + \sum_{i=1}^{3} T_{fP_i} \tag{7.56}$$

$$T_B = T_{BA} + T_{BD} + \sum_{i=1}^{3} T_{BP_i} = B_{qA}\dot{\theta} + B_{qD}\dot{\theta} + \sum_{i=1}^{3} B_{qP_i}\dot{\theta} \tag{7.57}$$

$$T_J = T_{JA} + T_{JD} + \sum_{i=1}^{3} T_{JP_i} = J_{qA}\ddot{\theta} + J_{qD}\ddot{\theta} + \sum_{i=1}^{3} J_{qP_i}\ddot{\theta} \tag{7.58}$$

The driving system's torque balance equation allows us to obtain

$$T_m - T_1 = T_f + T_B + T_J \tag{7.59}$$

$$F_0 = T_1 r \tag{7.60}$$

The back electromotive force produced inside the motor will also have an impact on the cable drive unit's output force when it is operating. The DC motor torque output equation allows us to obtain

$$L\frac{di}{dt} + Ri = u - e \tag{7.61}$$

$$e = C_e\dot{\theta}_m \tag{7.62}$$

$$T_m = C_m i \tag{7.63}$$

The cable can currently be considered an equivalent spring-damping model because its moving speed is slow throughout the stretching operation, allowing us to achieve

$$F = K(y - y_0) + B(\dot{y} - \dot{y}_0) \tag{7.64}$$

$$y' = y - y_0 \tag{7.65}$$

T_m represents the motor's driving torque. The motor's load torque is shown by T_1, while the cable drive unit's static friction is indicated by T_f. The motor rotor, cable drum, and three fixed pulleys' moments of inertia are shown by the symbols J_{qA}, J_{qD}, and J_{qP_i}. The viscous friction coefficients of the motor rotor, cable drum, and three fixed pulleys are indicated by the symbols B_{qA}, B_{qD}, and B_{qP_i}. The flexible cable drum's corner is indicated by θ_m, and its radius is shown by r. The motor torque is indicated by C_m, constant, where i and u stand for the motor's armature current and voltage, respectively; R stands for the armature circuit's resistance, L for its inductance, e for the motor's back EMF, C_e for the back EMF coefficient, and y for the drive end of the cable's displacement. The displacement of the cable's end is shown by y_0, and its elongation is indicated by y'. For the spring-damping model, F stands for the output tension at the cable's end in its actual condition, F_0 for its ideal state, K for the elastic coefficient, and B for the damping coefficient.

When Laplace modifies the aforementioned Eqs. (7.59)–(7.65), we obtain the system model's open-loop block diagram, which is displayed in [32]. The system model indicates that an extra torque T_d will be produced when the cable's end moves. When the cable drive unit's input command is fixed, the additional force produced by the cable end's movement is referred to as the redundant force. There are two different definitions of redundant force: broad and restricted. In a broader sense, the redundant force is any force that results in a loading error; in a more specific sense, the redundant force is the disturbance brought on by the motion of the load-bearing object [34]. The role of the static friction torque T_f in the feedback channel is disregarded during modeling since it is constant and has a negligible effect on the system as a whole. In conclusion, the cable drive system's mathematical model is

$$F(s) = P_1 I(s) - P_2 y_0(s) - f \tag{7.66}$$

Among them

$$P_1 = \frac{BLrC_m s^2 + (BR + KL)\, rC_m s + KRrC_m}{J_q Ls^3 + \left(BLr^2 + J_q R + B_q L\right) s^2 + \left(BRr^2 + KLr^2 + C_e C_m + B_q R\right) s + KRr^2} \tag{7.67}$$

$$P_2 = \frac{(Bs + K)\left[J_q Ls^3 + \left(J_q R + B_q L\right) s^2 + \left(C_e C_m + B_q R\right) s\right]}{J_q Ls^3 + \left(BLr^2 + J_q R + B_q L\right) s^2 + \left(BRr^2 + KLr^2 + C_e C_m + B_q R\right) s + KRr^2} \tag{7.68}$$

B_q represents the system equivalent viscous friction coefficient in Eqs. (7.67) and (7.68): $B_q = B_{qA} + B_{qD} + \sum_{i=1}^{3} B_{qP_i}$; J_q represents the system equivalent moment of inertia. $J_q = J_{qA} + J_{qD} + \sum_{i=1}^{3} J_{qP_i}$; P_1 represents the forward channel transfer function, which is the object model from output current $I(s)$ to output force $F(s)$;

P_2 represents the redundant force transmission channel, which is the object model from input displacement $y_0(s)$ to output force $F(s)$. The static friction torque T_f in the system produces the static friction force, denoted by f. The transfer function clearly exhibits differential characteristics when calculating the redundant force according to the cable end displacement $Y_0(s)$, as can be seen by looking at the expression in Eq. (7.68), where the highest order of the numerator is fourth order and the highest order of the denominator is third order. Furthermore, the simulation and experiment physics are subpar. Consequently, take into account figuring out the redundant force based on the cable end speed $V_0(s)$,

$$V_0(s) = sY_0(s) \tag{7.69}$$

$$P_2^* = \frac{(Bs+K)\left[J_qLs^2 + (J_qR + B_qL)s + (C_eC_m + B_qR)\right]}{J_qLs^3 + (BLr^2 + J_qR + B_qL)s^2 + (BRr^2 + KLr^2 + C_eC_m + B_qR)s + KRr^2} \tag{7.70}$$

When calculating the redundant force based on the cable end speed $V_0(s)$, the highest order of the denominator and numerator is the same on the third order. It is more feasible and offers more convenience for the controller's later design and choices. The completed system's new transfer function is

$$F(s) = P_1 \cdot I(s) - P_2^* \cdot V_0(s) - f \tag{7.71}$$

The object model from the output force $F(s)$ to the cable end speed $V_0(s)$ is P_2^*. We must limit the effect of excess force on the output force accuracy in addition to taking loading accuracy and speed into account.

7.4.2.2 Control Strategy Design

For the system control strategy diagram, refer to [32]. The main controller makes use of a PID controller; adding differential terms might enhance the system's dynamic properties, but introducing high-frequency signals is also simple, particularly when there are unexpected error disturbances. Because of the short length and excessive sensitivity of the differential, oscillation is easily caused. In light of this circumstance, the PID algorithm's differential term is enhanced, and a low-pass filter is included to eliminate the high-frequency signals that the differential term introduces [35]. Thus, in order to enhance the system's stability and dynamic quality, the primary controller implements partial differential PID correction,

$$G_m(s) = K_p\left(1 + \frac{K_I}{s} + \frac{K_D s}{T_f s + 1}\right) \tag{7.72}$$

K_P, K_I, K_D denote the PID control parameters, and T_f denotes the time coefficient of the incomplete derivative term.

The disturbance feed-forward compensation method based on the structural invariance is utilized because the precision of the cable driving force is decreased due to the interference of the redundant torque induced by the cable's end movement. Creating a feed-forward compensation controller based on disturbances is the fundamental idea behind this compensation technique. Dual-channel control can be utilized to remove superfluous force interference brought on by preexisting disturbances. This compensation plan has a straightforward structure, good real-time performance, and is simple to apply [36, 37]. The schematic diagram of structural invariance compensation is shown in [32]. $N(s)$ represents the disturbance, $R(s)$ represents the input, and $Y(s)$ represents the output. The purpose of the control compensator $G_c(s)$ is to remove the impact of the disturbance on the output. Consequently, it must fulfill

$$N(s) \cdot G_c(s) \cdot G_1(s) \cdot G_2(s) = N(s) \cdot G_3(s) \cdot G_2(s) \tag{7.73}$$

After calculation

$$G_c(s) = \frac{G_3(s)}{G_1(s)} \tag{7.74}$$

Based on Eq. (7.74), the principle of invariance, and the Compensation Controller's transfer function

$$G_v(s) = \frac{P_2(s)}{P_1(s)} = \frac{(Bs + K)\left[J_q Ls^2 + (J_q R + B_q L)s + (C_e C_m + B_q R)\right]}{BLrC_m s^2 + (BR + KL)rC_m s + KRrC_m} \tag{7.75}$$

The compensating controller has distinct features since the numerator's order is higher than the denominator's. An approximation compensation controller can be obtained by configuring two poles on the denominator while taking feasibility into consideration,

$$G_v^*(s) = \frac{P_2(s)}{P_1(s)} = \frac{(Bs + K)\left[J_q Ls^2 + (J_q R + B_q L)s + (C_e C_m + B_q R)\right]}{\left[BLrC_m s^2 + (BR + KL)rC_m s + KRrC_m\right](T_\tau s + 1)} \tag{7.76}$$

T_τ is the filter time constant.

Lastly, the cable driving force control block diagram of the overall system model is shown in [32], where ξ is the disturbance noise; $G_b(s)$ is filtered by the double-T band stop filter, and the band stop frequency is 45–55 Hz [38].

7.4.3 Control and Monitoring Program Design

This section lays the groundwork for the future control and monitoring of the entire cable-driven robot by designing the experimental platform LABVIEW monitoring system to confirm the efficacy of the control method and collect experimental records, including various sensor data. The particular flowchart for monitoring is shown in [32].

The experimental platform's displacement and tension sensors are connected to the PLC's AD analog module. LabVIEW retrieves sensor data from the PLC using the shared variables in OPC. The PLC's I/O serial port is linked to the fixed pulley module motor driver, and the PLC sends pulses to regulate the fixed pulley's position. To accomplish the control goal, LabVIEW utilizes the shared variables in OPC and, via program editing, writes the appropriate pulse number and frequency to the PLC register.

The experimental platform's main control unit is the torque motor and its driver. The industrial computer and the two torque motor drivers communicate via RS232 serial. By regulating the current, the torque motor may regulate the output torque. To achieve torque motor control, LabVIEW uses the control algorithm calculation (developed in the previous section) to determine the corresponding current value. Then, using the VISA serial communication function, it writes the command frame to the appropriate driver. By transmitting the instruction frame to read the drive status, LabVIEW may also acquire the motor's related operational characteristics. For the motor's program code, please refer to [32].

Generally speaking, the operator can use the LabVIEW monitoring program to calibrate the experimental platform's sensor, set the motor's operating and controller parameters, control the fixed pulley module's position, and start or stop each module. Throughout the experimental platform's operation, it can continuously record and monitor a variety of sensor readings and the motor's running state. It can then use algorithms to analyze the monitoring data, identify anomalous situations within the platform, and halt operations in an emergency. For the control and monitoring interface, refer to [32].

7.4.4 Test and Verification Experiment

This part will carry out platform experiments to confirm the efficacy of the control strategy and the overall functionality of the monitoring system. The experiment's methodology: set the desired speed at one motor to drag the cable drive unit, the expected output force from the motor's cable drive unit at one end, and the time required to finish the pulling motion's reciprocating cycle in the monitoring system. The experiment shows a discrepancy between the cable drive unit's actual output force and its expected force, although the other monitoring features of the system operate normally. Each parameter's initial value during the experiment is as follows:

Then we get the monitoring screen of LabVIEW during the experiment. The monitoring interface shows a number of parameters in real time. The monitoring system produces the desired outcome. Figure 7.33 shows the expected and actual output forces of the cable drive unit when one of the torque motors produces driving force on the cable (as drive force). We have gathered data for three cycles in

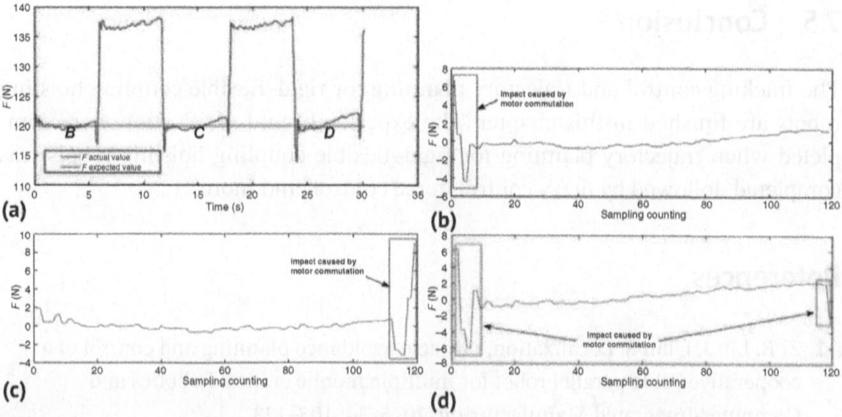

Figure 7.33 Cable force control experiment results. (a) Comparison of cable tension and expected output force in three cycles. (b) The error in Segment B. (c) The error in Segment C. (d) The error in Segment D cable tension and expected output force.

Figure 7.33a. Additionally, we have compared and analyzed errors. The inaccuracy of the output force is between 0 and 1.8 N, excluding the effect of the force produced during motor commutation. The accuracy of the test apparatus and system fault make this inevitable. The cable's output force control has produced the intended result.

7.4.5 Conclusion

A cable force control experiment platform was constructed based on the structural features of the cable-driven robot. The control strategy design of the cable force control technique has been accomplished through the use of the principle of structural invariance, imperfect differential PID correction, and an analysis of the experimental platform's mathematical model and structure. The cable force control experimental platform is controlled and monitored by program editing and data exchange between OPC and PLC. It builds the groundwork for the future cable traction robot's entire machine monitoring design technique and offers a convenient control process, as well as easy-to-understand data display and acquisition for further trial operations. According to experimental data, the control panel can fully control the experimental bench, and the surface monitoring system can produce the anticipated monitoring effect. The man–machine interface is user-friendly, and the monitoring data is effective in real time. Furthermore, the applied cable force control strategy can successfully guarantee the correctness of the cable output force by monitoring data analysis, providing a basis for the subsequent phase of cable force control strategy optimization.

7.5 Conclusion

The tracking control and trajectory planning for rigid–flexible coupling hoisting robots are finished in this chapter. The experiment and verification were completed when trajectory planning for rigid–flexible coupling hoisting robots was completed, followed by fuzzy control, force control, and more.

References

1 Zi B, Lin J, Qian S. Localization, obstacle avoidance planning and control of a cooperative cable parallel robot for multiple mobile cranes. Robotics and Computer-Integrated Manufacturing, 2015, 34: 105–123.

2 Scalera L, Gallina P, Seriani S, et al. Cable-based robotic crane (CBRC): design and implementation of overhead traveling cranes based on variable radius drums. IEEE Transactions on Robotics, 2018, 34(2): 474–485.

3 Yao R, Tang X Q, Wang J S, et al. Dimensional optimization design of the four-cable-driven parallel manipulator in FAST. IEEE/ASME Transactions on Mechatronics, 2010, 15(6): 932–941.

4 Tang X Q, Yao R. Dimensional design on the six-cable driven parallel manipulator of FAST. Journal of Mechanical Design, 2011, 133(11): 111012.

5 Kawamura S, Kino H, Won C. High-speed manipulation by using parallel wire-driven robots. Robotica, 2000, 18(1): 13–21.

6 Wang X G, Ma S Y, Lin Q. Hybrid pose/tension control based on stiffness optimization of cable-driven parallel mechanism in wind tunnel test. Proceedings of the 2016 2nd International Conference on Control, Automation and Robotics (ICCAR), Hong Kong, China, April 28–30, 2016: 75–79.

7 Wang L, Sheng Y, Qiu Y Y, et al. Design and implementation of the control device for a cable-driven camera robot. Applied Mechanics and Materials, 2013, 373–375: 225–230.

8 Barnett E, Gosselin C. Large-scale 3D printing with a cable-suspended robot. Additive Manufacturing, 2015, 7: 27–44.

9 Qian S, Bao K L, Zi B, et al. Kinematic calibration of a cable-driven parallel robot for 3D printing. Sensors, 2018, 18(9): 1–22.

10 Zi B, Wang N, Qian S, et al. Design, stiffness analysis and experimental study of a cable-driven parallel 3D printer. Mechanism and Machine Theory, 2019, 132: 207–222.

11 Xiong H, Diao X. A review of cable-driven rehabilitation devices. Disability and rehabilitation: Assistive technology, 2019, 15: 1–13.

12 Chen Q, Zi B, Sun Z, et al. Design and development of a new cable-driven parallel robot for waist rehabilitation. IEEE/ASME Transactions on Mechatronics, 2019, 24(4): 1497–1507.

13 Bosscher P, Williams R L, Bryson L S, et al. Cable-suspended robotic contour crafting system. Automation in Construction, 2007, 17(1): 45–55.

14 Izard J-B, Dubor A, Hervé P-E, et al. Large-scale 3D printing with cable-driven parallel robots. Construction Robotics, 2017, 1 (1–4): 69–76.

15 Jadhao K S, Lambert P, Bruckmann T, et al. Design and analysis of a novel cable-driven haptic master device for planar grasping. In Cable-Driven Parallel Robots, Cham: Springer, 2017: 307–318.

16 Pott A, Bruckmann T. Cable-Driven Parallel Robots. Cham: Springer, 2019.

17 Longval J M, Gosselin C. Dynamic trajectory planning and geometric analysis of a two-degree-of-freedom translational cable-suspended planar parallel robot using a parallelogram cable loop. Journal of Mechanisms and Robotics, 2019, 11(2): 020903.

18 Jiang X L, Barnett E, Gosselin C. Periodic trajectory planning beyond the static workspace for 6-DOF cable-suspended parallel robots. IEEE Transactions on Robotics, 2018, 34(4): 1128–1140.

19 Dion-Gauvin P, Gosselin C. Dynamic point-to-point trajectory planning of a three-DOF cable-suspended mechanism using the hypocycloid curve. IEEE/ASME Transactions on Mechatronics, 2018, 23(4): 1964–1972.

20 Sharma R, Rana K P S, Kumar V. Performance analysis of fractional order fuzzy PID controllers applied to a robotic manipulator. Expert Systems with Applications, 2014, 41(9): 4274–4289.

21 Liang X, Wang H, Liu Y H. Calibration-free image-based trajectory tracking control of mobile robots with an overhead camera. IEEE Transactions on Automation Science and Engineering, 2019, 17(2): 933–946.

22 Qian S, Zi B, Shang WW, et al. A review on cable-driven parallel robots. Chinese Journal of Mechanical Engineering, 2018, 31: 66.

23 Di Massa G, Pagano S, Strano S, et al. A comparison between linear and nonlinear modelling of a wire rope seismic isolator. International Review on Modeling & Simulations, 2013, 6(4): 124–128.

24 Zou Y, Lixun Z, Ma H, et al. Hybrid force control of astronaut rehabilitative training robot under active loading mode. Journal of Central South University, 2014, 21: 4121–4132.

25 Xing F, Li S. Design of control system of electric load simulator. Journal of Northwestern Polytechnical University, 2009, 26(5): 3596–3598 (in Chinese).

26 Zi B, Zhang L, Zhang D, et al. Modeling, analysis, and co-simulation of cable parallel manipulators for multiple cranes[J]. Proceedings of the Institution of Mechanical Engineers, Part C: Journal of Mechanical Engineering Science, 2015, 229(9): 1693–1707.

27 Biagiotti L, Melchiorri C. Trajectory Planning for Automatic Machines and Robots. Berlin: Springer, 2008.

28 Yang C, Jiang Y M, Na J, et al. Finite-time convergence adaptive fuzzy control for dual-arm robot with unknown kinematics and dynamics. IEEE Transactions on Fuzzy Systems, 2019, 27(3): 574–588.

29 Zhang S, Wu Y, He X, et al. Cooperative fault-tolerant control for a mobile dual flexible manipulator with output constraints. IEEE Transactions on Automation Science and Engineering, 2021, 19(4): 2689–2698.

30 Sun N, Fu Y, Yang T, et al. Nonlinear motion control of complicated dual rotary crane systems without velocity feedback: design, analysis, and hardware experiments. IEEE Transactions on Automation Science and Engineering, 2020, 17(2): 1017–1029.

31 Zhou B, Wang Y H, Zi B, et al. Fuzzy adaptive whale optimization control algorithm for trajectory tracking of a cable-driven parallel robot. IEEE Transactions on Automation Science and Engineering, 2024, 21(4): 5149–5160.

32 Liu W H, Wang Z Y, Zi B, et al. Control and monitoring of a double cable driven system. Proceedings of the 2021 6th International Conference on Intelligent Computing and Signal Processing (ICSP), IEEE, 2021.

33 Wang X, Qing H, Zongxia J, et al. Friction force and its compensation control in load simulator. China Mechanical Engineering, 2013, 14(6): 67–70.

34 Haifeng Z, Li S. Research on aircraft rudder motor servo loading system. Journal of System Simulation, 2004, 16(4): 657–659.

35 Lianghong W, Yaonan W, Shaowu Z, et al. Design of PID controller with incomplete derivation based on differential evolution algorithm. Journal of Systems Engineering and Electronics, 2013, 19(3): 578–583.

36 Fu W, Li S. Design and model-building of motor-driven load simulator with large torque outputs. Journal of System Simulation, 2009, 21(12): 3596–3598.

37 Nam Y, Lee J, Hong S K. Force control system design for aerodynamic load simulator. Proceedings of the American Control Conference, Chicago, Illinois, 2000: 3043–3047.

38 Shitao Q, Liming T, Min W, et al. Design of power frequency interference trap based on Matlab. Chinese Medical Equipment Journal, 2005, 26(8): 3–4.

8

Platform Development and Application for Rigid–Flexible Coupling Hoisting Robots

8.1 Preamble

This chapter addresses the design, development, and experimental validation of robotic platforms for rigid–flexible coupling hoisting systems, a critical domain where the integration of structural rigidity and cable-driven flexibility presents unique challenges and opportunities [1–4]. Unlike traditional rigid-link or purely flexible robotic systems, rigid–flexible coupling hoisting robots (RFCHRs) must balance the stiffness required for precise load positioning with the adaptability of flexible cables for extended workspace and payload variation [5, 6]. These systems are inherently characterized by nonlinear dynamics arising from cable elasticity, pulley friction, and kinematic coupling, which complicates motion control and trajectory tracking in applications such as heavy-load manipulation, automated stacking, and precision hoisting.

Existing research on rigid–flexible coupling robots has often focused on either simplified rigid-body models or purely flexible systems, leaving a gap in the integration of rigid–flexible dynamics for robust industrial applications. Mechanical decoupling [7], a key requirement for multi-degree-of-freedom (DOF) articulated robots, has been primarily addressed through algorithmic approaches, while hardware-based solutions capable of mitigating interjoint interference at the physical level remain underdeveloped. Additionally, under realistic operational conditions, the validation of complex trajectories, including cable winding variability and dynamic load shifts, has posed significant challenges for both theoretical modeling and experimental verification [8–11].

Against this backdrop, two representative rigid–flexible coupling robotic platforms are presented. First, a 7-DOF RFCHR is systematically engineered with novel mechanical decoupling mechanisms, such as discretized pulley systems and 3D-printed components, to reduce interjoint interference. Complementing this hardware design is a fuzzy PID control strategy [12], tailored to address the nonlinear dynamics of cable-driven systems and enable closed-loop position control. Experimental validation focuses on quantifying decoupling performance, trajectory tracking accuracy, and the impact of mechanical innovations on reducing kinematic coupling errors, providing a foundational framework for articulated hoisting robots [13].

Second, a specialized robot for yard operations is developed to demonstrate the translation of rigid–flexible coupling principles into real-world logistics tasks. This platform integrates high-precision motion capture systems, tension sensors, and a servo-driven cable architecture to validate complex trajectories under realistic conditions. By systematically analyzing linear, spiral, and task-space stacking motions, the study evaluates the robot's stability, positioning accuracy, and resilience to challenges such as cable winding superposition, measurement limitations, and prototype tolerances. The results highlight the interplay between mechanical design, sensor integration, and control strategy in enabling reliable performance for material handling and container stacking.

A core contribution of this work lies in its holistic approach to bridging theoretical modeling with physical prototyping. Both platforms are rigorously tested using custom experimental setups that incorporate aluminum alloy structures, 3D-printed components, and advanced sensing systems, offering actionable insights into the fabrication, calibration, and validation of rigid–flexible coupling robots.

The major contribution of this chapter lies in the systematic development and experimental validation of rigid–flexible coupling robotic platforms, establishing a foundational framework for addressing mechanical decoupling, nonlinear dynamics, and complex trajectory execution in cable-driven hoisting systems. By integrating novel mechanical designs, such as discretized pulleys and 3D-printed decoupling modules, with adaptive control strategies like fuzzy PID, the research provides practical solutions for enhancing motion precision and stability, which are critical for advancing industrial automation. The rest of the chapter is structured as follows. Section 8.2 focuses on the RFCHR for container yard operations in Chapter 2, presenting the construction of its experimental platform, integration of high-precision sensing systems, and validation of linear, spiral, and task-space stacking trajectories under realistic operational conditions. In Section 8.3, the development of a 7-DOF RFCHR is detailed, including the design of its physical prototype, control system, and experimental verification of decoupling performance through single-joint and trajectory-tracking experiments. Finally, Section 8.4 provides concluding remarks and thoughts on future work. These sections collectively demonstrate the technical viability and practical applications of the proposed robotic platforms, laying the groundwork for future advancements in rigid–flexible coupling robotics for heavy-load manipulation and automated logistics.

8.2 Platform Development and Performance Verification of an RFCHR for Yard Operations

In this section, the rigid–flexible coupling robot that can be used for yard operations in Chapter 2 is introduced. The following introduces the physical prototype development and motion performance experiments for the robot [14].

8.2.1 Physical Prototype Development

Considering the proportionality and rationality of the experimental prototype's construction, the robot's stationary platform uses European-standard 4040L aluminum profiles with a wall thickness of 2 mm and a unit length weight of 1.3 kg. The moving platform adopts European-standard 2020L aluminum profiles with a wall thickness of 1.5 mm and a unit length weight of 0.47 kg. The profiles are connected using right-angle corner brackets and 45° corner brackets. The experimental platform is shown in Figure 8.1.

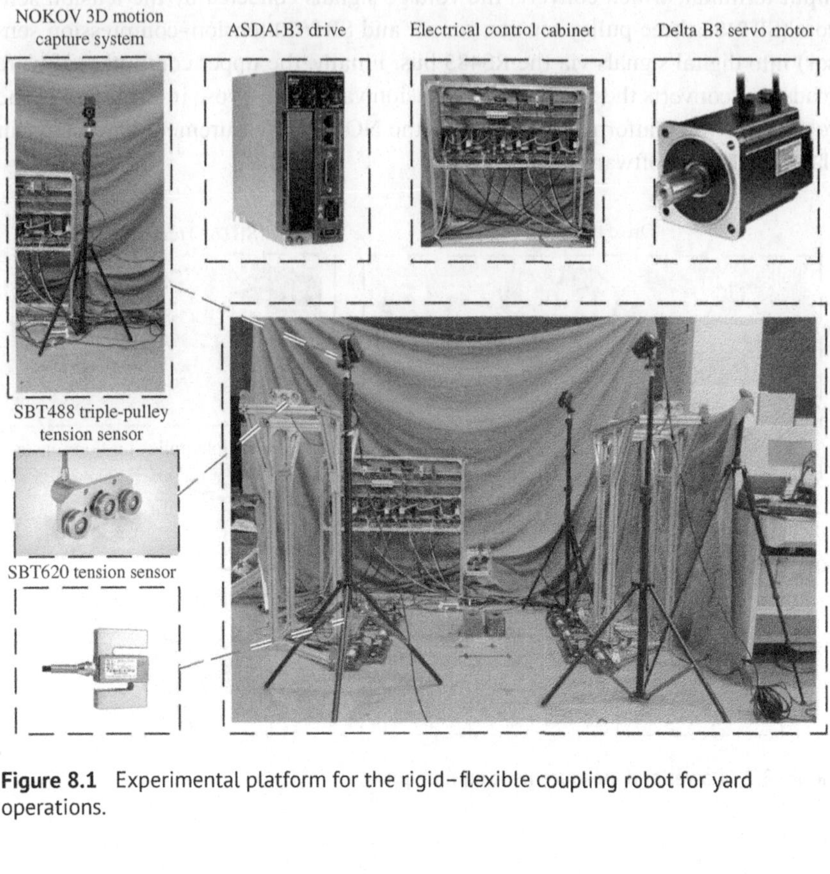

Figure 8.1 Experimental platform for the rigid–flexible coupling robot for yard operations.

In the experimental platform, a NOKOV 3D motion capture system is used to acquire in real time the spatial position of the robot's moving platform. When the calibration results are good, this device can, in real time, obtain the precise positioning of the moving platform. Its principle is to use an optical lens to emit infrared light, which is reflected back to the sensor matrix on the lens after irradiating the reflective marker points. The pose information of the target object in space can be calculated when three or more sensor cameras receive the reflection signal simultaneously.

As shown in Figure 8.2, the drive module consists of a servo motor connected to a planetary gear reducer, which is then connected to workpieces such as a coupling drum as the power output. A lightweight flexible cable is introduced from one side of the drum and connected to the fixed anchor points of the moving platform through multiple sets of pulleys, thus building the robot's power system. In the stacking experiment, a 3D-printed container shell model is used as the workload, and a remote-control relay is employed to control the H3017 electromagnet to simulate the empty container stacking process. For the acquisition of flexible cable tension data, an SBT908D digital signal amplifier is used as the data acquisition input terminal, which converts the voltage signals collected by the tension sensors (SBT488 three-pulley tension sensor and SBT620 tension–compression sensor) into digital signals via the RS485 bus. Finally, the upper computer software reads and converts these signals into tension values. The pose information of the robot's moving platform is obtained by the NOKOV measurement equipment in the XINGYING software.

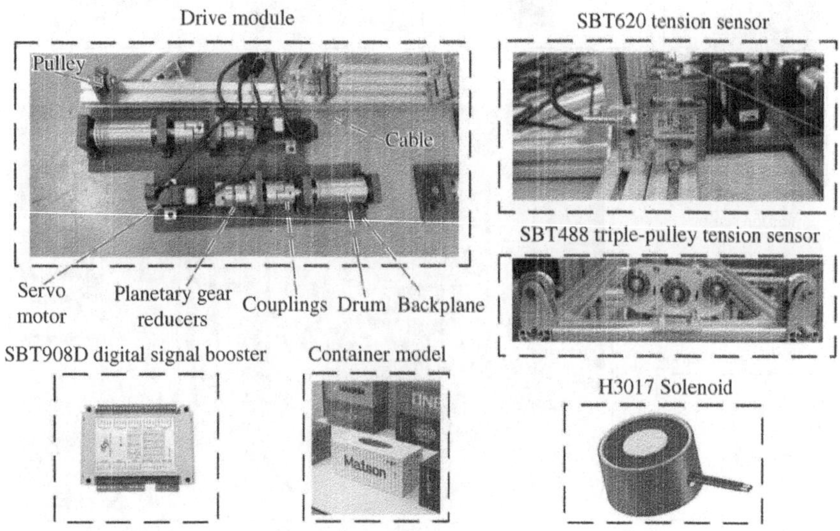

Figure 8.2 Power and measurement equipment.

8.2.2 Robot Motion Performance Experiment

First, a spatial linear trajectory is used to verify whether the pose change of the moving platform conforms to kinematic laws. The spatial linear trajectory path is as follows: initial position $A\,(0.52, 0.74, 0)$ m → intermediate position $B\,(0.52, 0.74, 0.3)$ m → linear starting point $C\,(0.7, 0.9, 0)$ m → linear ending point D (0.3, 0.5, 0.6) m. The experiment is controlled using the PT motion mode, and the experimental process is shown in Figure 8.3.

During the experiment, it can be intuitively observed that the posture of the robot's moving platform remains stable, without severe fluctuations or "cable detachment" phenomena. Due to the large number of experimental points captured by the motion capture system, a recording method of taking a set of effective experimental points every 60 points was adopted. The actual operation trajectory and theoretical operation trajectory of the moving platform are plotted as shown in Figure 8.4a. In the figure, it can be seen that near the middle section of the linear motion trajectory, individual points exhibit extreme values and deviate from the motion trajectory, which may be related to the singularity at these positions. To further analyze the trajectory error trend, error waterfall plots in three directions are plotted as shown in Figure 8.4b. It can be found from the figure that except for the error in the Y-axis at the extreme points, the average motion error of the trajectory in the X-axis is 22.17 mm, the average error in the Y-axis is 9.21 mm, and the average error in the Z-axis is 9.32 mm. All errors are within the allowable error range.

To verify whether the robot can ensure stable operation under more complex conditions, a spatial spiral trajectory is used to test the stability of the moving

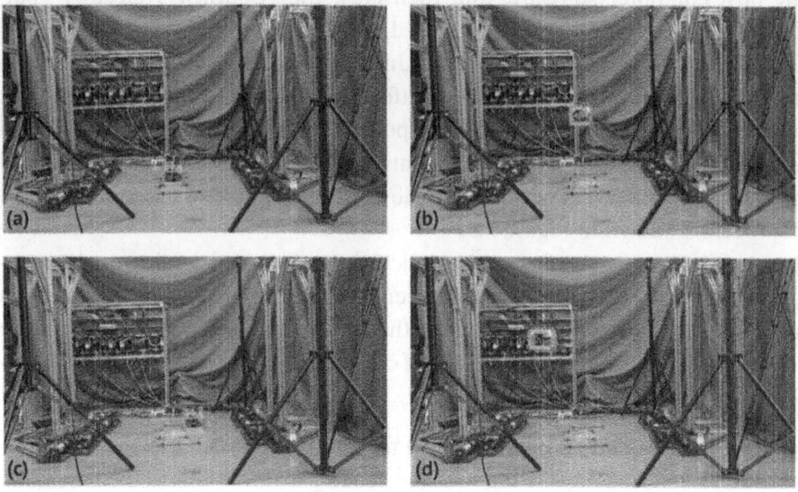

Figure 8.3 Linear trajectory motion process.

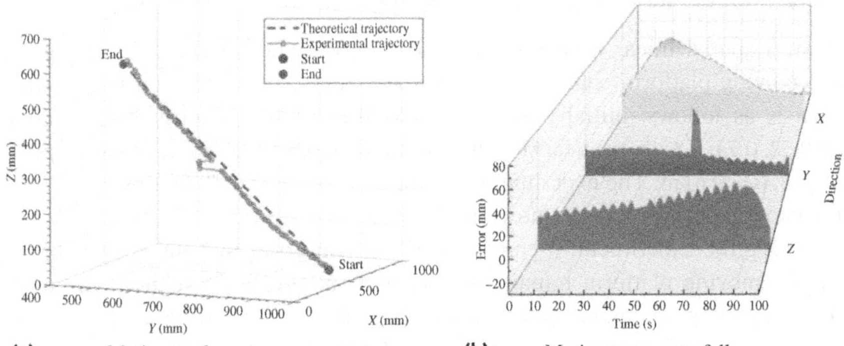

(a) Motion track (b) Motion error waterfall

Figure 8.4 Straight-line trajectory.

platform. The rule for using the spatial spiral trajectory here is to perform sinusoidal and cosine motions with the same amplitude in the *X*- and *Y*-directions, and a linear upward movement in the *Z*-axis direction. The trajectory equation of the spatial spiral is as follows:

$$\begin{cases} x = 0.52 + 0.15 \sin(2\pi/5 * t) \\ y = 0.64 + 0.15 \cos(2\pi/5 * t),\ t \in [0, 10] \text{ sec} \\ z = 0.15 + 0.04t \end{cases} \tag{8.1}$$

To complete the spatial spiral trajectory movement, it is necessary to input the corresponding parametric equations into the trajectory planning interface of the host computer software, and then click the "Confirm Trajectory" button. After the planning data is calculated, clicking the "Run Start Point" button initiates the trajectory movement. Under the spatial spiral trajectory, the operation path of the robot's moving platform can be briefly expressed as: start point A (0.52, 0.79, 0.15) m → left spatial point B (0.52, 0.49, 0.25) m → right spatial point C (0.52, 0.79, 0.35) m → left spatial point D (0.52, 0.49, 0.45) m → end point E (0.52, 0.79, 0.55) m → initial position F (A)(0.52, 0.79, 0.15) m. The movement process is shown in Figure 8.5.

The actual movement path and theoretical movement path of the moving platform are shown in Figure 8.6. As can be seen from Figure 8.6a, when the precision is in centimeters, the movement trend of the actual trajectory matches that of the theoretical trajectory, but there is a lack of coincidence. There are three main reasons for this phenomenon:

1. **Cable winding superposition:** The most significant factor is that the flexible cables may overlap during winding, leading to deviations in the kinematic model.

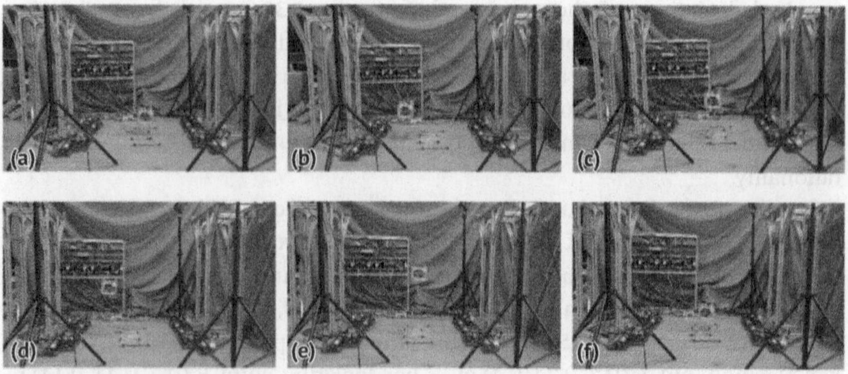

Figure 8.5 Spiral trajectory motion process.

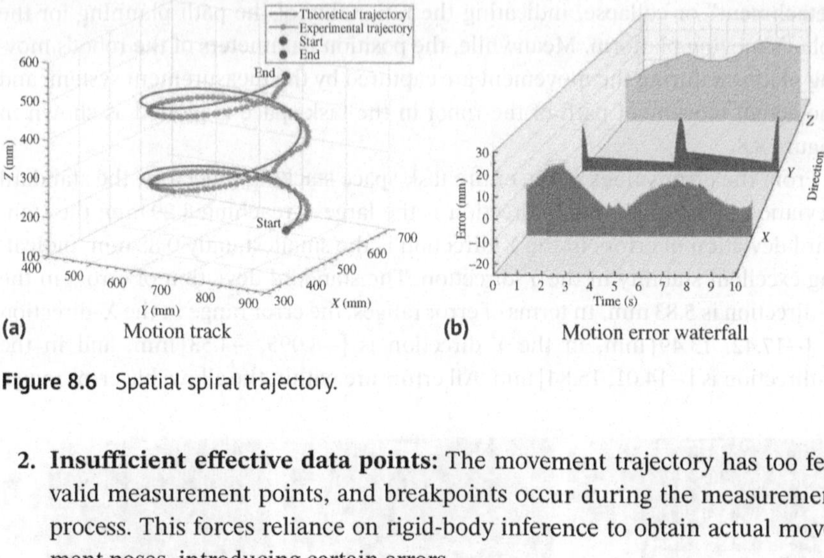

Figure 8.6 Spatial spiral trajectory.

2. **Insufficient effective data points:** The movement trajectory has too few valid measurement points, and breakpoints occur during the measurement process. This forces reliance on rigid-body inference to obtain actual movement poses, introducing certain errors.
3. **Prototype construction errors:** Measurement errors arise from significant construction inaccuracies in the experimental prototype, primarily constrained by the experimental site and setup conditions.

Figure 8.6b presents the error waterfall plots for each direction under this trajectory. Briefly, the error band of the moving platform in the Y-axis direction is relatively uniform and small, but larger errors occur at the starting position of each phase, which is caused by significant jitter of the moving platform at the starting points of the phases, thereby increasing the errors. In terms of data, the error standard deviations of the moving platform are 6.38 mm in the X-direction, 5.67 mm in the Y-direction, and 6.48 mm in the Z-direction. In terms of error ranges, the error ranges are [−13.94, 11.77] mm in the X-direction, [−20.13, 5.64] mm in the Y-direction, and [−0.98, 23.46] mm in the Z-direction, indicating that the robot's movement under complex trajectories has a certain degree of rationality.

Finally, the motion path is designed using a task-space stacking trajectory. Combining the requirements of the box-lifting segment, box-moving segment, and box-placing segment, the full-process handling path is planned as follows: box-lifting point A (0.3, 0.74, 0.125) m → box-moving point B (0.3, 0.74, 0.4) m → stable operation point C (0.4, 0.74, 0.5) m → stable termination point D (0.6, 0.74, 0.5) m → box-placing point E (0.7, 0.74, 0.4) m → box-lowering point F (0.7, 0.74, 0.14) m. The experimental process is shown in Figure 8.7.

The entire stacking process is smooth and gentle, without occurrences of "cable detachment" or collapse, indicating the rationality of the path planning for the robot's moving platform. Meanwhile, the position parameters of the robot's moving platform during the movement are captured by the measurement system, and the actual movement path of the robot in the task space is plotted as shown in Figure 8.8.

From the error values of the entire task-space stacking trajectory, the standard deviation of errors in the X-direction is the largest, reaching 8.29 mm; the standard deviation of errors in the Y-direction is the smallest, only 0.88 mm, indicating excellent stability in the Y-direction. The standard deviation of errors in the Z-direction is 5.83 mm. In terms of error ranges, the error range in the X-direction is [−17.42, 13.49] mm, in the Y-direction is [−8.095, −4.53] mm, and in the Z-direction is [−14.01, 15.84] mm. All errors are within the allowable error range.

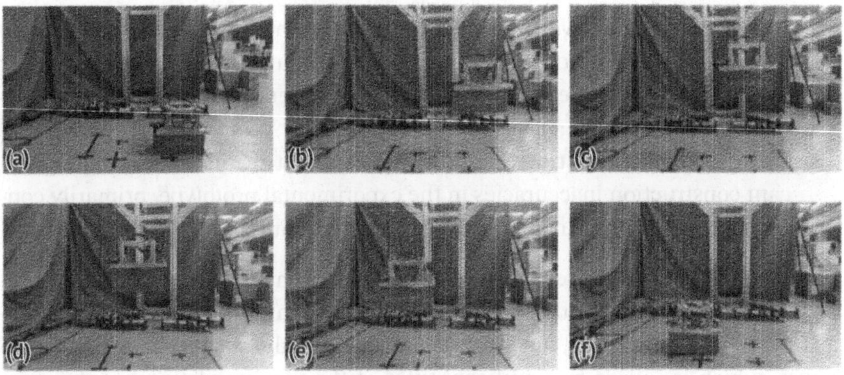

Figure 8.7 Task-space stacking trajectory motion process.

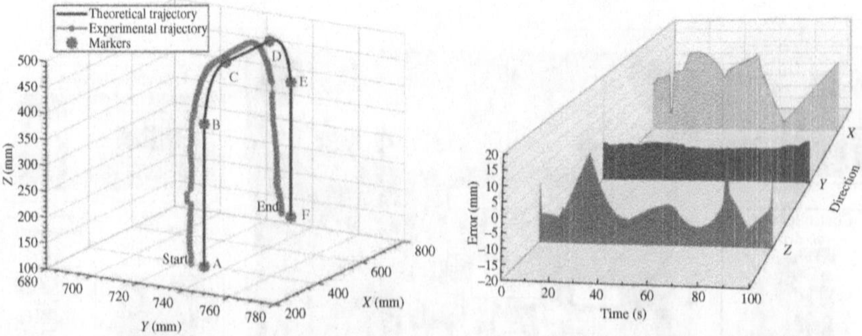

(a) Motion track (b) Motion error waterfall

Figure 8.8 Task-space stacking trajectory.

8.3 Platform Development and Performance Verification for the 7-DOF RFCHR

In this section, the system platform of the 7-DOF RFCHR introduced in Chapter 3 is designed, including the physical prototype, control system, and experimental verification of the decoupling performance for the robot based on the former two.

8.3.1 Physical Prototype Development

To validate the motion performance and decoupling capability of the 7-DOF RFCHR, a physical prototype was constructed, as shown in Figure 8.9. The robot is built from aluminum alloy components, while the decoupling modules are fabricated using 3D printing technology. To reduce excessive friction that may occur when the drive cables of joints 2–7 act on the decoupling wheel of joint 1, the decoupling wheel was discretized within the mechanism, effectively dividing it into multiple smaller pulleys, as highlighted in the detailed view of the decoupling mechanism in Figure 8.9. The driving system consists of seven servomotors integrated into the base, transmitting force to each joint via high-strength polyethylene fiber cables. Multiturn absolute encoders are installed at each joint to enable real-time acquisition of rotation angles for closed-loop position control. Additionally, a multi-camera visual motion capture system was used to record end-effector trajectories, allowing verification of the robot's motion performance.

The robot control system, illustrated in Figure 8.10, consists of an IPC, a control and operating system, a motion controller, a servo cable-driven system, and data acquisition and feedback modules. Communication between the software and hardware components is facilitated through the EtherCAT and RS-485 protocols.

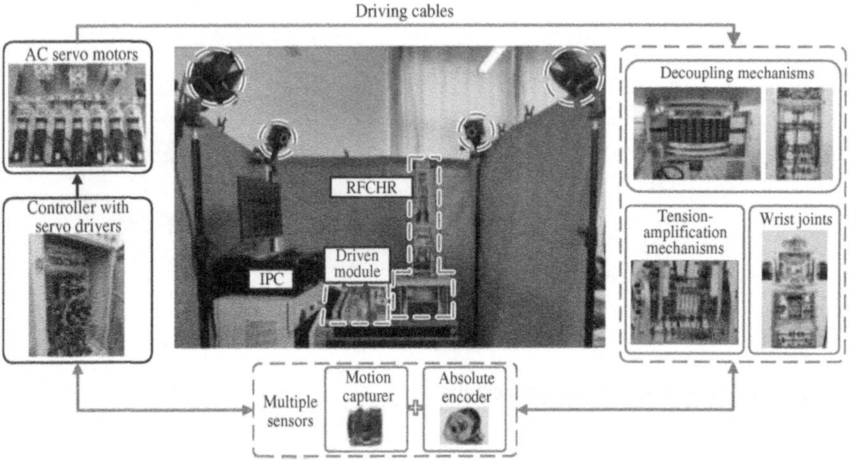

Figure 8.9 Experimental scaled prototype of 7-DOF RFCHR. *Source*: [14]/with permission of Elsevier.

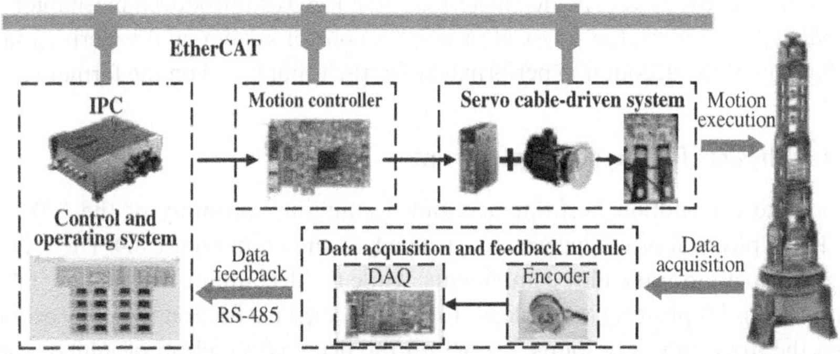

Figure 8.10 Diagram of control system for 7-DOF RFCHR. *Source*: [14]/with permission of Elsevier.

The RFCHR is a nonlinear system characterized by its rigid–flexible coupling and relatively low stiffness. Its motion performance is affected by factors such as the elastic–plastic deformation of cables, friction and damping from pulleys, and the decoupling modules, making accurate system modeling challenging. By applying the principles of fuzzy mathematics, dynamic systems can be simplified and generalized, achieving effective control even for complex or hard-to-model systems. Fuzzy control is widely used in engineering applications. In this study, a fuzzy control algorithm was employed to tune the PID controller parameters, thereby improving the overall performance and applicability of the control system, as shown in Figure 8.11.

8.3.2 Verification of Decoupling Performance

To verify the motion-decoupling performance of the decoupling mechanisms, an experiment was conducted focusing on single-joint decoupled motion. A 16-cable mechanism was used to decouple the rolling motion of joint 1, while a six-cable mechanism was applied to decouple the pitching motion of joint 4. During the decoupled joint movements, coupling effects influenced only the motion of the posterior joints. Therefore, experimental data collection concentrated solely on the angular displacements of the posterior segments of the decoupled joints:

(a) Decoupling performance of joint 1: All motors were mounted on the base, with the installation sequence illustrated in Figure 8.12a. Joint 1 was moved from 0 to $\pi/2$ and then back from $\pi/2$ to 0, as depicted in Figure 8.12.

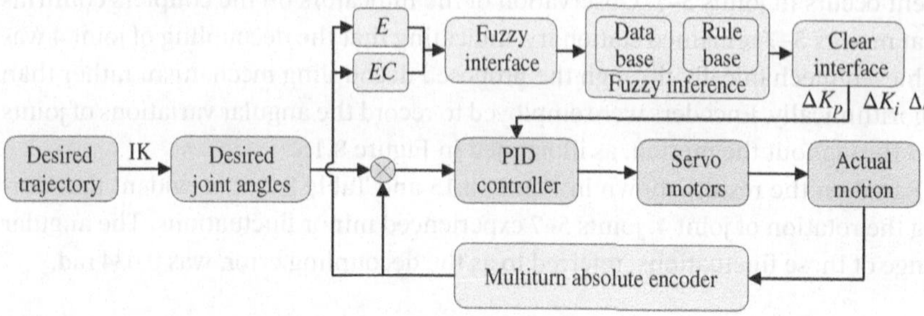

Figure 8.11 Control strategy for 7-DOF RFCHR.

Figure 8.12 Rotational compensation-based motion decoupling experiment for joint 1. (a) Time: 0 s; (b) Time: 5.8 s; (c) Time: 10.3 s; (d) Time: 14.7 s. *Source*: [15]/with permission of Elsevier.

As shown in Figure 8.12, during the motion of joint 1, no noticeable movement occurs in joints 2–7. Examination of the indicators on the couplers confirms that motors 2–7 remained stationary, demonstrating that the decoupling of joint 1 was achieved through the proposed mechanical decoupling mechanism rather than an algorithmic approach. Encoders were employed to record the angular variations of joints 2–7 during the motion, as illustrated in Figure 8.13.

Based on the results shown in Figure 8.13 and Table 8.1, it is clear that during the rotation of joint 1, joints 2–7 experienced slight fluctuations. The angular range of these fluctuations, defined as the decoupling error, was 0.035 rad. As the rotation decoupling mechanism employs a multi-pulley approximation, larger rotation angles of joint 1 lead to greater fluctuations in the posterior joints. This fluctuation trend aligns with the theoretical predictions.

(b) Decoupling performance of joint 4: All motors were mounted on the base, with the installation sequence illustrated in Figure 8.14a. Joint 4 was moved from 0 to $\pi/2$ and then returned from $\pi/2$ to 0, as depicted in Figure 8.14.

As shown in Figure 8.14, during the motion of joint 4, no noticeable movement occurs in joints 5–7. Observation of the indicators on the couplers confirms that motors 5–7 remained stationary, indicating that the decoupling of joint 4 was achieved mechanically through the proposed decoupling mechanism rather than algorithmically. Encoders were employed to record the angular variations of joints 5–7 throughout the motion, as illustrated in Figure 8.15.

Based on the results shown in Figure 8.15 and Table 8.2, it is evident that during the rotation of joint 4, joints 5–7 experienced minor fluctuations. The angular range of these fluctuations, referred to as the decoupling error, was 0.034 rad.

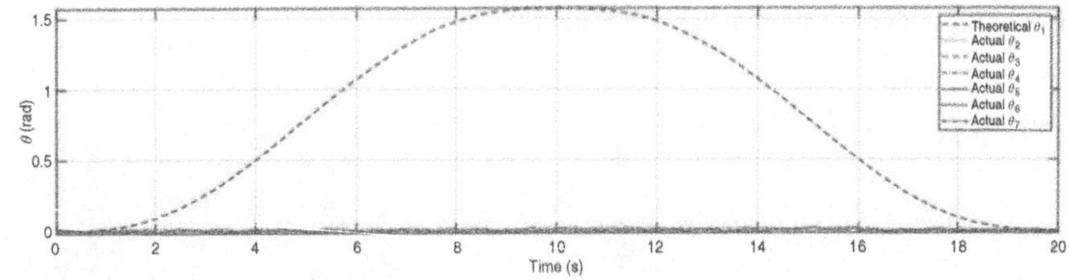

Figure 8.13 Actual angles of joints 2–7.

Table 8.1 Deviations of joints 2–7.

Joints	θ_2 (rad)	θ_3 (rad)	θ_4 (rad)	θ_5 (rad)	θ_6 (rad)	θ_7 (rad)
Lower	−0.033	−0.025	−0.024	−0.026	−0.027	−0.025
Upper	0.026	0.031	0.035	0.033	0.034	0.032

Figure 8.14 Translational compensation-based motion decoupling experiment for joint 4. (a) Time: 0 s; (b) Time: 5.3 s; (c) Time: 9.6 s; (d) Time: 15.5 s. *Source*: [15]/with permission of Elsevier.

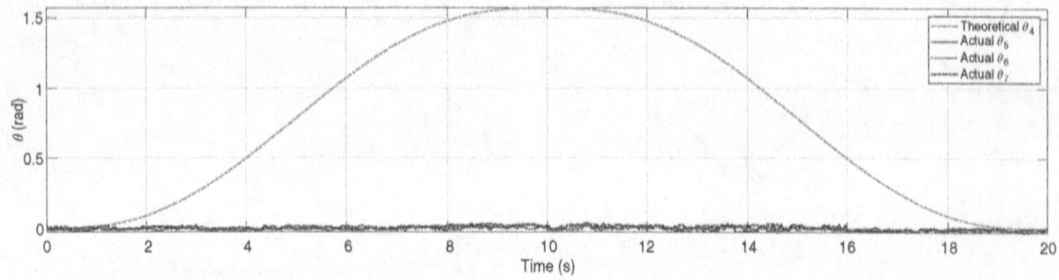

Figure 8.15 Actual angles of joints 5–7.

Table 8.2 Deviations of joints 5–7.

θ_5 (rad)		θ_6 (rad)		θ_7 (rad)	
Lower	Upper	Lower	Upper	Lower	Upper
−0.023	0.034	−0.017	0.031	−0.026	0.021

In conclusion, the proposed decoupling mechanism successfully enabled motion decoupling in the cable-driven robotic system, keeping decoupling errors below 0.035 rad.

8.3.3 Overall Performance

To evaluate the motion performance of the proposed cable-driven robotic system, a trajectory-tracking experiment was performed. Joint angles were measured using seven encoders, while the end-effector trajectories were recorded with a visual motion capture system. The ideal vs. actual joint angles are shown in Figure 8.16, and the corresponding ideal and actual end-effector trajectories are presented in Figure 8.17.

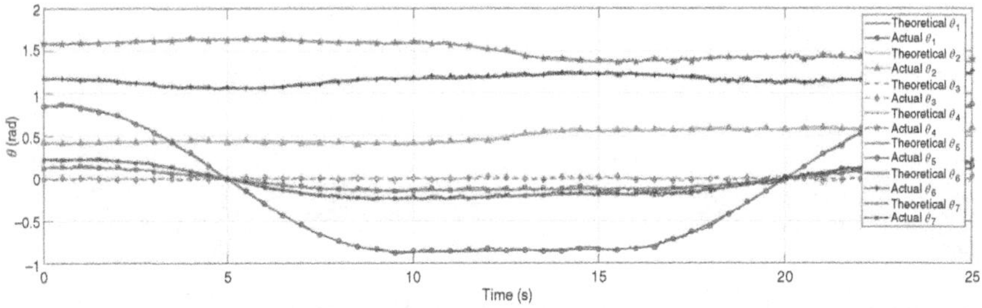

Figure 8.16 Theoretical and actual joint angles of the robot.

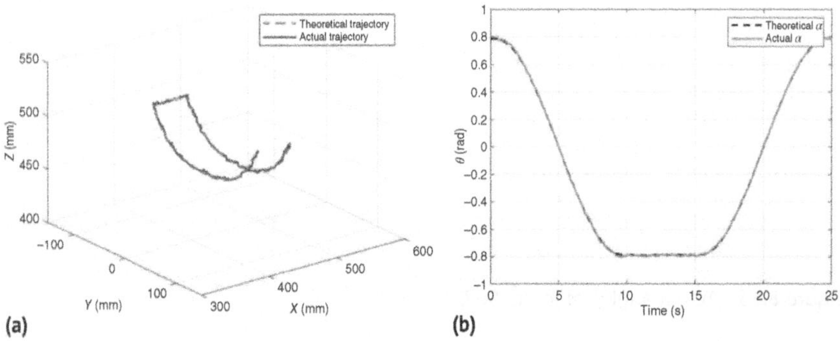

Figure 8.17 (a) Theoretical and actual trajectories. (b) Theoretical and actual α.

Table 8.3 Deviations of joints 1–7.

Joints	θ_1 (rad)	θ_2 (rad)	θ_3 (rad)	θ_4 (rad)	θ_5 (rad)	θ_6 (rad)	θ_7 (rad)
Lower	−0.0016	−0.0102	−0.0078	−0.0088	−0.0163	−0.0068	0.0021
Upper	0.0135	0.0181	0.0242	0.0251	0.0192	0.0287	0.0218

As shown in Figure 8.17 and Table 8.3, the deviations in joint angles remain below 0.0242 rad. Additionally, Figure 8.9 indicates that the overall motion of both the robot joints and the end-effector closely follow the ideal trajectories.

8.4 Conclusion

In this chapter, the systematic development and experimental validation of robotic platforms for rigid–flexible coupling hoisting systems, addressing critical challenges in mechanical decoupling, multi-DOF control, and complex trajectory execution, are introduced. For the RFCHR for container yard operations, the development of an experimental platform equipped with high-precision motion capture, tension sensing, and servo-driven cable systems enables rigorous validation of linear, spiral, and task-space stacking trajectories. Despite the challenges posed by cable winding dynamics, measurement limitations, and prototype tolerances, the robot's ability to maintain stable postures and follow complex paths highlights the effectiveness of its kinematic design and control architecture. Trajectory errors, consistently within allowable ranges across translational axes, underscore the system's adaptability to real-world operational demands, such as container handling and material stacking. For the 7-DOF rigid–flexible coupled crane robot, the integration of a hardware-based decoupling mechanism characterized by a discrete pulley system and 3D-printed components demonstrates a novel approach to minimizing interjoint interference. Complemented by a fuzzy PID control strategy, this effectively alleviates the nonlinear dynamics of the cable-driven system. Experimental results validate the feasibility of this design, showcasing precise motion decoupling and trajectory tracking with errors constrained within operationally acceptable margins, thereby establishing a robust framework for rigid–flexible interaction in articulated hoisting applications. Collectively, these investigations advance the state of rigid–flexible coupling robotics by bridging mechanical innovation with adaptive control, demonstrating practical solutions for precision hoisting and dynamic manipulation in industrial environments. The research not only validates the technical viability of such platforms but also underscores their potential to enhance efficiency and safety in logistics, manufacturing, and heavy-load operations. Future work may focus on optimizing cable management strategies, refining sensor fusion algorithms, and exploring scalability for larger-scale robotic systems, ensuring continued progress toward seamless integration of rigid–flexible dynamics in automated industrial applications.

References

1. Zhou B, Li S, Zi B, et al. Performance analysis and optimization design of a rigid-flexible coupling hoisting robot considering cable deformation and random loading. Mechanical Systems and Signal Processing, 2025, 223: 111903.
2. Khoshbin E, Otis M J D, Meziane R. Reconfigurable cable-driven parallel mechanism design: physical constraints and control. Robotica, 2024, 42(10): 3592–3638.
3. Liu F, Luo J, Mo J, et al. Modeling and analysis of rigid-flexible coupling dynamics of a cable-driven manipulator. Journal of Mechanical Science and Technology, 2024, 38(8): 4377–4384.
4. Zeng S, Zhang Y, Wang S, et al. Design and dynamic analysis of a bio-inspired rolling joint manipulator based on pulley-cable system. Journal of Mechanisms and Robotics, 2025, 17(8): 081006.

5 Jiang X, Zhang J, Zhang Y, et al. Rigid and flexible structures coupled system study and enhanced vision-aided control for a versatile cable winding robot. IEEE/ASME Transactions on Mechatronics, 2024, 29(6): 4618–4629.
6 Ye Z, Zheng L, Chen W, et al. Recent advances in bioinspired soft robots: fabrication, actuation, tracking, and applications. Advanced Materials Technologies, 2024, 9(21): 2301862.
7 Zeng W, Yan J, Cheng S S. Model-based design optimization for motion decoupling in dual-segment flexible robots. International Journal of Mechanical Sciences, 2024, 275: 109312.
8 da Veiga T, Chandler J H, Lloyd P, et al. Challenges of continuum robots in clinical context: a review. Progress in Biomedical Engineering, 2020, 2(3): 032003.
9 Cerik B C, Huang L. Recent advances in mechanical analysis and design of dynamic power cables for floating offshore wind turbines. Ocean Engineering, 2024, 311: 118810.
10 Zarebidoki M, Dhupia J S, Xu W. A review of cable-driven parallel robots: typical configurations, analysis techniques, and control methods. IEEE Robotics & Automation Magazine, 2022, 29(3): 89–106.
11 Goodarzi R, Korayem M H, Tourajizadeh H, et al. Nonlinear dynamic modeling of a mobile spatial cable-driven robot with flexible cables. Nonlinear Dynamics, 2022, 108(4): 3219–3245.
12 Zhang B, Zhao H, Zhang X. Adaptive variable domain fuzzy PID control strategy based on road excitation for semi-active suspension using CDC shock absorber. Journal of Vibration and Control, 2024, 30 (3–4): 860–875.
13 Lu K, Yang B, Wang B, et al. Decoupling skill learning from robotic control for generalizable object manipulation. 2023 IEEE International Conference on Robotics and Automation (ICRA). IEEE, 2023: 864–870.
14 Liu Y, Qian S, Zi B. Mechanical design and tension optimization of rigid-flexible hybrid robots for container yard operations. 2024 IEEE 2nd International Conference on Sensors, Electronics and Computer Engineering (ICSECE), IEEE, 2024.
15 Xu F, Zi B, Yu Z, et al. Design and implementation of a 7-DOF cable-driven serial spray-painting robot with motion-decoupling mechanisms. Mechanism and Machine Theory, 2024, 192: 105549.

Index

Note: Page numbers in *italics* and **bold** refers to figures and tables respectively.

a

Active Cord Mechanism (ACM) 2
active decoupling methods 52
adaptive adjustment inertia weight particle swarm optimization (AAIWPSO) algorithm 76, 83, **86**, 87–89, *88*
adaptive whale optimization algorithm (AWOA) 264, 269, 271, 274
  FPID-AWOA 243, 264, 269, *270–274*
algorithmic decoupling 51–52

c

cable coupling problems 51
cable deformation 45, 91
cable-driven continuous arms 4
cable-driven hoisting systems 286
cable-driven parallel robots (CDPRs) 5, 10, 11, 17–19, 22, 25, 114, 115, 118, 243, 283
cable-driven parallel robot (CDPR) 65, 77, 80, 81, *86*, 88, 143, 144, *145*, 146, 147, 149, 152, 155, 157, 160
  3D model *144*, *145*
  force analysis diagram 80
  IPAnema series 5, 10, *10*
  kinematic diagram *146*
  schematic diagram *145*
cable-driven RFCR 2
cable-driven robots 51, 53, 72, 243
cable driving force control method 275–279
  control strategy design 278–279
  mathematical models, establishment of 276–278
cable parallel manipulators for multiple mobile cranes (CPMMC) 117, 178, 238, 240
cable-suspended parallel robot (CSPRs) 241
incompletely restrained CPRs (IRCPRs) 64
compound function/fuzzy perturbation method (CFFPM) 19, 182–184, 185, 192, 194
  accuracy and efficiency 194–199, *195*, **196–197**, *198–200*
  static uncertainty analysis 183
continuum robotic system 4–5, *5*
control strategies 16
  design 278–279
coupling characteristic analysis 53–55, *53–54*
cycloid trajectory method 254–258, *255–257*

d

decoupling mechanism design
  rotational compensation 59–60, *60*
  translational compensation 58–59, *58*
decoupling performance verification 295–297, *295–297*, **296**, **297**
Delta software 272–273
Dempster–Shafer theory 118
Denavit–Hartenberg theory 17, 33
dense distance method 88–89
dexterity 82–83
double-S velocity curve method 258–260, *258–261*
dual automobile crane systems (DACS) 117, 125
  hybrid LAR analysis 163–164
  hybrid LAR field calculation 164–169
  hybrid random and interval parameters 160–173
  hybrid uncertainty, random and interval parameters 160–162
  LAR analysis with small uncertainty 162–164
  numerical examples 169–173, **169**, *170–174*, **175**, **176**, *176*, *177*
  random interval compound functional moment method 163
  random interval perturbation compound function-based method 162
dual crane system (DCS) 181
dual-machine collaborative rigid–flexible coupling hoisting robot (DCHR) 65
  dynamic modeling 69–71
  integrated mechanism design 65–71
  kinematic modeling 67–69, *67*
  mechanical design 66–67, *66*
dynamic analysis 181–183
  static uncertainty analysis, fuzzy parameters 183–201
  uncertainty with hybrid random and interval parameters 201–238
dynamic modeling 42, 246–249, *247*
  DCHR 69–71
  Lagrange method 42–45
  Newton–Euler method 45–49, *46*
  numerical solutions 265–266
dynamic uncertainty analysis
  hybrid LSOAAC response field calculation 207–215
  hybrid LSOAAC response field problem, random and interval parameters 216–233, **216**, *220–223*, *225*, **226–228**
  with hybrid random and interval parameters 201–238
  LSOAAC equilibrium equation 201–203, *202*
  LSOAAC response analysis 205–207
  LSOAAC response field problem, interval parameters 235–236, **236**, *236*
  LSOAAC response field problem, random parameters 233–234, **234**, *235*
  mathematical characteristics of 212–215
  MHUAM dynamic response 204–207
  monotonic analysis 207–212
  numerical examples 215–236, **216**
  random IPM and random interval moment method 204–205

e

end-effector (EE) 144
error 111
error transfer model 147–149, 151
evidence theory 143
  architecture 144, *144*, *145*
  error transfer model 147–149
  ETUAM 149–152, *150*
  kinematics 144–147, *145*, *146*
  kinematic uncertainty 154–160, *155*, *156*, **157**, *158*, *159*
  modeling methods, simulation of 152–154, **153**, *153–155*
  simulation and comparison 152–160

evidence theory-based uncertainty analysis method (ETUAM) 149–152, *150*, 157
exoskeleton robots 3–5

f

fifth-degree polynomial method 251–254, *252–254*
first-order Neumann series 183
first-order Taylor series 183
Five-hundred-meter Aperture Spherical radio Telescope (FAST) project 2, *3*, 9
flexible drive-rigid support robots 4
force control 275
  cable driving force control method 275–279
  control and monitoring program design 279–280
  control strategy design 278–279
  experimental platform, construction of 275
  mathematical models, establishment of 276–278
  test and verification experiment 280–281, *281*
fuzzy control 264
  adaptive whale optimization algorithm 267–269, *270*
  dynamic modeling, numerical solutions to 265–266
  experiment validation 272–274, *273–274*
  fuzzy PID controller 266–267, *267*
  fuzzy trajectory tracking control 265–269
  numerical simulations 270–271, *270–272*
fuzzy control strategy based on an adaptive whale optimization algorithm 267–269, *270*
fuzzy PID controllers 242–243, 266–267, *267*
fuzzy PID control strategy based on an adaptive whale optimization algorithm (FPID-AWOA) 243, 273–274
fuzzy PID control strategy based on particle swarm optimization (FPID-PSOA) 264
fuzzy PID control strategy based on whale optimization algorithm (FPID-WOA) 264, 273
fuzzy static equilibrium equation 184–185, *184*
fuzzy static response 192–194, *193, 194*
fuzzy theory 181
fuzzy trajectory tracking control 265
  dynamic modeling, numerical solutions to 265–266
  fuzzy control strategy based on an adaptive whale optimization algorithm 267–269, *270*
  fuzzy PID controller 266–267, *267*

g

genetic algorithm (GA) 105
global dexterity index 83

h

hybrid compound function/subinterval perturbation method (HCFSPM) 134–137
  subinterval perturbation method 134–135
hybrid LAR analysis 163–164
hybrid LAR field calculation 164–169
hybrid LSOAAC response field 207
  mathematical characteristics of 212–215
  monotonic analysis 207–212
  with random and interval parameters 216–233, **216**, *220–223*, *225*, **226–228**
hybrid MCM (HMCM) 201
hybrid perturbation compound function-based moment method 162
hybrid random and interval parameters 160–176
  hybrid LAR analysis of the DACS 163–164

hybrid random and interval parameters (*Cont.*)
  hybrid LAR field calculation of DACS 164–169
  hybrid uncertain DACS, random and interval parameters 160–162
  LAR analysis of DACS with small uncertainty 162–164
  numerical examples 169–173, **169**, *170–174*, **175**, **176**, *176*, *177*
  random interval compound functional moment method 163
  random interval perturbation compound function-based method 162
hybrid random-interval perturbation method 18
hybrid uncertain parameter model (HUPM) 118
hybrid uncertainty model 182–183
  LSOAAC equilibrium equation 201–203, *202*

i

incompletely restrained CPRs (IRCPRs) 64
integrated mechanism design of DCHR 65
  dynamic modeling 69–71
  kinematic modeling 67–69, *67*
  mechanical design 66–67, *66*
interval algorithm 117–118
interval and subinterval perturbation methods (IPM) 182
interval kinematic equilibrium equation 133
interval variables
  hybrid compound function/subinterval perturbation method 135–137
  interval kinematic equilibrium equation 133
  numerical examples 137–142, *138*, **139–143**, *141*, *142*
  subinterval perturbation method (SIPM) 134–135
inverse kinematic analysis 36–42, **36**, *36–38*, *40–42*, 120–122, *120*
inverse kinematic modeling 244–246, *245*

j

Jacobian equation 82
Jacobian matrix 79, 82, 83, 182–184, 186, 246
joint torque transmission mechanism 46

k

kinematic Jacobian matrix 184
kinematic modeling 76–81, *76*, *77*, *79–80*
  DCHR 67–69, *67*
  inverse kinematic analysis, low-load environment 36–42, **36**, *36–38*, *40–42*
  large-load environment 32–36, *32*, *34*
kinematics 144–147, *145*, *146*
kinematic uncertainty 154–160, *155*, *156*, **157**, *158*, *159*
kinematic uncertainty analysis 18, 117–119
  evidence theory 143–160
  hybrid random and interval parameters 160–176
  interval variables 133–142
  random parameters 119–132

l

LABVIEW monitoring system 279–280
Lagrange method 28
  dynamic modeling of RFCHR based on 42–45
Lagrangian dynamic formulation 45
luffing angular response (LAR) 118–119
  DACS with small uncertainty 162–164
  hybrid LAR analysis, DACS of 163–164
  PRCFM 123–124

random interval compound functional moment method 163
random interval perturbation compound function-based method 162
RVFMM 124–125
luffing system of an automobile crane (LSOAAC) 183
equilibrium equation 201–203, *202*
interval parameters 235–236, **236**, *236*
random parameters 233–234, **234**, *235*
response field problem 233–236

m

mathematical models, establishment of 276–278
mechanical decoupling 51, 285–286
mechanism design of rigid–flexible coupling hoisting robot (MDRFCHR) 29–32, 61–62
container yard operations 31–32, 41, 49, 50
design methodology 61–62
DOF rigid–flexible coupling hoisting robot 57–61
large-load environment 29–30, *29*
low-load environment 30–31, *30*
mechanical description 62, *62–64*
modular and reconfigurable mechanism design 61–65
typical configuration 64–65, *65*
modeling method simulation 152–154, **153**, *153–155*
modified compound function/fuzzy perturbation method (MCFFPM) 18, 182–184
accuracy and efficiency 194–199, *195*, **196–197**, *198–200*
static uncertainty analysis 188–192
modified hybrid random method (MHRM) 123, 127–133
LAR field, DACS narrow uncertainty 123–125
PRCFM for LAR analysis 123–124
RVFMM for LAR field calculation 124–125
modified hybrid uncertain analysis method (MHUAM) 19, 183, 201
Monte Carlo method (MCM) 97, 103, 125, 130, 156–157, 184
motion-decoupling method 52–57, *53–54*
motion-decoupling performance 295–297, *295–297*, **296**, **297**
multi-degree-of-freedom 16, 32, 285
multi-objective optimal design 83–89
AAIWPSO algorithm 87–89, *88*
multi-objective optimization model 83–84
PSO algorithm 84–85, *85*, **86**
multi-objective optimization design 101–108
dynamic response and optimization design method, analysis of 102–105, *103*, **103**, *104*
multi-objective optimization model 83–84, 101–102
optimization design method 105–108, **106**, *106*, **107**, *108*
reliability, workspace and stiffness 89–108
workspace and dexterity 76–89
multi-objective optimization model 75, 83–84, 101–102
multiple fuzzy variable (MFV) 182

n

Newton–Euler method 28
dynamic modeling of RFCHR based on 45–49, *46*
NOKOV 3D motion capture system 288
non-regularized method 94
normalization formula 94
novel winding mechanism 52

o

Octobot 7, *7*
optimization design method 75–76, 105–108, **106**, *106*, **107**, *108*

dynamic response analysis and 102–105, *103*, *104*
experiment and verification 109–113, *109–113*, **113**
multi-objective optimization design 76–103
OptiTrack system 263–264
origami anisotropic stiffness structure (OASS) 5

p

parallelogram mechanisms 31
Pareto optimal solution 88
particle swarm optimization (PSO) algorithm 76
performance indices 81–83, 89–100
dexterity 82–83
reliability 93–96, **95**, *95*, **96**
static stiffness 98–100, *98*
uncertainty quantification 89–93
workspace 81–82, 96–97, **97**
performance verification
7-DOF RFCHR 293–298
yard operations 286–293
perturbation-based random composite function method (PRCFM) 123
LAR analysis 123–124
physical prototype development 287–288, *287*, *288*, 293–295, *294*, *295*
planetary cable routing mechanism 52
platform development 285–286
and 7-DOF RFCHR performance verification 293–298
and yard operations performance verification 287–293
pneumatic honeycomb network structure 2
point-to-point trajectory planning 249–251, *249*
cycloid trajectory method 254–258, *255–257*
double-S velocity curve method 258–260, *258–261*
fifth-degree polynomial method 251–254, *252–254*
PSO algorithm 84–85, *85*, **86**, 105, 107–108

r

random interval compound functional moment method 163
random interval moment method (RIMM) 90, 119, 182, 183, 204–205
random interval perturbation compound function-based method 162
random interval perturbation method (RIPM) 76, 90, 102, 103, 204–205
random parameters, kinematic uncertainty analysis 119
DACS equilibrium equation, narrowly random model 122–123
inverse kinematics 120–122, *120*
mechanism description 119–120, *119*, *120*
MHRM for LAR of DACS with narrow uncertainty 123–125
numerical examples 125–132, **125**, **126**, *126*, **128–129**, *129–132*
PRCFM for LAR analysis 123–124
RVFMM for LAR field calculation 124–125
random theory 181
random variable functional moment method (RVFMM) 123
LAR field calculation 124–125
redundantly restrained CPRs (RRCPRs) 64
reliability 93–96, **95**, *95*, **96**
rigid actuation-flexible mechanism 1–3
rigid exoskeletons 4
rigid–flexible coupling hoisting robots (RFCHRs) 8, 75, 27, 97
construction 11–13, *12*
dynamic analysis 181–238
dynamic modeling 42–49
field applications 11–16
fuzzy control 264–274

history and development 8–11, *9–10*
kinematic analysis 117–177
kinematic modeling 32–42, 76–81, *76*, *77*, *79–80*
mechanism design 29–32
motion decoupling and reconfigurable design 51–71
ocean 13–14, *13*
optimization design 75–113
platform development and application 285–298
series manipulator 27–29, *28*
static modeling of 76–81, *76*, *77*, *79–80*
storage 14–16, *15*
trajectory planning and tracking control 241–281
rigid–flexible coupling robot (RFCR) 1, 27–28, 75, 76
evolution of 1–8, *3–5*, *7*
experimental platform *287*
flexible actuation-rigid support 3–6
physical prototype development 287–288, *287*, *288*
platform development and performance verification 287–292
rigid actuation-flexible mechanism 1–3
robot motion performance experiment 289–292, *289–293*
smart material-driven 6–8
yard operations 287–292
RoboCrane 4, 9, *9*
robot motion performance experiment
cable winding superposition 290
insufficient effective data points 291
linear trajectory motion process *289*
motion capture system 289
prototype construction errors 291
spatial linear trajectory 289
spatial spiral trajectory 289–290, *291*
spiral trajectory motion process *291*
straight-line trajectory *290*
task-space stacking trajectory 292, *292*, *293*
robot's power system 288, *288*
rotational compensation
decoupling mechanism design based on 59–60, *60*
motion decoupling based on 56–57, *57*

S

series elastic actuators 2
7-DOF cable-driven serial spray-painting robot 57
7-DOF rigid–flexible coupling hoisting robot 285–286
coupling characteristic analysis and 53–57, *53–54*
decoupling on rotational compensation 59–60, *60*
decoupling on translational compensation 58–59
decoupling performance verification 295–297, *295–297*, **296**, **297**
mechanical design 57–61, *58*
motion-decoupling design for 53–57
overall performance 297–298, **298**, *298*
performance verification 293–298
physical prototype development 293–295, *294*, *295*
rotational compensation 56–57, *57*, 59–60, *60*
tension-amplification mechanism design 60–61, *61*
translational compensation 55–56, *55*, 58–59, *58*
shape memory alloys (SMAs) 7
Sherman–Morrison–Woodbury (SMW) formula 182
six-degree-of-freedom (6-DOF) 241
hybrid palletizing robot 30–31, *30*
SkyCam sports camera 2, 9
Soft Exosuit 6
soft robotics 5
spatial spiral trajectory 289–290, *291*
static modeling 76–81, *76*, *77*, *79–80*
static stiffness 98–100, *98*
static uncertainty analysis

accuracy and efficiency, CFFPM and MCFFPM 194–199, *195*, **196–197**, *198–200*
CFFPM 185–188
fuzzy parameters 183–201
fuzzy static equilibrium equation 184–185, *184*
fuzzy static response 192–194, *193*, *194*
MCFFPM 188–192
numerical examples 192–199
structural coupling effect 91
subinterval perturbation method (SIPM) 134–135
surface generation method (SGM) 182

t

task-space stacking trajectory 292, *292*, *293*
tension-amplification mechanism design 60–61, *61*
Tensor Arm 4
3-DOF hybrid cable-actuated parallel mechanism 30–31
3-DOF DACS 120, *120*
total kinetic energy of payload 69
tracking control 241–244
force control for 275–281
fuzzy control for 264–274
trajectory planning for 244–264
trajectory planning 244
cycloid trajectory method 254–258, *255–257*
double-S velocity curve method 258–260, *258–261*
dynamic modeling 246–249, *247*
fifth-degree polynomial method 251–254, *252–254*
inverse kinematic modeling 244–246, *245*
numerical simulation and experiments 261–263, *262–264*
point-to-point trajectory planning 249–260, *249*
translational compensation
decoupling mechanism design based on 58–59, *58*
motion decoupling based on 55–56, *55*
Tri-Co Robot 5

u

uncertainty quantification (UQ) 89–93, 181
under-constrained cable-suspended parallel robot (UCPR) 17, 19, 242, 244–246
unit direction vector of cable 41
US National Institute of Standards and Technology (NIST) 4, 9, *9*

v

variable intervals 103–104
vector loop equation 39
vertex method 156–157
video tape recorder system 2, *3*

w

warehousing 14
whale optimization algorithm (WOA) 268–269
workspace 81–82, 96–97, **97**

x

XINGYING software 288

y

yard operations 286
performance verification 287–293
physical prototype development 287–288, *287*, *288*
robot motion performance experiment 289–293, *289–293*

Wiley The manufacturer's authorized representative according to the EU General Product Safety Regulation is Wiley-VCH GmbH, Boschstr. 12, 69469 Weinheim, Germany, e-mail: Product_Safety@wiley.com.

Printed and bound by CPI Group (UK) Ltd, Croydon, CR0 4YY
15/09/2026
02217457-0005